About the author

Morgan Downey is a successful commodities trader based in New York. In addition to the Americas, Morgan Downey has spent many years trading commodities while living in Asia, Australia, and Europe.

www.morgandowney.com

OIL 101

OIL 101

Morgan Downey

This book does not constitute any offer, recommendation, or solicitation to any person to enter into any transaction or adopt any hedging, trading or investment strategy, nor does it constitute any prediction of likely future movements in rates or prices or any representation that any such future movements will not exceed those shown in any text or illustration. Although the author and publisher used their best efforts in preparing this book, the author and publisher make no representations or warranties with respect to the accuracy or completeness of this book. The author and publisher specifically disclaim any implied warranties of merchantability or fitness for a particular purpose. Materials and processes described herein may be dangerous to health and property and you are not to rely on this book in any manner for proper storage, handling or use of materials and processes described herein. The author and publisher have no fiduciary duty towards you, and assume no responsibility to advise on and makes no representation as to the appropriateness or possible consequences of any action you may take with respect to any matter contained herein. You are to make your own independent judgment with respect to any matter contained herein and to seek your own independent professional advice where appropriate. Neither the author nor publisher shall be liable for any loss, loss of profit or any other damages, including but not limited to special, incidental, consequential, or other damages.

Wooden Table Press LLC

www.morgandowney.com

ISBN: 978-0-9820392-0-5 (hardcover)

First edition, Version 1.0

Table of Contents

CHAPTER ONE
A BRIEF HISTORY OF OIL

1859 – 1911: a new industry

The modern oil industry began as a result of a scarcity of whales. Artificial light, through all the ages of civilization, has served as the tool man has used to free himself from his dependency on daylight. Until 1859, most people obtained light by burning animal fats in the form of beeswax candles or whale oil. Whale oil shed the purest light of all available fuels, and became a luxury product. Overfishing lead to a decline in the whale population and a sharp increase in whale oil prices.

People have used oil obtained from the ground since at least 4,000BC. In the Middle East, crude oil that seeped to the surface was used to waterproof boats and as an adhesive in the construction of buildings and roads. Crude oil was also refined in minor quantities for lamp and heating oil in China around 1,000BC, though this technology never made it directly to modern times. By 600AD, the Byzantines used crude oil to produce a flame-throwing weapon known as Greek fire.

In the 1840s, a Pennsylvanian salt mine owner named Samuel Kier noticed his works were becoming fouled on the surface with an oily substance. The oil was initially dumped. Later, after noticing local Seneca Indians using the oil as liniment for skin ailments, Kier decided to try to commercially market this waste product of his salt business as medicinal Seneca Oil, or Rock Oil. Kier had little commercial success with his medicinal rock oils, but his prescience was later validated. Petroleum[1]-based oils are widely used today in the form of products such as Vaseline, developed by another entrepreneur in the 1870s.

In 1854, George Bissell and his business partners sent a sample of a crude oil skimmed from a surface pool in northwestern Pennsylvania to Professor Benjamin Silliman of Yale University for analysis. Bissell hoped that the crude oil could be distilled into a lamp fuel so that they could profit from high whale oil prices. Silliman confirmed that the sample could indeed be distilled to produce kerosene[2].

The fundamental process of distillation involves separating different products by heating them. The products have different boiling points, so they evaporate and are condensed separately. It is still the basic refining technique used today.

---

[1] Petroleum etymology: Latin *petra*, meaning rock, and *oleum*, meaning oil.
[2] Kerosene etymology: Greek *keros*, meaning wax.

Silliman's analysis was used to raise capital for the formation of the Pennsylvania Rock Oil Company in 1855. Now, all investors needed were large quantities of rock oil.

Bissell had heard about Kier's rock oil obtained as a byproduct of drilling for salt with derricks, and came up with the idea of setting up a derrick specifically to drill for oil. Drilling to produce oil was a far-fetched concept at the time. The Pennsylvania Rock Oil Company hired a railroad conductor named Colonel Edwin Drake to carry out the drilling (Fig. 1-1). Drake was never in the military, and his title was created by the Rock Oil Company so that he would impress locals. After a nervous few weeks in rural Pennsylvania, Drake struck oil on August 27, 1859. The first well was on a salt dome rock formation. The well was 69 feet deep and yielded 15 barrels a day. Others quickly followed Drake and drilling soon spread across the region.

The petroleum that flowed from the world's first wells in what became known as Oil Creek, near Titusville, Pennsylvania, started the modern oil industry we know today. Drake never patented any of his drilling equipment. His inventions included a stove pipe used to prevent the hole from collapsing and to stop ground water from leaking into the well. This is the same principle used in modern day casing. He died without ever achieving any great fortune. Drake's legacy is that Pennsylvania was, for a brief time, the largest commercial oil producing area in the world.

Fig. 1-1

"Colonel" Edwin Drake (right) and his good friend Peter Wilson, a Titusville pharmacist, in front of the location of the world's first oil well.
Source: Pennsylvania Historical and Museum Commission, Drake Well Collection, Titusville, PA

There were no pipelines or railway lines near the Pennsylvanian oil fields until several years after Drake's discovery. Crude oil was stored and transported to refineries in any readily available container. Wooden whiskey and wine barrels were the most common means of transporting liquids at the time and were requisitioned to haul crude oil. By the 1870s, railroad tanker cars and pipelines began to replace barrels as the preferred, and less expensive, methods of moving crude oil.

The original product that drove oil demand was kerosene. However, when refineries distilled crude oil, they produced other products that were either too light and volatile or too dense and smoky when burned to be used safely as lamp oil, including what are now known as gasoline and diesel.

Gasoline acquired its name from its tendency to vaporize and become a gas easily. The internal-combustion gasoline engine uses the heat of electrical sparks to ignite measured doses of gasoline and air in an enclosed space. The first practical gasoline internal combustion engine was developed in 1867 by Nikolaus August Otto of Germany. Diesel fuel, which contains slightly heavier hydrocarbon molecules, was named after another German, Rudolph Diesel. Diesel invented an engine in 1892 that could ignite this heavier fuel without a spark plug, by simply heating gas via compression.

In spite of the development of internal combustion engines, gasoline and diesel were seen as useless byproducts in the production of kerosene and burned or dumped near refineries until the early 20th century and the appearance of mass produced automobiles.

Given the extremely high cost of whale oil and the expedient alternative which kerosene presented, Drake's first well created a scramble reminiscent of the California gold rush of 1849. Oil became known as black gold. The price of a barrel of crude oil in January 1860 sold for $18 in the money of the time, or over $375 in today's terms. These prices were initially supported as kerosene was replacing hugely expensive whale oil. By the end of 1861, however, a barrel of crude collapsed to just 10 cents per barrel, or $2.60 in today's terms, due to rampant overproduction.

The Pennsylvania oil rush, followed by the careless depletion of the reservoirs, led many to call for regulation. In addition to causing extremely low oil prices, which discourages conservation, the drilling of too many wells on a single reservoir and depleting it too quickly can significantly reduce the total amount of recoverable oil.

Reservoir management techniques involve the judicious use of natural drive pressure from water aquifers underneath the oil so that it will lift the maximum number of barrels out of the ground. It is also important to utilize the hydrocarbon gas dissolved within or on top of crude oil. If used in a coordinated manner across a reservoir, natural drive pressures can lift more total crude oil out of the ground than if everyone independently tries to maximize their individual production from a single reservoir by drilling into it like a pin cushion and producing at a rapid rate. Once reservoir pressures fall, a large amount of crude oil becomes unrecoverable.

The new industry was gradually consolidated and monopolized by one man, John D. Rockefeller, and his Standard Oil Company. Rockefeller started as a produce seller during the Civil War. He purchased his first refinery in Cleveland in 1865. He formed The Standard Oil Company in 1870 and by 1890 controlled 90% of the US oil market. If Rockefeller couldn't buy a competitor, he simply drove it out of business by dropping prices or undercutting railroad oil transportation costs in that competitor's sales region. His control of the oil industry was to such an extent that oil companies Rockefeller did not control were known as independents.

Journalist Ida Tarbell brought the anti-competitive practices of Standard Oil to public attention in a series of investigative reports published from 1902 through 1904. Hastened by the Tarbell information, the Sherman Antitrust Act of 1890 was used in 1911 to split Standard Oil into several competing firms. Thus sprang forth 34 companies including Exxon, now known as ExxonMobil. To this day, ExxonMobil is known as ESSO outside of the United States, a throwback to the Standard Oil acronym. Among many other firms originating from the breakup of Standard Oil were Chevron and Texaco, now known together as Chevron; and Conoco, now part of ConocoPhillips. The split companies became far more valuable than the original Standard Oil and added further to Rockefeller's wealth.

Although whiskey and wine barrels were only briefly used to store and transport oil, the barrel remained the default volume measure in oil markets. In the early years of the industry a person could buy a barrel of oil and not be sure exactly how much oil the barrel would contain. Standard Oil standardized the volume of a barrel to be a Standard Oil Blue Barrel, or bbl (an acronym used to this day), which is 42 US gallons[3]. At the time, a 40-gallon barrel was used to carry whiskey and wine. The Standard Oil Blue Barrel, or bbl, was 42 gallons to allow for 2 gallons of evaporation and leaks during shipment of the crude oil to a

---

[3] There is some dispute whether the term bbl originated with Standard Oil as it had been used as an acronym for barrels before Standard Oil existed. However, Standard Oil's 42 gallon Blue Barrel size did set the default standard volume for the term bbl in the oil industry.

refinery. Since oil expands and contracts with temperature changes and evaporates easily, volume operational tolerances of one to five percent are often permitted even today in large transactions.

Oil volume conversions

Table 1-1

| 1 Oil Barrel = 42 US Gallons ≈ 159 liters |
| 1 US Gallon ≈   3.8 liters |

Early 20ᵗʰ century: boats, planes, trains and automobiles

Following the initial discovery of crude in Pennsylvania, additional small discoveries were made in Texas, Oklahoma, and California. Very few wells in the US produced more than 50 barrels per day. There were sufficient oil discoveries for the lamp oil industry, but not much else. So little oil was produced that Henry Ford initially designed the Model T to run on ethanol. Ethanol is alcohol made from fermenting corn and other starch-based crops. Rudolph Diesel originally designed his engine to run on etherized vegetable oils, now known as biodiesel. Then came a gusher named Spindletop (Fig. 1-2).

Spindletop, discovered in 1901 just south of Beaumont, Texas, produced what was then a gigantic amount of crude - over 50,000 barrels per day. The discovery, made by Patillo Higgins, came from the deepest well in the US at the time, at 1,100 feet underground. Although oil wells had been drilled for over forty years, Spindletop was the first large gusher in the US where oil actually shot out of the well into the sky. Texas tea, as it became known, gushed out of Spindletop, stunning local oilmen and forming a huge oil lake before the well could finally be controlled nine days later. This individual well produced 20% of daily US production at the time.

Spindletop, 1901

Fig. 1-2

Source: The American Petroleum Institute

In addition to large US discoveries such as Spindletop, major discoveries began to occur in other parts of the world. On the shores of the Caspian Sea, production in Baku, Russia, today part of modern day Azerbaijan, began in the 1870s. This development was led by the Nobels, of dynamite and prize fame, and the Rothschild banking family. Production in the Middle East commenced in Persia when the UK government-controlled Anglo-Persian Oil Company, as BP was then known, struck oil in 1908. Royal Dutch discovered oil in the 1890s on the island of Sumatra, nowadays part of Indonesia.

Spindletop and other subsequent large discoveries created the needed cheap and ubiquitous supply of fuel to launch the automotive age. Kerosene for lamp fuel had until then been sold to end users in small cans from grocery and general supply stores. The advent of the mass produced automobile, led by Henry Ford's Model T in 1908, created demand for crude oil distillates other than kerosene. The quantity needed ultimately changed the principal marketing mechanism to dedicated oil service stations.

Up until the First World War, although economically important for lighting, oil was not of much strategic significance. This changed when Winston Churchill, then commander of the British Navy, decided to replace slow coal-powered vessels with rapid response oil-powered military ships in WWI. This became a conclusive factor in the outcome of the war. Ocean-going commercial and military ships continue to use residual fuel oil as their primary fuel to this day. Residual fuel oil is also called fuel oil, resid, or bunker fuel, a vestige of the days when ships had coal storage bunkers. Resid is one of the more inexpensive heavy oil products produced from crude.

The high energy density and rapid energy release of oil was essential to the development of powered flight. A gasoline engine powered the Wright Flyer in 1903. Gasoline continued to be the only fuel choice for powered aircraft until the 1940s.

The jet engine, or gas turbine was developed during the 1930's by Frank Whittle in the UK, and Hans von Ohain in Germany. Jet turbines rely on burning fuel to create a pressure differential resulting in air moving to spin a turbine. These types of engines generate a very high power output for their weight, which is why they are ideal for aircraft. The first jet flight took place in 1939 in Germany. As gas turbines are not very efficient at low speeds or handling constantly varying speeds, they are rarely used for ground transportation.

A gas turbine engine can in theory be configured to burn a wide range of oil products, such as gasoline, diesel and kerosene. Kerosene was chosen for use as jet fuel because gasoline and diesel were desperately needed for existing WWII military machinery and kerosene was plentiful since it was no longer used for lighting following the widespread adoption of the electric light bulb in the early 20th century.

By 1950, crude oil had completely transitioned from a source of lamp oil to a transportation fuel, with gasoline, diesel, residual fuel oil and jet fuel/kerosene accounting for about two thirds of crude oil consumption (Fig. 1-3).

The remaining third of oil consumption was used in a variety of ways to facilitate rapid economic growth. Bitumen, a very heavy and cheap product of crude oil could be combined with small rocks to make asphalt. This enabled the creation of vast, inexpensive road networks during the 20th century. Plastics and other petrochemicals derived from crude oil began to gain widespread acceptance in the 1930s and 1940s. Diesel and kerosene found additional use as heating oils in several countries. Except in unusual circumstances, oil has been rarely used to produce electricity. Coal, nuclear fuel, natural gas and hydro are all relatively cheaper in non-transportation roles.

Oil Statistics   Fig. 1-3

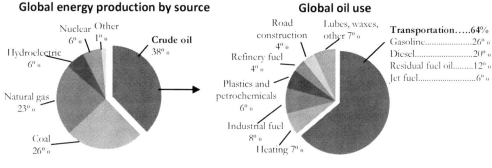

Global energy production by source

Nuclear — 6%
Other — 1%
Crude oil — 38%
Hydroelectric — 6%
Natural gas — 23%
Coal — 26%

Global oil use

Road construction — 4%
Lubes, waxes, other — 7%
Refinery fuel — 4%
Plastics and petrochemicals — 6%
Industrial fuel — 8%
Heating — 7%

Transportation.....64%
Gasoline...................26%
Diesel.........................20%
Residual fuel oil.........12%
Jet fuel..........................6%

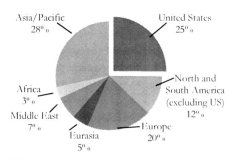

Global electricity generation by fuel

Crude oil — 7%
Other — 2%
Coal — 40%
Nuclear — 16%
Hydro — 16%
Natural gas — 19%

Global oil consumption by region

Asia/Pacific — 28%
United States — 25%
Africa — 3%
North and South America (excluding US) — 12%
Middle East — 7%
Europe — 20%
Eurasia — 5%

Sources: EIA and IEA 2007 data.

Mid to late 20th century: managing excess capacity

The success of the oil industry in finding oil became its greatest challenge. Oil prices remained low during long periods of unforeseen oversupply while demand steadily caught up.

In 1928, following large oil production increases from Russia as it recovered from the First World War, English and American oil companies became worried that the world was once again moving into a dangerously oversupplied situation. To address the glut, the heads of the most powerful oil companies in the world attended a grouse shoot at Achnacarry Castle in Inverness, Scotland. The shoot was hosted by Henri Deterding of Royal Dutch Shell and attended by John Cadman of Anglo-Persian Oil, now BP, Walter Teagle of Standard Oil of New Jersey, now ExxonMobil, William Mellon of Gulf Oil, now part of Chevron and BP, and Robert Stewart of Standard Oil of Indiana, which later became Amoco and is now part of BP.

The result of the grouse shoot was the As-Is Agreement, or Achnacarry Agreement. Under the Agreement the oilmen agreed that they would not compete against each other outside of the US and instead would act to ensure price and profit stability for each of them. The oil companies were required to compete within the US due to the anti-trust legislation that had been used to split Standard Oil. The respective market share of each company outside of the US was to be kept 'as-is', at 1928 ratios, with joint ventures and other alliances removing competition. The Soviet Union in 1929 agreed to participate in the Agreement.

Despite providing conspiracy theorists with plenty of fodder, the Achnacarry Agreement failed as the participants to the agreement, although large, did not hold sufficient market share to control supply and prices. More nimble independents could simply undercut the As-Is group's price by a very small amount to gain market share. Compliance with As-Is ratios was also difficult to verify, so cheating among participants could easily occur.

The need to stabilize prices at profitable levels in the face of oversupply was satisfied three years later by the US government. In late 1930, a wildcatter by the name of Dad Joiner discovered a massive oil field in East Texas. Oil prices that had been over a dollar a barrel in 1930, US$10.95 in today's terms, plummeted due to rampant production from this huge field to ten cents per barrel by mid-1931, $1.11 in today's money. In the midst of dealing with the Great Depression and trying to kick start US industry including the oil business, the US federal government required the Texas Railroad Commission (TRC) and similar but smaller organizations in other oil producing states to impose production restrictions to ration the amount of crude produced in each state.

TRC control of East Texas production spare capacity made it the arbiter of global prices from 1931 through 1971.

Until the 1970s, exploration and production outside the US was dominated by a group of large oil producers known as the Seven Sisters because of their close ties. The Seven Sisters, through mergers and acquisitions, are now four: ExxonMobil, Chevron, BP, and Royal Dutch Shell.

The Seven Sisters were:

> **Standard Oil Company of New Jersey**, referred to in the industry simply as 'Jersey', became Esso (S.O.), then renamed Exxon, now ExxonMobil.
> **Standard Oil Company of New York (SoCoNY)** became Mobil, which merged with Exxon to form ExxonMobil.
> **Standard Oil of California (SoCal)** became Chevron.
> **Texas Company** was renamed Texaco, which is now part of Chevron.
> **Royal Dutch Shell**, the Anglo-Dutch company which continues to have the same name.
> **Anglo-Persian Oil Company** became the Anglo-Iranian Oil Company (AIOC) and then, following the Abadan crisis, described shortly, was renamed British Petroleum, then BP Amoco following a merger with Amoco, formerly Standard Oil of Indiana. It is now simply named BP.
> **Gulf Oil**, most of which has now become part of Chevron and BP. An independently owned network of retail stations in the northeastern United States still carries the gulf oil brand name, but does not produce or refine crude oil.

The four remaining Sisters have been joined by two other large international oil corporations (IOCs), ConocoPhillips and TOTAL, to form a group known today as the six Majors (Fig. 1-4). Today, the Majors have lost their market dominance. Together, they control only 14% of global crude oil production, although they still own 24% of global refinery capacity.

The Majors   Fig. 1-4

ExxonMobil	Chevron	BP	Royal Dutch Shell	ConocoPhillips	Total

Much of the oil produced outside of the US in the twentieth century until the 1970s was carried out on the basis of concessions. A concession is a particular type of legal agreement between an IOC, most often a Major, and the government of the country in which the oil was being produced.

Saudi Arabian oil, for example, which was first discovered in 1938, was developed on a concession basis by a consortium of western oil companies led by Chevron.

Concessions negotiated by the Majors were as a rule on a 50/50 profit sharing basis. The oil Major would set what is known as the posted price – an arbitrary buying price the Major determined for crude oil. The oil Major would buy crude at this posted price and split net revenue after costs 50/50 with the country. The posted price was generally a US gulf coast posted oil price adjusted lower based on the prevailing freight rate from the country in which production took place. The posted price in each country took into account the freight charge from wherever the crude oil was located globally to the US gulf coast, even if the crude was never actually going to the US. This is a form of netback pricing.

Netback pricing links a commodity price in an obscure location to a more transparent market price. The posted price mechanism used freight netback pricing, which linked the price of crude at various locations around the world to the transparent US gulf coast price by simply adjusting for freight costs.

The Majors were subject to the TRC control of US spare capacity. Since the TRC controlled all US spare capacity, it effectively controlled US gulf coast crude oil posted prices. Consequently, the TRC controlled global crude prices via freight netback pricing. However, the TRC had no control over the retail price of refined products. The governments of oil producing nations often felt that this posted price for crude oil was set at a level too low relative to retail market prices of refined finished products, such as gasoline and diesel. The producing nation governments also felt disenfranchised as the posted prices were set unilaterally by the Majors.

Additionally, some producing country governments felt that a 50% share of revenue was insufficient, and the lack of a time limit on concessions was too generous, given the unusual production longevity some non-US oil fields showed.

Cracks in the 50/50 concession arrangements began to emerge in 1951 as Mohammed Mossadegh, the democratically elected prime minister of Iran, nationalized his country's oil industry. Mossadegh took possession of the British owned and operated Anglo-Iranian Oil Company (AIOC) production and refining facilities on behalf of the Iranian state in what became known as the Abadan Crisis.

The British government immediately imposed a trade embargo on Iranian oil. The sole asset of the Anglo-Iranian was the concession share of Iranian production and the Abadan refinery in Iran. The company took legal action to prevent refineries or tanker operators from buying Iranian oil over the following two years, which brought Iran close to bankruptcy. The embargo ended in

1953 when a British SIS-sponsored coup, with US CIA support, reinstalled the Shah of Iran. The Shah, Mohammed Reza Pahlavi, had fled Iran the previous year and lived in France during his exile. Once re-instated, the Shah permitted AIOC's Iranian oil operations to be divided among British, American, and French oil companies on the usual 50/50 concession basis.

Enrico Mattei, president of the Italian Company ENI, was denied an ownership stake in the new consortium controlling Iranian oil production following the reinstatement of the Shah. The shutout enraged Mattei such that he decided to allow AGIP, a subsidiary of ENI, to be the first large IOC to break with the 50/50 concession convention. In 1957, Mattei negotiated an oil concession with a 25% share of profits for ENI and 75% for Iran. Other breaches of the 50/50 arrangement occurred in 1957 between the Japanese Arabian Oil Company and Saudi Arabia and in 1958 between Standard Oil of Indiana and Iran.

These new agreements remained outliers and the 50/50 splits held elsewhere, until 1970. Colonel Muammar al-Qaddafi of Libya obtained an increased 55% share of oil revenues from Armand Hammer of the large US independent Occidental Petroleum, by threatening to nationalize his country's oil industry. This break in the long-standing and sacred 50/50 share quickly snowballed with every other producer nation seeking 55% or more.

Meanwhile, OPEC (the Organization of Petroleum Exporting Countries) was formed in 1960 in Baghdad. The organization was based in Vienna and modeled after the TRC. Five founding member countries: Saudi Arabia, Kuwait, Iran, Iraq, and Venezuela, were joined in later years by a further nine nations, the United Arab Emirates, Qatar, Libya, Algeria, Indonesia, Nigeria, Angola, Ecuador and Gabon. Gabon and Indonesia subsequently left OPEC. There are currently twelve remaining OPEC countries. Sudan is seeking to join OPEC, which would bring membership to thirteen.

During the 1960s, OPEC didn't really have any power: firstly, because western oil Majors controlled production in OPEC countries via concessions, and secondly, but more importantly, the TRC still controlled global pricing as the US had surplus production capacity since Dad Joiner discovered the East Texas fields in 1930. The TRC would add or subtract oil supply to manage global prices, as OPEC later would do.

The global pricing ability of the TRC disappeared in 1970 when US oil production peaked and began to steadily decline (Fig. 1-5 and Fig. 1-6). In 1971, facing declining US production, the TRC gave producers in Texas, previously the only global production area with excess capacity, free reign to produce as much oil as they could.

The US no longer had any spare production capacity. The world's spare capacity, and with it the balance of international oil pricing power was now in the hands of OPEC nations alone. In order to stem the growing dissent over the 50/50 concession sharing and to restore cohesion to international oil concessions, 22 oil companies, including the Majors, negotiated with OPEC in a serious manner for the first time in February 1971.

OPEC's negotiations yielded a higher posted price and incorporated a minimum of 55% participation in revenues for producing countries. The deal, known as the Tehran Agreement, which was to be fixed for five years, quickly fell apart.

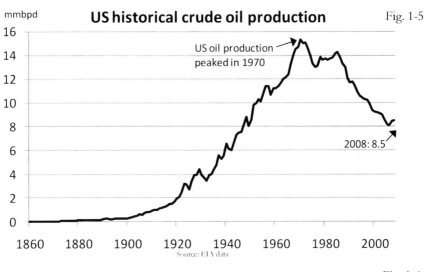

Fig. 1-5

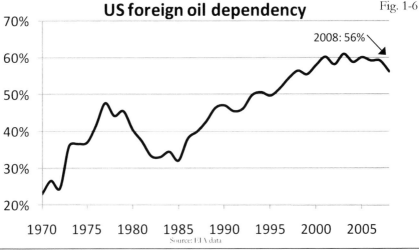

Fig. 1-6

Oil price control timeline Table 1-2

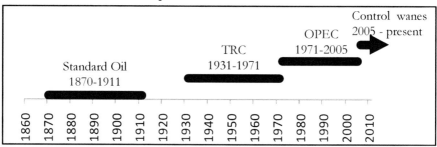

The purchasing power of the US dollar, which was and is the default currency for oil transactions, declined rapidly following the removal of the US currency from the gold standard by Richard Nixon in 1971. Dollars could no longer be exchanged for gold or any other metal held by the US government. Today, oil prices generally rally when the dollar weakens and fall when the dollar strengthens. However, in 1971 there was no transparent free floating global 24-hour market in oil as there is today. The US gulf coast posted price set by the Majors stood unchanging for months or years at a time as the reference price which all international oil trade negotiations referred. With the use of fixed posted prices in the Tehran Agreement, oil was stuck at a set US dollar exchange rate and couldn't rally as the dollar weakened. This put a huge strain on the posted price mechanism.

A surprise invasion of Israeli occupied territory by Arab states Egypt and Syria on Yom Kippur, October 6, 1973, was the final straw leading to the collapse of the Tehran Agreement.

The US military sent supplies to Israel to combat Soviet Union supplies the Egyptians had received. It was also as an attempt to prevent the Israelis having to resort to using their nuclear weapons. Arab nations at the time were mostly US allies, but were also being actively wooed by the Soviets. They viewed the US military re-supply of Israel as an overtly provocative anti-Arab gesture.

In support of Egypt and Syria, on October 16, 1973, Arab OPEC members Saudi Arabia, Libya, and four Gulf Arab states decided to stop supplying crude oil to the US and the Netherlands, the dominant oil trading hub for Europe at the mouth of the river Rhine. The Arab oil embargo cut world supply dramatically by between 5% and 10% overnight. The action was not an official OPEC embargo as non-Arab OPEC countries, such as Iran, (the British-reinstated Shah still in his throne), did not cut production.

The cut caused a consumer panic in the oil market. OPEC nations unilaterally began to ignore the Tehran Agreement's posted price and sell their oil for whatever price the market would bear. The price of crude increased almost fourfold between October 16, 1973 and Jan 1, 1974, from $14 to over $50 in today's terms. The embargo continued for six months until March 1974, when Arab oil producers, except Libya, announced its end. This was the first and only time to date the Arab oil weapon has been used. Arab nations appeared somewhat shocked themselves that they could now have such a large effect on oil prices.

With the Arab oil embargo, low and stable posted crude oil prices became a thing of the past.

The sudden increase in revenue resulting from the Arab oil embargo and a reluctance to continue providing such a large concession share to IOCs quickened the pace of resource nationalization in oil producing nations during the 1970s.

Nationalization, whereby IOC concession shares and production facilities in producing nations were either compulsorily purchased or seized without remuneration by a government owned National Oil Company (NOC) had begun in 1938 in Mexico as a result of anti-colonialist sentiment; in the Middle East as a result of Arab nationalism, led by Nasser in Egypt in 1952; and Persian nationalism led by Mossadegh in Iran in 1951.

Some nationalization processes were carried out abruptly, as in Algeria, whereas others took place gradually, as with the seven year transition of the concession owned Arabian American Oil Company (Aramco) to the state owned NOC Saudi Aramco during the 1970s. The NOCs of nations which engaged in a gradual transition and retained some services from the international oil companies tended to be able to better optimize their crude production following the handover as the departing company would, for a fee, provide training and transfer any important technical data relative to the oilfields before leaving.

In reaction to the energy crisis, the Organization for Economic Cooperation and Development (OECD) formed the International Energy Agency (IEA) in November 1974 to coordinate the response of developed nations to restrictions in supply. The IEA recommended minimum stockpile levels of oil to be created in consumer countries to enable petroleum consumers to shelter themselves better from such crises. In 1975, as part of the IEA effort, President Gerald Ford established a Strategic Petroleum Reserve (SPR) of crude oil to be used for US emergency purposes. The reserve was composed of crude oil because a long-term reserve of a finished product, such as gasoline, is very difficult to maintain. Refined finished products degrade relatively quickly, usually within a year of production.

Contrary to popular perception, there was no global shortage of oil in the 1970s. Government price controls in individual countries resulted in artificial and localized shortages. The US, for example, had lines outside gasoline stations not because oil was unavailable, but because the US government had set domestic oil prices below international oil prices. If the government did nothing, and allowed the free market to operate, then prices would have been higher, but there would have been no domestic shortage.

The first oil shock of 1973-74, caused by the Arab oil embargo, was followed by the second oil shock of 1978-81, as a result of different Middle Eastern events.

Beginning in October 1978, a strike by Iranian oil workers in defiance of the Shah led to a complete cessation of Iranian oil production by December of that year. In early 1979, the Shah was deposed in a populist revolution by a fundamentalist Islamic regime led by a Muslim Cleric, Ayatollah Khomeini. The Shah had warned successive US governments not to place spies in Iran. The US and other western nations, worried about upsetting a major oil supplier, complied. The net result was that Western oil consumers were caught by surprise with the success of the popular Iranian revolution and how easily the Shah gave up control. It was later discovered that the Shah knew he had terminal cancer in 1978, and he died in 1980.

The decline in Iranian production in 1978 resulting from the Iranian strike was compensated somewhat by a temporary increase in production from Saudi Arabia and other OPEC nations with spare capacity, but the suddenness of the oil supply decline was frightening for oil consumers who were yet again shocked at how dependent modern life had become on such unstable oil suppliers.

The overthrow of the Shah was quickly followed by US embassy staff in Iran being held hostage in Tehran for over a year in late 1979 and, in 1980, the outbreak of the Iran-Iraq war. With the loss of a large amount of production from both of these countries, the price of crude ran up to over $38 per barrel in 1980, just over $90 in today's terms (Fig. 1-8).

The Middle East appeared to be falling apart, and spending on oil by US consumers in 1980 climbed to over 8% of GDP. Global oil spending as a share of GDP was not far behind (Fig. 1-7).

Oil is essential to the modern way of life and demand is said to be inelastic, in that it is relatively invulnerable to price movements. However, there is a breaking point. Spending on oil at 4% of GDP is a point above which consumers have in the past become more efficient, in a process called demand destruction. Global demand for oil has traditionally been highly correlated to steady population growth of around 2% per year since the oil industry began.

However, due to high prices, demand fell due to efficiency gains in 1974-75 and between 1980 and 1984 (Fig. 1-9 to 1-12). Demand destruction was a novel concept in the oil industry because it had simply never occurred before.

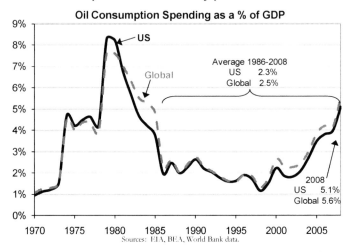

Fig. 1-7

Oil consumption efficiency, measured as the average number of barrels of oil consumed per average person per year, improved by advances in engine technology and lower horsepower engines. People began to drive lighter, less powerful vehicles.

Since 1987, engine technology has continued to progress, but these engine efficiency gains have been wiped out by the increased weight of vehicles, in particular due to the prevalence of large SUVs and more powerful engines. Although a less significant contributory factor in stagnating efficiency, the weight of passengers has also been steadily increasing (Table 1-3).

One positive way of looking at the decline in efficiency since 1987 is that the technology now exists to produce more efficient vehicles. Consumers must simply have an incentive to switch back to lighter lower horsepower vehicles rather than SUVs and light trucks.

Table 1-3

US non-Commercial Vehicle Statistics

	1975	1987	1998	2008
Fuel Economy (MPG)	13.1	22	20.1	20.8
Gallons per 10,000miles	763	455	498	481
Weight (lbs)	4060	3221	3744	4117
Horspower	137	118	171	222
0-to-60 time (sec)	14.1	13.1	10.9	9.6
Percent Truck*/SUV**	19%	28%	45%	48%
Percent Diesel	0.2%	0.3%	0.1%	0.1%
Weight of average adult (lbs)***	159	168	173	177

*trucks such as pickup trucks; **Sport Utility Vehicles
Sources: US EPA, ***CDC Survey data, author's calculations.

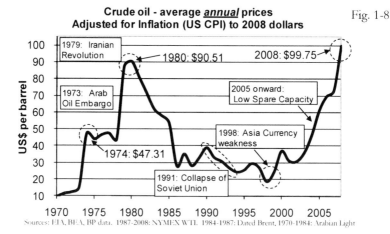

Fig. 1-8

Crude oil - average _annual_ prices
Adjusted for Inflation (US CPI) to 2008 dollars

Sources: EIA, BEA, BP data. 1987-2008: NYMEX WTI. 1984-1987: Dated Brent, 1970-1984: Arabian Light

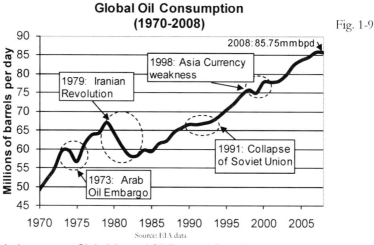

Fig. 1-9

Global Oil Consumption
(1970-2008)

Source: EIA data

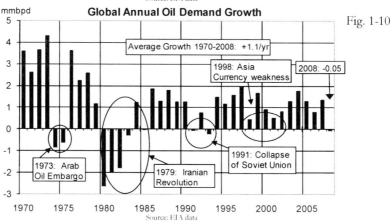

Fig. 1-10

Global Annual Oil Demand Growth

Source: EIA data

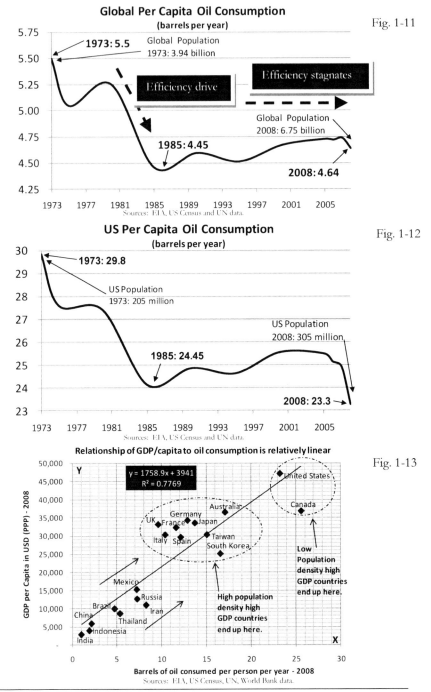

Global Per Capita Oil Consumption
(barrels per year)

Fig. 1-11

1973: 5.5
Global Population
1973: 3.94 billion

Efficiency stagnates

Efficiency drive

Global Population
2008: 6.75 billion

1985: 4.45

2008: 4.64

Sources: EIA, US Census and UN data.

US Per Capita Oil Consumption
(barrels per year)

Fig. 1-12

1973: 29.8

US Population
1973: 205 million

US Population
2008: 305 million

1985: 24.45

2008: 23.3

Sources: EIA, US Census and UN data.

Relationship of GDP/capita to oil consumption is relatively linear

Fig. 1-13

$y = 1758.9x + 3941$
$R^2 = 0.7769$

United States

Australia
Germany
UK France Japan
Italy Spain
Taiwan
South Korea

Canada

Low
Population
density high
GDP countries
end up here.

Mexico

Russia
Brazil
China Iran
Thailand
Indonesia
India

High population
density high
GDP countries
end up here.

GDP per Capita in USD (PPP) - 2008

Barrels of oil consumed per person per year - 2008
Sources: EIA, US Census, UN, World Bank data.

18

1980 – 1984: oil industry bust

In 1978, during the Iranian strike, Saudi production had rapidly and very briefly ramped up to 10.5 million bpd, well above its normal production pace. Total global oil production in 1978 was 67 million bpd. Subsequently, in order to keep prices from collapsing during the early 1980s as Iranian oil production resumed, and to bring production back to more optimal rates, Saudi Arabia cut its production back to 2 million bpd, which is a huge swing for any producer.

Most OPEC nations at this time had spare production capacity and they were supposed to hold oil off the market to support OPEC quotas. However, the more the Saudis cut back, the more other OPEC countries produced by cheating on their quotas. Iran and Iraq, now at war and requiring financing, were particularly lax in adhering to OPEC quotas.

Saudi Arabia became known as the swing producer for its singular attempts to manage prices. Tired of falling prices as a result of quota cheating by fellow OPEC member countries and the global efficiency drive by consumers, Saudi Arabia decided in August 1985 to stop selling oil at its official posted prices. The Saudis instead linked the sale price of their crude oil to open market prices for finished products such as gasoline and diesel, and increased their production from 2 to 5 million bpd. The new method of pricing, whereby crude oil is sold at a price linked to finished products like gasoline, was another form of netback pricing, linking an opaque (Saudi crude oil) price to another more transparent free-market price (retail gasoline and other product prices in the US and elsewhere).

The international market did not need the additional Saudi oil. Within a year, oil prices collapsed more than 70% to just over $15 by 1986, $28 in today's terms, and remained between $10 and $20 until 1990. The mid 1980s drop in prices effectively bankrupted the Soviet Union, as Russia was the number two oil producer behind Saudi Arabia, and nearly bankrupted the entire US oil industry. It has been alleged that the Saudis were instructed by the US government to increase supply in order to bankrupt the Soviets. However, the Saudis were genuinely hurting and were actually acting economically rationally in the short-term. Saudi Arabia was borrowing heavily to fund large welfare programs for a rapidly growing population and made a decision to increase production to save its economy. Oil prices halved, but Saudi production more than doubled, so total Saudi oil revenue increased.

The wild increases in Saudi production to 10.5 million bpd in 1978 during the Iranian revolution, and subsequent decrease to 2 million bpd a couple of years later led many to speculate that Saudi oil fields simply could not physically sustain high production rates for long periods of time.

If the Saudis can only sustain 10.5 million bpd output rates for short periods, then their current production, which once again reached these levels, is worth monitoring closely. Signs of strain may portend a sharp unavoidable pullback.

1984 – 1990: move to transparent oil pricing

Oil began to be traded on futures exchanges during the late 1970s. A futures exchange is a marketplace where one can buy or sell a commodity for delivery at a point in the future. Heating oil futures first traded on the New York Mercantile Exchange (NYMEX) in 1978, followed by gasoline futures in 1981, and crude oil futures in 1983, after US domestic crude oil prices were deregulated in 1981. In 1988 oil began to trade on the International Petroleum Exchange (IPE) in London. The IPE now trades electronically as the Atlanta-based ICE (IntercontinentalExchange).

The oil traded on these two exchanges (NYMEX and ICE) created price transparency between producers and consumers never witnessed before and finally removed oil price determination from the shady smoke-filled rooms where the somewhat artificial posted prices were set.

Heating oil, gasoline, and other finished product prices being openly quoted on futures exchanges enabled Saudi Arabia, followed by other large producers, to begin using refining margin netback pricing in 1984, linking the price at which they sold their crude oil to the price of finished products. For example, if a western oil refiner managed to sell gasoline, heating oil and the other products in its basket of finished products linked to futures prices for $50 per barrel, and the pre-agreed netback margin between the OPEC crude oil producer and the refiner is $10 per barrel, then the refiner pays $40 per barrel for the crude oil. In this way the revenue of the crude oil producer is more closely linked to the final market price for the refined petroleum products, and a refiner is guaranteed a profit margin.

By 1986 OPEC crude oil producers tired of guaranteeing western refineries a fixed profit margin at a time when there was a glut of spare refinery capacity. Refining margin netback pricing began to be replaced by crude oil formula netback pricing.

Table 1-4

International crude oil pricing mechanism timeline

Posted prices set by OPEC: 1973-1984	Formula netback pricing: 1986 →			
Posted prices set by Majors ended in 1973	Refinery margin netback pricing: 1984-1986			
1965	1975	1985	1995	2005

Crude oil formula netback pricing, which is the mechanism still in use today, links the price at which OPEC crude producers, and others, are willing to sell crude oil to an openly traded free market crude oil benchmark or a combination of benchmark prices.

Benchmark oil prices, also known as price markers, are oil prices set at the close of business each day on futures exchanges, such as the NYMEX or ICE futures exchanges. They also include prices assessed daily by oil trade journals Platts and Petroleum Argus, among others (Table 1-5).

Table 1-5

Oil benchmark sources for formula pricing

Futures exchanges	Trade journals
1. New York Mercantile Exchange (NYMEX)	1. Platts
2. IntercontinentalExchange (ICE)	2. Petroleum Argus

Platts and Petroleum Argus are the two most widely used oil trade journals. They assess prices during a window of time at the end of each business day for hundreds of grades of oil at various locations around the world based on spot market (immediate delivery) trading in physical oil at those locations. Oil traders and their brokers report to these journals in real time during the daily time window, via phone or internet instant messenger, the price and quantity of any trades they have transacted, or are willing to, transact.

Some of the prices the journals assess are more commonly referenced in oil contracts than others because they are based on spot physical oil markets in more actively traded locations such as Singapore, Dubai, the Netherlands, New York harbor, the US Gulf Coast, and Los Angeles.

As a real example of formula pricing using a benchmark price, the sales price for Saudi Arab Light crude is set to ICE Brent crude futures weighted average settlement plus or minus a differential. The differential used in the formula is set each month by Saudi Aramco, the national oil company of Saudi Arabia. Formula pricing, when used by an NOC is referred to as an Official Selling Price, or OSP. By linking their sales price to a benchmark price, the Saudis, and other crude oil producers, ensure that their oil is competitively priced against openly traded international market prices and they can sell all of their production quickly.

Formula netback pricing using benchmarks is used in the wholesale international market between large buyers and sellers of oil. The subsequent price at a retail pump then differs around the world mostly due to government taxation and subsidies (Table 1-6).

Table 1-6

	Retail Gasoline Prices		
	(January 2009)		
	US$ per Liter	US$ per US Gallon	Population Density (Pop per km²)
Portugal	1.77	6.69	114
Netherlands	1.74	6.59	392
Belgium	1.61	6.11	341
Finland	1.60	6.06	16
Germany	1.57	5.95	232
Italy	1.57	5.95	193
United Kingdom	1.55	5.88	246
France	1.55	5.85	110
Denmark	1.54	5.85	126
Norway	1.53	5.78	12
Slovakia	1.49	5.65	110
Ireland	1.41	5.32	59
Sweden	1.33	5.02	20
Austria	1.31	4.95	98
Switzerland	1.29	4.90	176
Poland	1.28	4.83	123
Czech Republic	1.27	4.82	130
Luxembourg	1.27	4.80	180
Hungary	1.25	4.74	109
Spain	1.24	4.69	88
Japan	1.21	4.56	339
Greece	1.20	4.53	84
Slovenia	1.16	4.37	97
India	1.15	4.35	336
Estonia	1.13	4.26	29
Latvia	1.12	4.22	36
Lithuania	1.08	4.10	53
China	0.80	3.02	137
New Zealand	0.78	2.97	15
Australia	0.75	2.84	98
Russia	0.68	2.59	8
Nigeria	0.60	2.27	142
Canada	0.59	2.24	3
Mexico	0.55	2.07	55
United States	0.44	1.66	31
NYMEX Gasoline	0.30	1.14	0
NYMEX Crude	0.30	1.13	0
Saudi Arabia	0.12	0.45	11
Venezuela	0.03	0.11	29

In developed countries higher population density means public has a greater ability to use mass transit which lowers dependence on oil. Governments then find it easier to tax oil in these densely populated nations.

Sources: U.S. Department of Energy, Japanese Oil Information Centre, U.K. Automobile Association,
Australian Institute of Petroleum, Mexican Banco de Información Económica,
The Oil Information Center Japan, Moscow Fuel Association, CNN, Reuters.
Population Density from the United Nations.

1990 - Today: excess capacity disappears

The invasion of Kuwait by Iraq in 1990 caused crude oil prices to more than double in a few days from $15 to $33 per barrel. The release of crude from the US SPR in addition to an extremely quick end to the war caused the price to collapse just as quickly to under $20.

Apart from a brief period in the middle of the 1990s, crude oil prices remained subdued for most of the decade with OPEC spare capacity (Fig. 1-14) keeping a lid on the market. The low point of around $10 per barrel in 1997 due to a fall off in demand during the Asian financial crisis was followed by OPEC production cuts, in addition to cuts by non-OPEC nations Mexico and Norway, to try to rescue prices.

By 2005, demand finally caught up with OPEC's 35 years of managing excess capacity and, apart from a small amount of difficult to refine low quality heavy crude, for the first time since the oil industry began in 1859 there was effectively no buffer against unexpected crude oil supply outages.

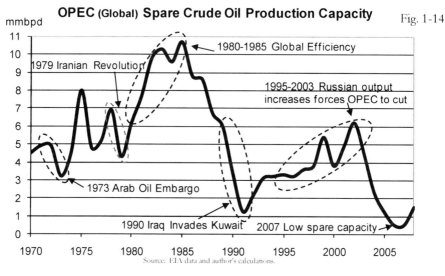

OPEC (Global) Spare Crude Oil Production Capacity   Fig. 1-14

Source: EIA data and author's calculations.

Despite a huge increase in exploration and drilling programs between 2000 and 2008, OPEC seemed unable to increase production capacity. Instead of asking whether OPEC could increase production to keep pace with increasing demand, the question became: could OPEC producers, particularly its leading member Saudi Arabia, even manage to sustain current production levels?

Saudi Arabia's large fields, purported spare capacity, and ability to increase production were almost entirely based on dry land. However, between 2004 and 2008 offshore dayrates for jackup drilling rigs suddenly soared from

US$30,000 per day to over US$150,000 per day due to demand from Saudi Arabia. The only time a crude oil producer goes offshore is when onshore production opportunities have been maximized. Offshore exploration is much more risky and production costs are four to five times higher than onshore. The Saudis going offshore in a rapid manner indicated that they may be struggling to maintain current onshore production rates, and with it, the world's only remaining spare capacity.

A lack of spare capacity does three things. Firstly, individuals and organizations build a security premium into the price of oil as supply disruptions are now more likely. Secondly, the volatility of oil price movements increases as a disruption to supply cannot be compensated for by any crude oil producer. Thirdly, and most importantly, it means that crude oil producers cannot respond to high prices by increasing production and inventories.

Fortunately for consumers, a strong resumption of non-OPEC Russian oil production, which had collapsed after the fall of the Soviet Union, managed to temporarily make up the difference which other non-OPEC and OPEC producers could not meet. However, Russia is also now struggling to increase daily production capacity.

The 1970s oil shocks were driven by temporary events which oil consumers believed and hoped would be over quickly. What is so worrying about today and beyond is that the lack of spare crude oil production capacity is not a temporary concern resulting from an embargo or war; it is instead due to the inability of oil market supply to keep pace with steady demand growth. Oil has simply not been found to increase production.

A side effect of the price increases beginning in 2003 caused by the lack of spare capacity was a renewed trend toward resource nationalism which had lain dormant since the 1970s. The governments of Russia, Venezuela, Peru and several other nations began to increase taxes and seize IOC facilities or at least make it much more difficult for them to operate by suddenly levying heavy fines. Nationalized oil companies have traditionally been worse at spending oil revenue on exploration compared with private industry and this will exacerbate the long term decline in discoveries.

In addition to a lack of spare crude oil production capacity, in the early 2000s, there was only a small amount of spare refinery capacity globally (Fig. 1-15). The reason very little refinery capacity had been added by the oil industry was that a huge glut of refining capacity had existed since the 1970s and because of this refiners struggled to make a profit.

After twenty years of low profitability, it was only between 2002 and 2007 that refineries made decent profits as demand finally caught up with available worldwide refinery capacity.

Fig. 1-15

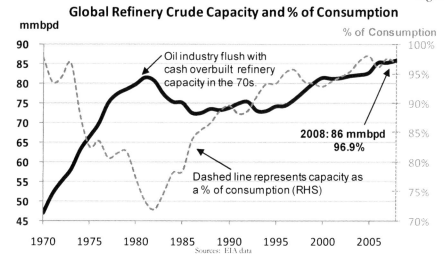

Global Refinery Crude Capacity and % of Consumption

Sources: EIA data

A large number of new refineries built around the world are now coming online, in addition to capacity and complexity expansion at existing plants, creating spare refinery capacity once again.

While the issue of a lack of spare refinery capacity quickly disappeared, the inability of the oil industry to increase crude production may not.

The Future: what next?

Annual global crude oil discovery rates peaked in the 1960s and have been steadily declining since in spite of huge advances in exploration technology. The decline in discoveries has been put succinctly by veteran oil analyst Charley Maxwell:

> "In 1930, we found 10 billion new barrels of oil in the world, and we used 1.5 billion. We reached a peak in 1964, when we found 48 billion barrels and used approximately 12 billion. In 1988, we found 23 billion barrels and used 23 billion barrels. That was the crossover when we started finding less than we were using. In 2007, we found perhaps 6-7 billion, and we used 31 billion. These numbers are just overwhelming."

As a result of this long-term slide in discoveries, global daily production rates of conventional crude oil began to struggle to meet demand in 2005. We are witnessing the beginning of the end for oil. Even in the most extreme optimistic scenario, conventional oil production will effectively cease to exist well before the end of this century - a fact even the most optimistic oil company agrees with.

As oil prices rally in the coming years, free market innovation and efficiency will occur. Governments and electorates without an understanding of oil markets and history will call for taxation, price controls, and consumption rationing, forgetting the painful lessons of those countries imposing such policies in the 1970s.

The two major challenges with finding a replacement for oil are energy storage and sourcing energy to store. As oil is primarily a transportation fuel, with very little used to generate electricity, the goal of any replacement has to be to produce and safely store energy in a lightweight form which can reasonably fit on a vehicle.

Storage of energy for transportation may involve hydrogen. The trouble with hydrogen is that it is very difficult to compress large volumes of the element into a form which is as energy dense as liquid oil. Storage technology is in its infancy and considerable efforts are going into finding solutions.

While hydrogen may be used as a store of energy, it doesn't naturally exist by itself in significant volumes. Hydrogen is usually found bonded to elements, such as with oxygen to form water. Large amounts of energy are required to break, by splitting water for example, hydrogen's bond with other elements. That energy has to come from somewhere.

Current nuclear technology, wind, and solar power are frequently mentioned as sources of energy with the potential to replace oil. However, none of them, even all together as a portfolio, can come close to replacing more than a small portion of the energy in our current rate of oil use, which was created by millions of years of solar energy.

The scale of oil consumption today is truly massive. For instance, in order to produce sufficient nuclear energy to completely offset modern oil use, one would have to build an additional 4,000 1.5 gigawatt (GW) nuclear power stations globally. Today there are approximately 440 reactors around the world with a combined capacity of 363GW. With 4,000 new nuclear reactors using current technology, all known uranium reserves would be depleted in just over

10 years. Breeder reactors, which are a more efficient method of using uranium, are not practical at present.

Wind and solar generation face the same scale problem as nuclear power. The number of wind turbines and solar arrays one would need to build makes them impractical except as a minor part of a portfolio of solutions.

Oil sands, oil shale, ethanol and biodiesel are also often mentioned as alternative sources of energy to replace oil, but unfortunately none of these, with current technology, can realistically become large scale alternatives for conventional oil.

Oil sands are produced by separating extremely heavy crude oil from sand in a process similar to mining. Even with maximum investment in Canada and Venezuela, the two major global sources of oil sands, it will only account for 4 to 5 million bpd (under 3% of forecasted oil consumption) of oil by 2025. Five million bpd coming from oil sands by 2025, while nothing to be scoffed at, is immaterial as a long-term global oil supply solution. Oil sands production also requires inexpensive natural gas or nuclear energy to provide heat to melt oil out of the sand. If oil is expensive and scarce in the future then so will natural gas and nuclear energy.

The amount of energy consumed to produce energy is called the energy return on investment (EROI). In other words, the energy spent on drilling, transporting, refining, storing and retailing oil versus the energy the resulting oil actually contains. Conventional petroleum currently has an estimated EROI of around 15 to 1 – such that for every unit of energy required to get oil to a consumer, the resulting oil contains 15 units of energy. In the 1970s the EROI was over 25 as the more obvious land-based oil fields were discovered first, leaving the vastly more expensive deep offshore and arctic areas for today's oil prospectors. The EROI on oil sands is less than 4 to 1, because of all the energy, primarily in the form of natural gas, required to melt the oil out of the sand. The closer one gets to an EROI of 1, the closer one gets to net wasting energy to produce useable energy.

Oil shale, which is organic material known as kerogen never buried deep enough to have formed oil, is known to be impractical. It takes more energy to produce oil from shale than the energy the produced oil ultimately contains – the process has an EROI of less than 1. Oil shale is a fundamental non-starter, and the science, which is quite simple, is well known, and no amount of technology can make it practical.

Ethanol and biodiesel are crop-based fuels produced in the US from corn and soybeans, respectively. Brazil, because of its favorable climate and cleared rich rainforest soil, produces ethanol from inexpensive sugarcane. No other country can replicate Brazil's sugar crop on a similar scale because of its unique climate and soil.

Crop-based fuels face serious challenges as a replacement for oil. Corn is the number one source of food calories in the US. If every bushel of corn grown in the US were converted into ethanol, it would only offset 6% of the energy in current total US oil consumption. Soybeans used to produce bio-diesel face a similar limit when production quantities are scaled up. The crop-based fuels debate brings to mind a quote from Ludwig von Mises, an Austrian economist:

> "It may sometimes be expedient for a man to heat the stove with his furniture. But he should not delude himself by believing that he has discovered a wonderful new method of heating his premises."

Second generation fuels produced from non-food crops including switch grass and other cellulosic materials are being researched. The hope is that in the future these may reduce reliance on food crops to produce fuels.

Natural gas could potentially offset conventional oil production rate declines for a brief time. Global natural gas production rates are not expected to fall for at least ten years. The technology to use natural gas as a transportation fuel is well known and natural gas is burned to power vehicles in many urban areas around the world. Shipping natural gas and storing it is currently much more expensive than conventional oil, but at least it is an available proven technology.

As global natural gas production rates are expected to begin to decline within twenty years, it can only be a temporary solution to offsetting decreasing conventional oil production rates.

Producing natural gas from an unconventional source called methane hydrates, which are possibly buried under the earth in large quantities, could potentially be a large source of energy. However, the realm of methane hydrates is highly speculative at the moment and also potentially environmentally disastrous as methane, if accidentally released into the atmosphere (i.e. not burned first) is a much worse greenhouse gas than carbon dioxide.

Global coal reserves are quite large. Producing liquid hydrocarbon fuel from coal is a known but very expensive and highly polluting technology made infamous in a desperate way by Nazi Germany and Apartheid South Africa. The EROI on producing liquid oil from coal is under 1, however due to the sheer abundance of coal, many will think it convenient to waste energy to produce liquid oil from coal so it can be used in transportation. Coal to liquids

(CTL) technologies could, therefore, possibly offset declines in oil production for several decades if crude oil were, all other costs being unchanging, to rally to several hundred dollars per barrel on a sustained basis in order to cover the high economic and environmental costs of production.

There are no easy solutions to replacing oil as a liquid transportation fuel. A portfolio of solutions will likely be implemented, but each of them has major disadvantages and limitations. Eventually we need to make a leap to another, as of yet untamed, source of energy.

Long-term money, in the form of billions from US and EU taxpayers, among others, is betting on thermonuclear fusion which is the process occurring within the sun, not to be confused with current nuclear reactor fission. The road to practical nuclear fusion is full of technological obstacles. The technology may not leave research labs until the next century, if ever. One common quip about the promise of commercial nuclear fusion is that it has been 40 years in the future in each of the past 60 years. If this power source becomes practical, then it could be used to generate electricity which could be used to split sea water to produce hydrogen which could be burned as a fuel in transportation engines.

One way or the other, in the 22nd century there will be transportation energy. It may be much more expensive, perhaps not. It is almost certain that the source of that energy will not be conventional petroleum.

CHAPTER TWO
A CRUDE OIL ASSAY

Crude oil is literally crude. To be used, it must be processed in a refinery to separate out individual finished products including gasoline, diesel, heating oil, jet, and residual fuel.

There are several hundred grades of crude oil produced today. Crude is not always black, and can be straw colored or brown. The viscosity of crude can vary from water-like to a near-solid. Crude oil from different reservoirs can contain varying ratios of undesirable elements such as sulfur, nitrogen, water, metals and sediment, and when refined, different crudes may yield vastly differing quantities of finished products.

The characteristics of an individual grade of crude oil are defined in an analysis called an assay prepared on behalf of a crude producer for sales purposes. An assay outlines properties of a crude oil important to a refinery, particularly the expected yields of various finished products, density, sulfur content, acidity and viscosity (Table 2-1).

Crude oil which is not within its tight range of assay specifications is known as off spec crude, and can result in expensive delays when delivered to a refinery. Off spec crude may have to be shipped to another refinery that can accept those specifications, or may have to be blended with other crudes to bring it within acceptable specs. A tanker containing oil which has no ready buyer is referred to as a distressed cargo, and often trades at a steep discount.

The first thing an assay will generally mention is the reservoir, oilfield or location from which the crude originated.

A crude oil reservoir does not resemble an underground lake; instead, an accumulation of crude oil is contained between grains of sand or within tiny pores inside an otherwise solid rock matrix, like a rigid finely perforated sponge.

An oilfield is an area containing a single reservoir or group of reservoirs related to the same geological structural feature. Crude oil can come from a single well which taps into a field or more typically from a group of wells drilled into the field. Crude oil from an individual well tends to have relatively stable properties, although these may change very slowly over time.

Sample crude oil assay
(Bryan Mound Sweet Crude)

Table 2-1

Specific Gravity, 60/60° F	0.8454	Ni, ppm	3.41	RVP, psi @ 100° F	5.28
API Gravity	35.9	V, ppm	4.12	Acid number, mg KOH/g	0.10
Sulfur, Wt. %	0.33	Fe, ppm	0.822	Mercaptan Sulfur, ppm	7.26
Nitrogen, Wt. %	0.111			H₂S Sulfur, ppm	5
Micro Car. Res., Wt. %	2.21	Org. Cl, ppm	0.3	Viscosity: 77° F 6.99 cSt	
Pour Point, °F	25	UOP "K"	11.96	100° F 4.666 cSt	

Fraction	Gas C₂-C₄	1	2	3	4	5	6	Residuum	Residuum
Cut Temp	C₂-C₄	C5-175°F	175°-250°F	250°-375°F	375°-530°F	530°-650°F	650°-1050°F	650°F+	1050°F+
Vol. %	1.9	7.0	8.2	14.1	16.8	12.5	28.8	39.6	10.8
Vol. Sum %	1.9	8.8	17.0	31.1	47.9	60.4	89.2	100.0	100.0
Wt. %	1.3	5.5	7.2	12.9	16.5	12.7	31.1	43.9	12.8
Wt. Sum %	1.3	6.8	14.0	26.9	43.4	56.1	87.2	100.0	100.0
Specific Gravity, 60/60° F		0.6747	0.7391	0.7774	0.8275	0.8604	0.9143	0.9371	0.998
API Gravity		78.2	60.0	50.5	39.5	33.0	23.3	19.5	10.4
Sulfur, Wt. %		0.0013	0.0018	0.0113	0.07	0.25	0.51	0.65	0.98
Molecular Weight		96	111	134	185	245	403		
Hydrogen, Wt. %		15.88	14.73	na				12.91	10.82
Mercaptan Sulfur, ppm		3.6	8.8	27.8	19.4				
H₂S Sulfur, ppm		<0.1	<0.1	<0.1	<0.1				
Organic Cl, ppm		4.1	1.0	0.1	<0.1				
Research Octane Number*		69.9	62.4	46.7					
Motor Octane Number*		67.5	60.0	44.8					
Flash Point, °F				77	172	246	301		
Aniline Point, °F			123.0	143.2	163.0	194.1			
Acid Number, mg KOH/g					0.04	0.10			
Cetane Index					45.5	51.0			
Diesel Index				62.1	56.6	53.7			
Naphthalenes, Vol %					4.83	10.24			
Smoke point, mm					19.9	15.6			
Nitrogen, Wt. %					0.0006	0.010	0.154	0.276	0.572
Viscosity, cSt 77°F					2.537				
100°F					1.990	5.691			
130°F						3.814	39.07	109.5	
180°F							14.77	32.12	2923
210°F									920.6
275°F									143.5
Freezing Point, °F					-28.1				
Cloud Point, °F						31.1	105		
Pour Point, °F						27.0	101	85	
Ni, ppm								7.66	25.8
V, ppm								9.29	31.4
Fe, ppm								6.41	21.6
Micro Car. Res., Wt. %								5.00	17.25

Source: DOE

Most oilfields produce less than 100,000 barrels per day (bpd) and of these a large number are marginal fields producing just 10 bpd or less. Approximately 60% of daily global production is concentrated in just 317 super giant, and giant, elephant fields (Table 2-2).

Global oilfield statistics

Table 2-2

		Reserves*	Number of fields
Elephant Fields	Super Giant	> 5 billion barrels	54
	Giant	0.5-5 billion barrels	263
	Large	50-500 million barrels	481
	Others	< 50 million barrels	70,000+

Source: IEA World Energy Outlook 2008 at pages 224-226; * 2P reserves.

Production from wells on an individual oilfield is combined for pipeline transportation to form a stream of crude (Fig. 2-1). Subsequently, streams from different pipelines are often combined to create a blend. In addition to blending to save on pipeline transportation costs, crude oil streams are blended to reduce extreme characteristics of individual crude streams such as high sulfur content or acidity.

Fig. 2-1

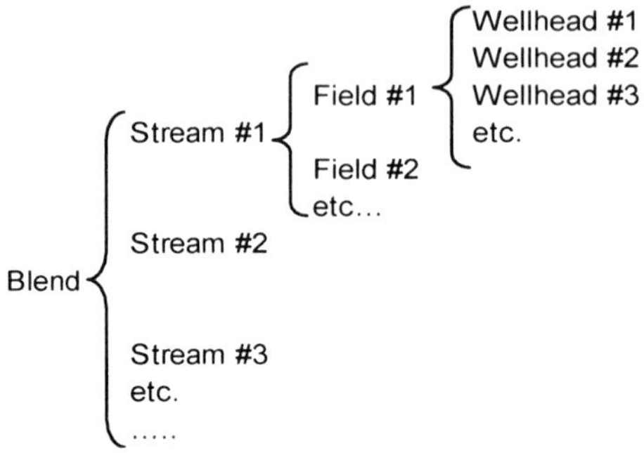

```
                                                    ┌ Wellhead #1
                                                    │ Wellhead #2
                                      ┌ Field #1 ┤ Wellhead #3
                         ┌ Stream #1 ┤           └ etc.
                         │            │ Field #2
                         │            └ etc...
                         │ Stream #2
                 Blend ┤
                         │ Stream #3
                         │ etc.
                         └ .....
```

When marketing a crude oil to refineries, producers frequently give it a name which references one of the component fields of a blend, or a nearby location. For example, Brent Blend is from the Brent stream, among others, in the North Sea and Kirkuk Crude is named after the nearby Kirkuk City, Iraq.

The most well known oilfield in the world is the Ghawar super-giant oilfield in Saudi Arabia, discovered in 1948. At its height it produced close to a whopping 6 million bpd. Although the field has been drained for such an unusually long period of time, it continues to be the largest in production today by a hefty margin.

Density is the most important physical characteristic of a crude oil mentioned in an assay as it gives an indication of the hydrocarbon molecules the crude oil contains and thus the products the crude oil will yield when refined.

Heavy crudes are denser because they contain larger hydrocarbon molecules containing more atoms than light crudes. In general, less dense, or lighter, crude is more valuable as it will readily yield more high value lighter products such as gasoline.

Density of crude oil varies with temperature and pressure, which, in addition to being relevant for indicating the potential refinery yield of finished products, has significant implications for storage and transportation. For example, crude oil loaded on a tanker in a cold climate, although it weighs the same, will occupy a larger volume within the same tanker when it arrives in a warmer climate, as it has become less dense.

The three measurements of density used for crude oil are metric density, specific gravity, and API gravity (Table 2-3).

Table 2-3

Comparison of density measurements (at 60°F and 1atm)

API Gravity °API	Specific Gravity (relative density)	Metric Density Kg/M³	Barrels per Metric Tonne
0	1.076	1076	5.93
10	1.000	1000	6.35
20	0.934	934	6.77
30	0.876	876	7.19
40	0.825	825	7.64
50	0.780	780	8.06
60	0.739	739	8.48
70	0.702	702	8.90
80	0.669	669	9.32
90	0.639	639	9.74
100	0.611	611	10.16
110	0.586	586	10.58
120	0.563	563	11.00
130	0.541	541	11.42
140	0.521	521	11.84
150	0.503	503	12.26
160	0.485	485	12.68

Density of water.

API gravity is the density measurement used most often for oil.

Metric density of oil is expressed in kilograms per cubic meter (kg/m³) measured at 15°C and 1 atmosphere of pressure.

Specific gravity and API gravity are more commonly used than metric density for crude oil and measure density of the oil in relation to the density of water using a simple instrument called a hydrometer (Fig. 2-2).

A hydrometer, in a practical application of the Archimedes principle, is used to measure the density of oil by the amount of water displaced by an oil sample. The hydrometer is made of glass with lead shot at the bottom to weigh it down when inserted into the oil sample. A thermometer is often an integral part of the measuring device, in which case it is called a thermohydrometer. The density of the oil can simply be read from lines on the hydrometer and a table can be referred to adjust for any deviation in the sample temperature from 60°F, which is the most common temperature at which crude oil density is measured. Hydrometers are occasionally carried by motorists and pilots to test fuel in countries where unscrupulous suppliers may add inexpensive materials to expensive petroleum products such as jet fuel and gasoline, which have precise densities that can be measured to confirm the fuel has not been tampered with.

Fig. 2-2

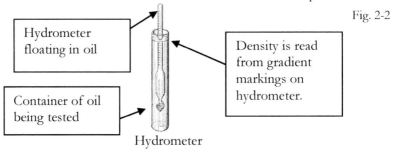

Hydrometer floating in oil

Container of oil being tested

Density is read from gradient markings on hydrometer.

Hydrometer

Specific gravity, also known as relative density, is the density of oil divided by the density of water. Water at 60°F and 1 atmosphere of pressure is assigned a specific gravity of 1. If oil has a specific gravity of less than one then it will float in water; otherwise, it will sink.

API gravity is simply specific gravity set to an index and is the most commonly used measure of crude oil density (Tables 2-3, 2-4 and 2-5). In 1921, the American Petroleum Institute (API) created the index ranging from 0° to 100°, 0° being the heaviest and 100° the lightest. Although API gravity was developed for crude oil, API gravities can be applied to very light hydrocarbon products of crude oil, such as propane, which can be above 100° API.

Table 2-4

Degrees API Gravity   = (141.5/Specific Gravity) – 131.5
example (WTI crude): 39.6° API = (141.5 / 0.827) - 131.5

Water has a density of 10° API. Bitumen, a heavy product of crude oil used for road paving, has an API gravity of between 5° and 9° API and will sink in 60°F water. Gasoline has a density of around 50° API which will float in 60°F water.

Crude oils with densities in the 30° to 39° API range are the most commonly produced and in the highest demand, as most refineries are configured to handle these grades of crude. Refining higher API crude oils will more easily generate a larger proportion of gasoline and other high value products, and very little heavy products such as residual fuel oil and bitumen.

West Texas Intermediate (WTI) crude, the grade traded on the New York Mercantile Exchange (NYMEX), has a density of 39.6 degrees API and is referred to as an intermediate, or medium, density crude. Brent Blend crude, the grade traded on the London based ICE futures exchange, is also an intermediate crude. Brent deliverable on the ICE has a slightly heavier density of 38.3° API, which is falling as a stream from the heavy North Sea Buzzard field is added to the blend. Some grades of oil are called "light" as a marketing tool, but are actually medium density. For example, Saudi Arabia's Arab Light crude, most of which comes from the Ghawar super-giant oilfield, has a density of 34° API.

Table 2-5

API density classifications

Crude oil density classification	API gravity
1. Condensate/Extra-light	>50°
2. Light	40-50°
3. Intermediate/Medium	**30 - 39°**
4. Medium-heavy	25-29°
5. Heavy	< 25°
6. Extra-heavy	<10°

Intermediate density crudes are most commonly produced.

Distillation profile is closely related to density, and shows the ratios, known as cuts, or fractions, of products which crude evaporates into at various True Boiling Points (TBP) ranges. The assayed volume evaporating in each TBP range will give refineries an idea as to the amount of each finished product the crude oil will yield.

An assay will show results of tests performed on a sample of the crude oil as a whole in addition to specific tests applied to individual distillation profile TBP ranges.

Sulfur content lowers the value of crude oil. Sulfur reduces the energy content of crude oil by displacing hydrocarbon molecules. Sulfur also corrodes metal piping and tanks in producer and refining facilities. It becomes a pollutant when burned, and sulfur in tailpipe exhaust damages catalytic converters.

Crude oil can be referred to as being sweet, low in sulfur, or sour, high in sulfur (Table 2-6). The term sour is in reference to the distinct rotten-egg smell of sulfur-laden crudes.

Crude oil sulfur content (% by weight)

Table 2-6

Sour	> 1.5
Medium sour	0.5-1.5
Sweet	< 0.5

At extremes, crudes tend to be either light-sweet or heavy-sour as sulfur binds more easily to heavy complex hydrocarbon molecules, and is therefore found in heavier crudes moreso than lighter crudes (Fig. 2-3). For the same reason, products with smaller molecules, such as gasoline, diesel and jet fuel have lower levels of sulfur than heavier products such as residual fuel oil and bitumen.

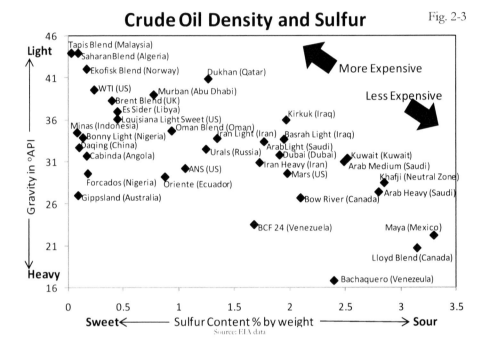

Crude Oil Density and Sulfur   Fig. 2-3

Source: EIA data

Two crudes often referred to as being sweet are WTI crude, which has a sulfur content of around 0.24%, and Brent Blend crude which has a slightly higher 0.37% sulfur content by weight. Other examples of sweet crudes are Nigerian Bonny Light, Saudi Arab Light and Malaysian Tapis crudes. Examples of sour crudes are UAE Dubai, Saudi Arab Heavy, US Mars, US Poseidon and Mexican Maya crudes.

The proportion of sour crude production in the world has been increasing over recent years as a larger ratio of less valuable heavy crude is being produced (Fig. 2-4). The US in particular has been able to cope with the larger amount of heavy sour crude as it has more complex refineries capable of processing such crude compared with the rest of the world.

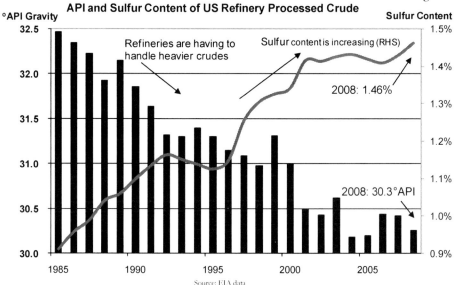

Due to environmental controls, refineries have to remove sulfur from finished products using expensive hydrotreater units. Refineries that lack hydrotreater capacity may not be able to handle crude above a certain sulfur content.

The sulfur compounds found in crude oil include mercaptans, hydrogen sulfide, and thiophenes. Although refineries attempt to reduce and remove all sulfur compounds from finished products, mercaptans, chemically known as thiols, receive special additional treatment as they have a distinct rotten egg odor which is unpleasant to consumers. Mercaptans in finished products such as gasoline are neutralized in a refinery unit known as a sweetener.

The smell of mercaptans is so powerful that very small amounts are used as a safety additive to normally odorless natural gas so that gas leaks can be easily detected by consumers.

Acids accumulate as a waste product of the biodegradation of crude oil by bacteria in a reservoir. Since bacteria metabolize lighter oil molecules more easily, acidic crudes tend to consist of the remaining heavy hydrocarbon molecules bacteria find more challenging to digest.

In addition to generally being heavier, highly acidic crude oils are cheaper than less acidic crude oils. The acid must be neutralized before it corrodes steel pipes used in transportation and refineries.

Acid content in crude is measured by a Total Acid Number (TAN), referred to as the Neutralization Number, equal to the milligram weight of potassium hydroxide (KOH) needed to neutralize a gram of crude oil. Most refineries are set up to run crude with a TAN under 0.5. Highly acidic crude oils are those with a TAN greater than 0.7.

Table 2-7

> Highly acidic crude oil → TAN > 0.7

Examples of highly acidic crude oils are Chad's Doba, Angolan Hungo Stream and Angolan deepwater Kuito as well as crude from the North Sea Alba field. Chad's Doba crude, for example, has a TAN above 2.0. In addition to using potassium hydroxide, refineries sometimes blend high TAN crudes with lower, more neutral, TAN crudes to reduce the risk of corrosion.

Viscosity defines how easily a liquid resists flowing. For example, honey has high viscosity and water low viscosity. In general, highly viscous material is comprised of large molecules which tangle as they pass each other. Heavy crude oil, which is comprised of larger hydrocarbon molecules is, therefore, highly viscous.

When oil is heated, long hydrocarbon molecules which tangled at lower temperatures stretch out and do not catch on each other as much. Because viscosity changes with temperature, it is frequently measured at both 40°C and 100°C.

Highly viscous crude oil is commonly heated or blended with less viscous crude oil, or low viscosity finished products such as kerosene in order to flow through a pipeline.

Viscosity of crude oil in a reservoir is measured in poise (P). A poise is a unit of absolute viscosity – absolute, in that no other force, such as gravity or capillary action, is involved in the measure. Absolute viscosity is also referred to as dynamic viscosity.

Outside of reservoir conditions, most viscosity measuring devices rely on timing oil falling due to the force of gravity – known as kinematic viscosity - through a calibrated hole in a testing device.

The most commonly used kinematic viscosity measurement unit for crude oil once it has been removed from a reservoir is the centistoke (cS or cSt), which is also used as a key reference for heavy oil products, such as residual fuel oil. A stoke is a kinematic viscosity of $1 cm^2$/second.

Crude oil, for example, with a gravity of 36° API may have, for example, a kinematic viscosity of 3.5 cSt at 40°C and a kinematic viscosity of 1.50 cSt at

100°C. Water at 20°C (68°F) has a kinematic viscosity of approximately 1 cSt and 0.65 cSt when heated to 40°C (104°F).

As kinematic viscosity is the ratio of absolute viscosity to density, if one knows the specific gravity of the oil being tested, one can flip back and forth between kinematic viscosity and absolute viscosity using the formula: Kinematic Viscosity (cSt) = Absolute Viscosity (cP) / Density (specific gravity).

Kinematic viscosity is measured with a viscometer. There are generally three types of viscometers used in the oil market: Redwood, Engler, and Saybolt. Each of these viscometers is used to measure amount of time oil takes to pass through a calibrated hole.

Redwood viscometers are used to measure the number of seconds it takes for 50ml of oil to pass through the measuring device. The result is shown in units called redwood seconds.

Engler viscometers measure the time it takes 200 ml to pass though a measuring device. The time is then compared to the time it takes the same amount of water to pass through the device. The result is shown in engler degrees.

Saybolt viscometers measure the number of seconds it takes for a 60 cc oil sample to move through the device. There are two types of Saybolt viscometer:

Saybolt universal viscometers are used for light low viscosity oils. Results are referred to as Saybolt Universal Seconds (SUS) or Saybolt Seconds Universal (SSU.)

Saybolt fural viscometers are used for heavy high viscosity oil products, such as motor oil and transmission oil with viscosities greater than 1,000 SUS. The word Fural was created from FUel and RoAd oiL. The results are referred to as Saybolt Seconds Fural (SSF)

Pour point is related to viscosity and is the lowest temperature at which crude oil behaves as a fluid and thus can be pumped easily. Any colder than the pour point and oil will not flow. Pour point is often measured at 5°F above the temperature at which the oil shows no surface movement when inclined for 5 seconds.

Characterization (K) factors use specific gravity and average boiling points to roughly characterize a crude oil as either paraffinic or aromatic in a single number. Paraffins and aromatics are the two most common types of hydrocarbon molecular structures found in crude oil. Two slightly different measurements known as the Watson Characterization (K) factor and the UOP Characterization (K) factor are used. A K-factor less than or equal to 10 indicates an aromatic crude. A K-factor greater than 12 indicates paraffinic crude, which is more valuable than aromatic.

Vapor pressure, also known as volatility, indicates how readily the entire crude oil and certain products of the crude evaporate. The volatility measure used for oil is Reid Vapor Pressure (RVP) in pounds per square inch (psi.) RVP quantifies the pressure which vapors from the oil exert. Crude oil with a high RVP indicates that the crude should produce a larger amount of light valuable products, such as gasoline and diesel. Although a higher RVP for crude oil is desirable, an RVP which is too high is not, as RVP of gasoline is limited by environmental regulations. Evaporating gasoline is a major cause of smog.

Nitrogen content of crude oil is important for refineries producing products meeting environmental restrictions. Nitrogen Oxides (NOx), produced when oil products burn, are a key pollutant which environmental regulators monitor and limit. Nitrogen can also poison catalysts used in refinery processes.

Carbon content is an indicator of a crude oil's suitability for coke production. Coke is a solid coal-like product. Three tests, the Ramsbotton test, the Conradson Carbon Residue (CCR) test and the Microcarbon test are used to define the amount of carbon by weight in a crude oil. Conradson carbon values between 0.5 and 1.0 are typically acceptable to a refinery.

Asphaltenes are the heaviest aromatic hydrocarbon molecules contained in oil and are responsible for the dark color of crude as the large molecules absorb light. High levels of asphaltene in crude oil can clog pipes and cause pumping difficulties. A high level of asphaltenes also indicates that the crude oil will produce large amounts of undesirable shot coke in a refinery coking unit.

Salt in crude can lead to corrosion of steel piping in a refinery and storage tanks. Crude must usually be de-salted before it is processed by a refinery.

Metals and other elements commonly found in crude oil are nickel, iron, vanadium, silver, mercury, sodium, and calcium. These elements, even in very small quantities, can interfere with catalytic reactions at a refinery.

Basic Sediment and Water, referred to as BS&W, is a catch all category for any water, dirt or junk brought up with the crude. Less than 1% BS&W by weight is desirable for most refineries.

Crude oil almost always contains water when produced. Crude can comprise well over 10 barrels of water, called the water cut, for every barrel of crude produced before dewatering. Water is removed in a dewatering plant close to the producing well, as it is uneconomical to haul the water along with crude oil in an oil tanker or using valuable pipeline capacity.

CHAPTER THREE
COMPONENTS OF OIL LIQUIDS

Total worldwide oil liquids production is currently approximately 85 million barrels per day (mmbpd) and is comprised of several components.

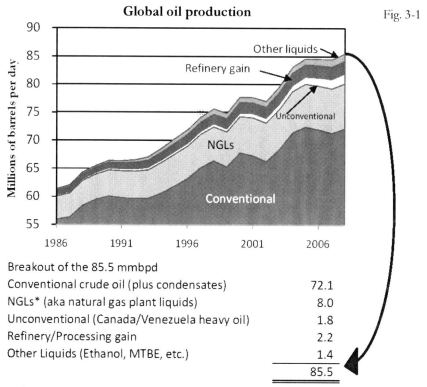

Global oil production   Fig. 3-1

Breakout of the 85.5 mmbpd	
Conventional crude oil (plus condensates)	72.1
NGLs* (aka natural gas plant liquids)	8.0
Unconventional (Canada/Venezuela heavy oil)	1.8
Refinery/Processing gain	2.2
Other Liquids (Ethanol, MTBE, etc.)	1.4
	85.5

*NGLs (Natural Gas Liquids) = ethane, propane, butanes, natural gasoline
Source: EIA 1986-2008 data, authors calculations.

Conventional crude oil is what most people think of when the word crude oil is mentioned. When removed from a reservoir, conventional crude is a hydrocarbon liquid with a density between 10°-50° API. Refineries are typically configured to optimally run a more narrowly defined density crude, 30°-35° API for example, within the 10°-50° API conventional range.

Condensates are very light hydrocarbons liquids with a density above 50°API, and thus are more valuable crude oils. Some have gone as far as to call condensates the champagne of crude oils because they are so easily refined into high value products such as gasoline. Condensates are so called because they are a gas at reservoir temperatures and pressures and then condense to become a liquid at surface temperatures and pressures. Condensates are usually added, as a relatively minor component, to conventional crude oils for production analyses on a global basis, and are rarely broken out in data.

Condensates are often produced alongside natural gas. The condensates turn to a liquid on the surface, whereas methane and other components of natural gas remain as gases. Separation of condensates and gases is carried out in one of three plants: lease separators; field separators; and plant separators.

If the separator is specific to a single well, or lease, then it is known as a lease separator and the result is known as lease condensate. Sometimes hydrocarbons are gathered from various wells and put through a field separator, which essentially does the same job as a lease separator, but on larger collected volumes. Condensates produced in a lease or field separator are called field condensates. Condensates can also be produced from the distillation of raw natural gas in a gas fractionating plant, also known as a plant separator. These plants handle flows from many different fields. Condensates produced in a gas fractionating plant are known as plant condensates.

As condensates are very light hydrocarbon molecules with a very low sulfur content, expensive secondary refinery treatments, such as crackers, cokers, reformers or hydrotreaters, may not be required to produce high value products. A condensate splitter is a very basic refinery which takes advantage of the ease of producing light straight-run products from condensates. The splitter is basically a distillation tower similar to those found in most refineries. The feed to a splitter is condensate and the yields of a condensate splitter are typically NGLs, naphtha, and middle distillates.

NGLs, or Natural Gas Liquids, are very light hydrocarbon molecules which are gases at standard atmospheric conditions but which can easily be compressed or cooled into liquid form. NGLs typically include ethane, propane, and butanes, but do not include methane, the most basic hydrocarbon molecule and the primary component of natural gas. In many oil company financial statements, hydrocarbon production is divided into liquids and gas. The gas referred to is most often methane.

Refinery gain is also known as processing gain. The oil output from a refinery occupies more space than inputs although weight is unchanged. Heavy dense oil molecules are cracked with heat from refinery processes into a larger number of light molecules which occupy more space. Think of refinery gain as analogous to making popcorn. Refinery gain adds to the total volume, but not weight, of oil liquids which can be sold to consumers.

Other liquids are components of finished oil products not naturally occurring in large quantities and may not even be hydrocarbons. Other liquids include MTBE, which is added to gasoline in some parts of the world, in addition to ethanol, and bio-diesel.

Unconventional crude oil is any hydrocarbon which has to go through special processing before it can be run through a refinery as a 10°-50° API conventional crude oil. Due to the additional processing involved, unconventional crude oil is much more expensive to produce than conventional crude oil.

Currently, there are three main sources of unconventional crude oil: oil sands; methane and coal-based syncrude; and shale oil.

Oil sands

Oil sands are the most common source of unconventional crude. Virtually all of the unconventional crude oil produced today is from Canadian and Venezuelan oil sands.

Oil sand, also known as tar sand, bitumous sand, or natural bitumen, is very heavy crude oil mixed with grains of sand and other inorganic material. The heavy crude is sometimes referred to simply as bitumen (a finished product of crude oil used for road surfacing) and it has a high asphaltene (a very large, and thus heavy, type of aromatic hydrocarbon molecule) content. Oil sands typically contain at least 10-15% bitumen by weight.

Large deposits of oil sands are found in the Athabasca oil sands deposit in Alberta, Canada, the Orinoco oil belt of Venezuela, and in the Olenek oil sands of Siberia, Russia. The gravity of crude oil produced from oil sands is around 8°-12° API and would thus be classified as extra-heavy.

Two methods of unconventional oil recovery are applied to oil sands: surface mining and in-situ production.

Surface mining involves open faced mining of the oil sands which are most often found within a few hundred feet from the surface. These mines are much shallower than conventional oilfields. The proximity to the surface is a major reason oil sands are so heavy, as light volatile components have had more opportunity here to evaporate, oxidize, or be metabolized by bacteria.

The surface mined sands must be transported to a separation facility, where the sands are treated with very hot water to detach the petroleum from the sand. Transporting oil sands to the separation facility can involve crushing the mined material and mixing it with water to convert it into slurry so it can be easily moved using a pipeline. Surface mined oil sands are also transported using huge trucks and dragline scoops.

Fig. 3-2

Massive oil sands truck in Alberta, Canada
Source: EIA

In-situ production involves removing the bitumen from a well in liquid form. There are six methods of in-situ recovery: CHOPS, CSS, SAGD, solvent-based recovery, hybrid thermal/solvent recovery, and air injection recovery.

Cold heavy oil production with sand (CHOPS) is used in situations where the oil is fluid enough to actually move through a reservoir to a wellhead, albeit still mixed with a large amount of sand. CHOPS is a somewhat conventional method apart from the fact that a large amount of sand is produced with the oil.

Cyclic steam stimulation (CSS), nicknamed 'huff and puff', involves injecting steam into the heavy oil reservoir and allowing it to soak for a period of time. The heat reduces the viscosity of the heavy oil and allows it to flow to the same wellhead the steam was pumped through. The cycle of huff and puff is repeated over and over.

Steam assisted gravity drainage (SAGD) is a more recent development and involves drilling two wells into the reservoir. One well is used to inject steam into the deposit. A second well is bored to a lower depth and as the heavy oil becomes less viscous and falls to the level of the second well it flows out.

CSS and SAGD require natural gas to heat water to produce steam. As natural gas has become more expensive an alternative fuel called Multi-phase Superfine Atomized Residue (MSAR) which is a mixture of bitumen, water and detergent is sometimes used to produce heat. In addition to heating water inexpensively, the availability of water in large quantities is a significant challenge with CSS and SAGD thermal recovery processes.

Solvent-based recovery, also known as vapor extraction (VAPEX) involves injecting a vaporized solvent such as propane into the well to reduce the oils viscosity and loosen the heavy oil molecules from the sand which allows it to flow to the well.

Hybrid thermal/solvent recovery is sometimes called a solvent aided process (SAP) because a small quantity of solvent is added to the steam.

Air injection recovery involves injecting air into one well to displace oil in the reservoir which is recovered from a second well.

Once the oil sands have been recovered, the bitumen must be separated from the sand. This is most often carried out at separation facilities using heat, water and centrifuges.

After it has been separated from sand, the next major challenge in dealing with heavy and extra heavy crude (bitumen) produced from oil sands is that it must be either upgraded to synthetic crude oil (SCO), or blended with lighter products to a gravity of 20-23° API or higher, so as to enable it to be transported to, and processed in, a conventional refinery.

Upgrading bitumen involves using refinery processes including distillation, coking (thermal cracking), catalytic cracking, and hydroprocessing (a sulfur removal process.) Treatment occurs in an upgrading plant, which is very similar to a refinery. Upgrading plants cost several billion dollars to construct and require vast amounts of energy to run. The upgraded oil, referred to as synthetic crude oil, can be, depending on the processes used, upgraded to a light sweet crude which can then be sent to a conventional refinery.

Bitumen blending is a less expensive process than upgrading and simply involves blending the bitumen with light products so that it can be pumped through pipelines to a refinery which can handle heavy oil. The diluent which is blended with the bitumen can be condensate, which is very light crude oil, or light products of crude oil in the kerosene-naphtha boiling point range which have APIs in the mid 50s or higher. Synthetic crude oil can also be blended with the bitumen.

Depending on the capability of the refinery to which the blend is being sent, the final crude oil can be tailored. Commonly produced unconventional crude oils from oil sands are:

> **DilBit** (Diluted Bitumen blend): 30% diluent (such as condensate) + 70% bitumen
> **SynBit** (SynBit Blend): 50% Synthetic crude oil + 50% Bitumen
> **SynDilBit** = SynBit + Diluent
> **SynSynBit** = Synthetic crude oil + SynBit

Oil sands processing summary

Fig. 3-3

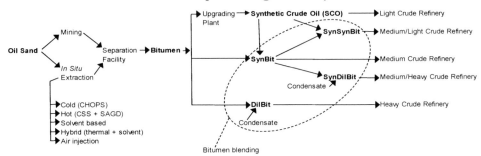

The bitumen from oil sands is heavy and therefore sells at a steep discount to light crude. Furthermore, the recovery, separation, upgrading and blending processes are expensive and require large quantities of water to produce steam in addition to natural gas or other fuels to generate heat. For these reasons, oil sands production requires relatively high crude prices and low natural gas prices to remain economical.

Total resources of oil contained in sands are estimated to be 3.5 trillion barrels (compared with conventional reserves estimates of around 1 trillion barrels.) However, at current 2008 prices and technology, the estimated proven reserves from oil sands are 174 billion barrels.

Presently, crude oil production in Canada from oil sands is 1 million bpd which is forecasted to rise to approximately 5 million per day by 2025. Venezuela produces roughly 570,000 bpd of crude oil from oil sands. The capital investment, energy consumption and construction lead times required in oil sands are enormous, starting in the low tens of billions of dollars and take 10 or more years to bring online.

Natural gas may increasingly become a transportation fuel over the next 25 years, which will drag natural gas prices steadily higher to converge with oil prices. This could mean that oil sands production may only see a very brief window of profitability, if at all, over the next ten to fifteen years before the natural gas used to produce heat for oil sands production becomes too expensive.

Methane and coal-based syncrude

Creating a conventional 10°-50° API hydrocarbon liquid out of coal or methane, is a proven, but expensive technology, made infamous by Nazi Germany and Apartheid South Africa. They both had to resort to producing oil from coal because of a lack of access to conventional crude oil.

Both methane and coal are hydrocarbons. Coal is comprised of hydrocarbon molecules with a lot of carbon atoms bonded to very few hydrogen atoms. Methane is the other extreme with lots of hydrogen atoms relative to carbon atoms.

Coal is treated in a two stage process to produce synthetic crude: the first stage involves gasification and the second stage involves liquefaction. Conversion of methane to liquid hydrocarbon products involves the liquefaction stage alone.

Gasification involves carbonizing coal by heating it under pressure until a combustible hydrocarbon gas is released. Basic coal gas, or town gas as it was known, was initially produced by heating coal under pressure in the absence of air. Coal gas technology was developed in the 1790s, well before the modern oil industry began. The use of coal gas spread throughout the 19th century and was used for street lighting, known as gas lights, in addition to domestic cooking and heating. Coal gasification plants were often located in poorer parts of towns as the process produced a lot of pollution. When this basic technique is applied, only 20% of the coal hydrocarbons are released as a hydrocarbon gas – the remainder is coke, which is a dense carbon-laden material, and slag, which is a waste product. Coke is today often burned in heavy industry such as in the manufacture of steel or cement. Slag, an inorganic material which contains no oxidization energy potential, is commonly used as filler in road construction.

As the basic gasification process leaves a lot of energy in the coke, more advanced forms of gasification were developed in the late 1800s.

The water gas method, which creates what is sometimes called syngas, was invented in 1873. This method increased the energy yield of the coal gas to 350-500 btu/ft³, compared with 150 btu/ft³ coal gas produced using the earlier method. The production of water gas involves heating coal to a very high temperature, of around 800°C, with steam as the source of hydrogen, and oxygen, not regular air, present. The process produces hydrogen gas and carbon monoxide as byproducts. Water gas, which has an energy content of 300-500 btu/ft³, is called a medium btu gas as it compares to the approximately 1,030 btu/ft³ energy content of natural gas.

Once one has a gaseous hydrocarbon, the next stage is to convert it into a liquid hydrocarbon. Producing liquid petroleum crude or petroleum products from coal can be done via two methods: indirect liquefaction, or direct liquefaction.

Indirect liquefaction is so named because coal is first converted to syngas and then to liquid petroleum. The use of coal gas for lighting became less common following the development of the electric light bulb and this, along with the

German need for a domestic source of aircraft and automotive fuel, led to the development of a process called Fischer-Tropsch synthesis in 1923. This process is used to convert syngas into liquid paraffinic hydrocarbon molecules which can in turn be used to produce gasoline, kerosene, diesel, and other liquid petroleum products. Nazi Germany used the Fischer-Tropsch synthesis process, which involves using heat, pressure and a catalyst, to produce 200 million gallons (13,046 bpd) of fuel per year during WWII. South African oil companies also used this process to produce liquid petroleum products during Apartheid. It goes without saying that the process is very expensive and highly polluting. More energy is consumed in the production of oil than the oil contains. Although energy is wasted producing the fuel, the benefit is that the finished product can be used in aircraft and automotive engines which burn liquid petroleum fuels.

Methane can also be converted into heavier liquid hydrocarbon molecules using indirect liquification/Fischer-Tropsch synthesis. The conversion of gas to liquid is often referred to as GTL technology.

Direct liquefaction involves producing synthetic crude, or syncrude, directly from coal without the intermediate step of producing a coal gas first. Producing liquid petroleum directly from coal uses what is called Coal to Liquids (CTL) technology. Direct liquefaction is much more difficult than indirect liquefaction and involves temperatures over 400°C, hydrogen, pressure of over 100 atmospheres and a catalyst. Germany produced only 1 million gallons of fuel per year during WWII with direct liquefaction. The process has not come a long way since the 1940s and even today direct liquefaction is not economically viable.

While producing petroleum from methane or coal is vastly more expensive than producing oil conventionally, methane-based and coal-based syncrude are in a different, more technologically realistic league than shale oil.

Shale oil
Oil shale is a fine grained sedimentary rock rich in kerogen, the organic precursor to crude oil. Kerogen is organic material which only reached the initial stage of transformation to crude oil and was never buried to the depths required to crack very heavy hydrocarbon molecules into the lighter hydrocarbon molecules of crude oil.

When heated to 450°C in the absence of oxygen, a process called pyrolysis (chemical change of organic material due to heat alone), kerogen can be separated from rock to produce an oily substance called kerogen oil. Pyrolysis takes place in a large kiln called a retort.

Kerogen oil is comprised of very long chain (heavy) hydrocarbon molecules. By breaking some of the heavy molecules in the presence of additional hydrogen, shorter, lighter hydrocarbon molecules, which comprise crude oil, can be created. The additional hydrogen (H) atoms are usually obtained by splitting water (H_2O) in a procedure which requires large amount of energy and water.

Lean shale contains approximately 5% kerogen oil while rich shale can contain up to 40% kerogen oil by weight. Lean shale produces 6 gallons of crude oil per metric tonne of shale and rich shale about 50 gallons of crude oil per metric tonne.

Kerogen laden shale is strip mined and then sent for processing at a plant near the deposit. The heating process causes the shale rock remaining after the kerogen oil has been removed to expand by around 30%, which is called the pop-corning effect. The rubble remaining after the heating process has no economic value. Because it is too large to be put back into the same hole the result is a mountain of rubble.

Another method of producing oil from shale involves dropping heaters into shafts drilled into the deposits and pumping up the produced oil, a process called in-situ converting. This avoids the issue of disposing of rubble. One challenge with this process is that the liquid kerogen oil may simply flow away underground, and possibly contaminate groundwater used for human consumption as there are no petroleum reservoir or trap rocks (which are essential components of an oil reservoir.) One way of creating a temporary trap is to freeze the rock in an area around the shafts such that any produced oil is contained.

The process of producing shale oil has been carried out since the 1850s, but is vastly more costly and energy intensive than producing liquid oil conventionally. During the 1970s, when oil spending as a percentage of GDP became very high, the development of shale oil deposits came back into vogue. Investors, including the US government, witnessing the decline in US oil production, poured billions of dollars into shale oil research. Fortunes were lost as everyone realized the technical and economic futility of the process and investment in shale oil dwindled such that there is no significant production today.

Instead of producing oil from the kerogen-laden shale, a few countries, notably Estonia, simply burn the unprocessed mined rock as a very low-grade alternative to coal.

The US Naval Petroleum and Oil Shale Reserves (NPOSR), a part of the US Department of Energy which manages a tract of land in the US containing oil shale, had until recently referred to themselves as the least well known strategic reserve. You now know why nobody wants to hear about it.

It has been estimated that globally there are 1.66 trillion barrels of crude oil locked in shale oil of which 1.2 trillion are in the US, compared with total global conventional oil reserves estimated to be around 1 trillion barrels. The largest known oil shale deposits in the world are located at the Green River shale deposits in Colorado, Utah and Wyoming. Oil produced from the Green River deposits has a density of around 12°API and would produce 30 gallons of crude oil per ton of shale, if you had the water supply and the energy to waste producing those 30 gallons.

Shale oil has been found in Canada, Sweden, Estonia, Scotland, Spain, China, Russia, South Africa, Australia, Brazil, and France.

The costs of mining, transport, crushing, heating and adding hydrogen, which requires huge quantities of water and energy, along with the disposal of the large quantities of waste material, make shale oil production prohibitively expensive and logistically a waste of time.

In the US, there is ongoing small-scale research by Shell into making shale oil recovery profitable.

The science is very well known. In short, although conventional oil can technically be created out of oil shale, one would be better off simply using any hydrogen created from water as a fuel in itself and leaving the low energy oil shale rock in the ground.

CHAPTER FOUR
CHEMISTRY OF OIL

An understanding of the chemical structure of crude oil forms a basis for appreciating refinery processes, why crude oil and finished oil product values differ, and petrochemicals. Although it covers all key aspects of crude oil chemistry, this chapter assumes no previous knowledge of the subject.

A molecule is made up of two or more atoms held together by chemical bonds. A molecule of water, for example, contains two hydrogen (H) atoms and one oxygen (O) atom – hence the chemical name for water – H_2O.

Typically, crude oil consists of 84-87% carbon atoms, 11-14% hydrogen atoms, 0%-6% sulfur atoms, and less than 1% nitrogen atoms, oxygen atoms, metals, and salts.

The large amount of carbon and hydrogen in crude oil combine to produce molecules known as hydrocarbons. A barrel of crude oil can contain thousands of different types of hydrocarbon molecules with carbon and hydrogen atoms arranged in many different ways.

The primary factor which makes oil so valuable is that hydrocarbon molecules release a large amount of energy when combined with oxygen in a process called oxidization, more commonly referred to as combustion or burning (Fig. 4-1). The oxidization reaction of hydrocarbon molecules and oxygen is exothermic, in that it net releases heat. The results of a complete reaction are carbon dioxide (CO_2) and water (H_2O). For example, methane, the simplest hydrocarbon molecule, is oxidized to produce carbon dioxide and water:

Fig. 4-1

$$\underbrace{CH_4}_{methane} + \underbrace{2O_2}_{oxygen} \xrightarrow{\text{heat}} \underbrace{CO_2}_{carbon\ dioxide} + \underbrace{2H_2O}_{water}$$

As one burns more complex hydrocarbon molecules, the results will still include carbon dioxide and water. An incomplete reaction, where the hydrocarbon molecule is not fully oxidized, may result in carbon monoxide (CO), carbon (C) and other molecules. Incomplete reactions may occur because the particular hydrocarbon molecules involved are very complex (heavy with a lot of atoms) or the reaction temperature doesn't get high enough for oxidization to occur, or perhaps there is insufficient oxygen for all the hydrocarbon molecules to react with.

Since air contains mostly nitrogen and because sulfur is sometimes found bound to hydrocarbon molecules, a number of side-reactions can result in nitrogen oxides (NOx) – where x is the number of oxygen atoms in the molecule – and sulfur oxides (SOx), both of which are pollutants. Adding oxygen, raising the reaction temperature, and removing sulfur from petroleum products can reduce the amount of pollutants produced.

At normal atmospheric pressure and room temperature, hydrocarbon molecules with 1-4 carbon atoms are typically gases, those with 5-24 carbon atoms are liquid and those with 25 or more are usually solid (Table 4-1).

The simplest hydrocarbon molecule is methane (CH_4), which has only one carbon atom attached to 4 hydrogen atoms. Methane is the primary component of natural gas. Methane is followed in complexity by ethane (C_2H_6), propane (C_3H_8), and butane (C_4H_{10}), with 2, 3 and 4 carbon atoms respectively. Increasing the number of carbon atoms in a hydrocarbon molecule makes it heavier – from gases to liquids and finally solids at room temperature and pressure. Thus, a heavy crude oil contains a lot of hydrocarbon molecules with many carbon atoms.

Refining is a process which separates crude oil into various products based on the number of carbon atoms. Hydrocarbon molecules have different boiling points depending on the number of carbon atoms in their molecular structure. Hydrocarbon molecules with fewer carbon atoms boil (turn into a gas) at lower temperatures.

There are a relatively small possible number of hydrocarbon molecular structures with five or less carbon atoms. For example, the only hydrocarbon in the petroleum complex with one carbon atom in its molecular structure is methane. However, once hydrocarbon molecules begin to contain six or more carbon atoms the number of different possible molecular structures grows rapidly. Most of the hydrocarbon molecules in crude oil contain less than twenty carbon atoms; but that allows for thousands of different types of hydrocarbon molecules.

Breakdown of crude oil

Table 4-1

Petroleum fraction	Petroleum product	Number of carbon atoms in molecules	Physical state*	Approx. boiling temp.	Primary uses
Petroleum gases	Methane	1	Gas	-161.6° C	Heating/cooking, electrical power.
	Ethane	2	Gas	-88.6° C	Petrochemicals, plastics.
	Propane	3	Gas	-42.1° C	Propane and butane together are called LPG when pressure liquefied for consumer use.
	Butane	4	Gas	-11.7° C	
Light ends	Naphtha	5-11**	Liquid	70°-200° C	Petrochemicals, plastics, solvents, blending for gasoline.
	Gasoline	7-10**	Liquid	100°-150° C	Transportation fuel.
Middle distillates	Kerosene	11-18**	Liquid	200°-300° C	Jet fuel, lighting, cooking, heating.
	Distillate fuel oil/ Gas oil	11-18**	Liquid	200°-300° C	Diesel fuel, home heating oil.
Heavy ends	Lubricating oil	18-25**	Liquid	300°-400° C	Motor oil, transmission oil.
	Residual fuel oil	20-27**	Liquid	350°-450° C	Marine shipping fuel, electrical power, industrial fuel.
	Greases & Wax	25-30**	Solid	400°-500° C	Lubricants, candles and coating fruit.
	Bitumen	35+**	Solid	500° + C	Road paving & roofing.
	Coke	50+**	Solid	600° + C	Industrial fuel for steel production.

*at standard atmospheric pressure and room temperature; **approximate range

Hydrocarbons are assigned names based on the number of carbon atoms in their molecular structure:

Number of carbon atoms	Prefix	Paraffin (alkane) example	Mono-olefin (alkene) example	Table 4-2
1	meth–	methane	–	
2	eth–	ethane	ethene	
3	prop–	propane	propene	
4	but–	butane	butene	
5	pent–	pentane	pentene	
6	hex–	hexane	hexene	
7	hept–	heptane	heptene	
8	oct–	octane	octene	
9	non–	nonane	nonene	
10	dec–	decane	decene	

The terms paraffin, olefin, mono-olefin, and diolefin, although still commonly used in the oil industry, have been replaced in modern chemistry nomenclature by the terms alkane, polyene, alkene, and diene, respectively. Also, the alkenes, ethene, propene, and butene are still referred to in the oil industry as ethylene, propylene and butylene, respectively. The simpler names, without the 'yl', were created by the International Union of Pure and Applied Chemistry (IUPAC.)

When using basic distillation, which involves boiling and condensing hydrocarbon molecules in order to separate them by molecular weight, a refinery is at the mercy of the range of hydrocarbons which nature placed in a particular grade of crude oil. However, by using heat, pressure, catalysts, and other chemicals, a refinery can crack and break down hydrocarbon molecules or conversely combine hydrocarbon molecules to form desired hydrocarbon molecules. Hydrocarbons can be cracked such that a molecule of residual fuel oil which contains 20 carbon molecules can be broken into two molecules which contain 10 carbon atoms each, which are in the gasoline range. The process can work the other way around also, such that molecules containing few carbon atoms, such as butane range molecules, can be combined to produce heavier molecules, such as those in the gasoline range.

This cracking and combining process allows a refinery to increase the amount of gasoline from perhaps 20% of a barrel with basic distillation to 55% or more by cracking and combining of hydrocarbon molecules.

Hydrocarbon structures

As mentioned above, crude oil contains thousands of different types of hydrocarbon molecules. To analyze the full molecular composition of each grade of crude oil every time it arrives into a refinery would take too much time and is unnecessary. Instead, it is useful to categorize crude oil by the ratio of hydrocarbon molecules with similar structures. This is because hydrocarbon molecules with similar arrangements of carbon and hydrogen atoms have similar physical properties.

Despite the wide variety of hydrocarbon molecules found in crude oil, each of them will have one of only three molecular structures: paraffinic, naphthenic, or aromatic. Rarely, a fourth molecular structure called an olefin is found in crude oil, although it is usually too unstable to exist for long in crude oil in a reservoir and is more commonly created from paraffinic molecules in crude oil at a refinery.

A refinery will have a good idea of the products they can produce from crude oil given the ratio of each of the hydrocarbon molecular structures in the crude oil. For example, a crude high in paraffinic molecular structure hydrocarbons will produce a lot of gasoline whereas a crude high in aromatic hydrocarbons will tend to produce less gasoline and more heavy residual fuel oil.

Crude oils are often characterized with reference to the four molecular structures by the acronym **PONA** (Paraffinic-Olefinic-Naphthenic-Aromatic). Crude is most valuable when it is paraffinic because gasoline contains a large amount of paraffins and, also, aromatics content is limited in gasoline by environmental regulations. The higher the API gravity, the lighter the crude and, usually, the higher the paraffinic content of the crude.

We will firstly discuss the molecules most commonly found in crude: paraffins, naphthenes and aromatics; and then move onto refinery created molecular structures, olefins.

Each carbon atom, due to its valency, seeks to bond with four other atoms. If carbon atoms in a hydrocarbon molecule do not have not enough hydrogen atoms to bond with, the carbon atoms form a double bond, which is a single bond, doubled back on itself, between carbon atoms in order to satisfy carbon's need for a total of four bonds. Double bonds are weaker than single bonds and make the molecule relatively unstable in that it reacts more easily with other molecules. Triple bonds are also possible and unstable. Hydrogen atoms have a valency requiring only one bond and thus can only link to one atom at a time.

Hydrocarbons are said to be *saturated* if all of the carbons bonds are satisfied – in other words, each carbon atom is bonded to four other atoms with single bonds (no unstable double or triple bonds.) Paraffinic molecules are saturated and this saturation makes paraffins chemically very stable. Naphthenic molecules are also saturated, and thus also relatively stable.

Aromatic and olefinic hydrocarbon molecules, however, are *unsaturated*. They have double, and even triple, bonds between some carbon atoms, which makes those molecules unstable and more reactive, which is useful for petrochemical production but more dangerous to the environment.

Carbon atoms can bond together in straight chains, branched chains, or rings (cyclics.) Aliphatic is the term used to describe non-cyclic structures, i.e. straight chains or branched chains. Paraffins and olefins are aliphatics.

Aliphatics (paraffins and olefins)

Table 4-3

Paraffins (Alkanes) = Saturated aliphatics with single bonds, present in crude. **Olefins** (Polyenes): Unsaturated aliphatics created from paraffins in refinery processes: 1. **Mono-olefins** (Alkenes) = Unsaturated aliphatics with at least **one double** carbon-carbon bond.   2. **Diolefins** (Dienes) = Unsaturated aliphatics with **two double** carbon-carbon bonds.   3. **Alkynes** = Unsaturated aliphatics with at least **one triple** carbon-carbon bond.

Paraffins (alkanes)

General formula: C_nH_{2n+2}
Molecular structures: 1. Straight chain (normal); 2. Branched chain (isomer)
Bonds: Saturated (i.e. single bonds only)
Examples: Methane, ethane, propane, butane, pentane and hexane

Paraffinic hydrocarbons have the general formula C_nH_{2n+2}. The molecular structure of a paraffin can be either straight-chained, called normal, or branch-chained, which is called an isomer. An isomer molecule contains the same number of atoms as a normal molecule but, because of the different arrangement of atoms, isomer and normal molecules differ in chemical and thus physical properties. Normal paraffins use the prefix n-, and iso-paraffins use the prefix iso-, or simply, i-.

Straight chained paraffins tend to have lower octane ratings than naphthenes and aromatics. Examples of straight-chain paraffinic molecules are methane, ethane, propane, n-butane, n-pentane and n-hexane, with 1 to 6 carbon atoms, respectively. Oil traders refer to these simply as C1, C2, C3, nC4, nC5 and nC6.

Branched-chain paraffinic isomer molecules are typically found in heavier hydrocarbon fractions and have a higher octane rating, important for gasoline, than straight-chained normal molecules.

Paraffinic molecules with very long chains, and thus a large numbers of carbon molecules, tend to have high melting points. Because of this they form waxes even at room temperature. A particular type of finished product called petroleum wax is sometimes called paraffin wax, as it is comprised mostly of very long chain paraffinic molecules.

Fig. 4-2

Paraffinic structure examples		
Methane (CH$_4$) Simplest chain molecule	Normal Butane (nC$_4$H$_{10}$) Straight chain molecule (note the chain of carbon)	Isobutane (iC$_4$H$_{10}$)* Branched chain (Isomer) molecule (note the branch off the straight chain)
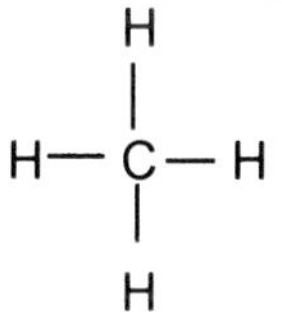 (structure)	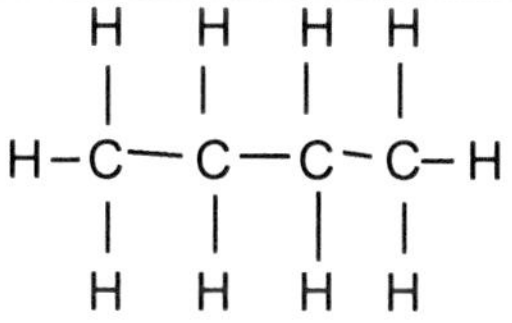 (structure)	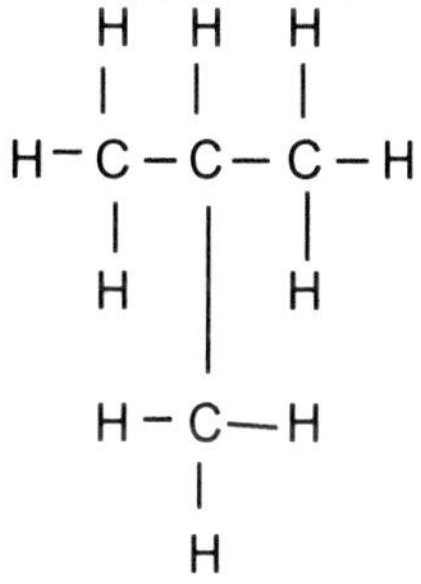 (structure)

*normal-butane and isobutane both have the same chemical formula (number of carbon and hydrogen atoms) but as they have different molecular structures they have differing chemical and physical properties.

Naphthenes

General formula:	C$_n$H$_{2n}$ (for single ring structures)
Molecular structures:	1. Ring (cyclic); 2. Branched ring; 3. Fused rings
Bonds:	Saturated (i.e. single bonds only)
Examples:	Cyclohexane; Ethylcyclohexane

Naphthenes have closed ring, cyclic, molecular structures, with the general formula C$_n$H$_{2n}$. Naphthenes can either have a single ring, to form monocycloparaffins, or a double ring, to form dicycloparaffins. Naphthenic hydrocarbon molecules are saturated and are chemically relatively stable with properties similar to paraffins.

Fig. 4-3

Naphthenic structure examples	
Cyclohexane (C_6H_{12}) Carbon ring Molecule (Note the ring of carbon) 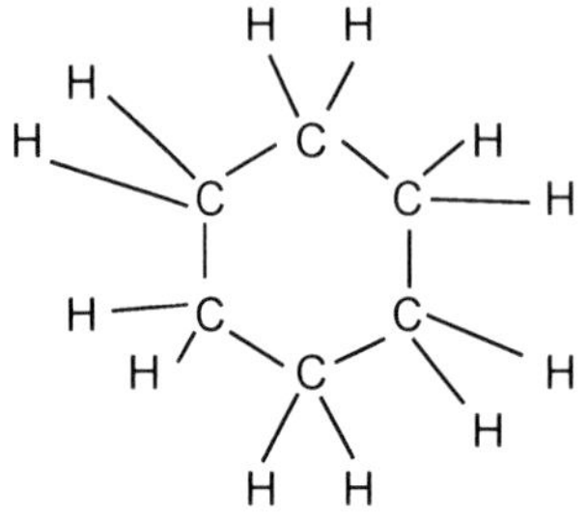	**Dimethyl Cyclohexane (C_7H_{14})** Branched carbon ring molecule (Note the ring of carbon) 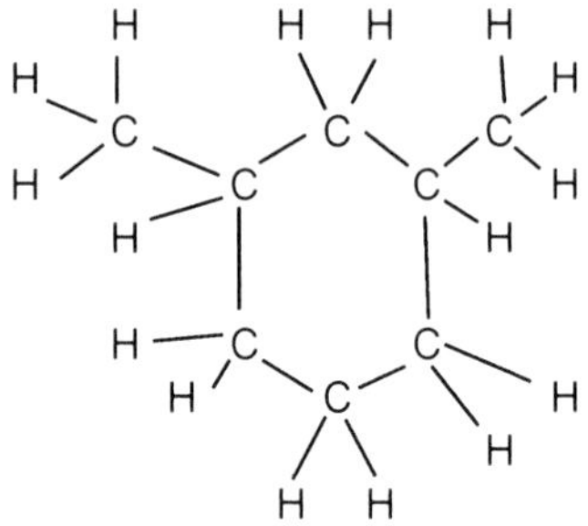

Aromatics

General formula: C_nH_{2n-6} (for single ring structures)
Molecular structures: 1. Ring (cyclic); 2. Branched rings; 3. Fused rings
Bonds: Unsaturated (i.e. double bonds exist)
Examples: BTX (Benzene, Toluene, Xylenes)

Similar to naphthenes, aromatic hydrocarbons are ring-type, cyclic structures. As aromatic hydrocarbons are unsaturated, lacking sufficient hydrogen atoms to form single bonds, they are chemically more unstable than paraffins or naphthenes. The instability of aromatics makes them useful for petrochemicals where the goal is to create tailored molecules by breaking apart existing ones and combining the atoms in new ways.

Aromatics occur naturally in crude oil and can also be created in refinery processes.

Aromatics all contain at least one ring of the molecule benzene as part of their structure. A benzene ring has 6 carbon atoms, three double bonds and three alternating single bonds, i.e. every other bond is single. Examples of aromatic molecules are benzene, toluene and xylenes – known sometimes in a group as BTX.

Compounds with two or more benzene rings are known as Polycyclic Aromatic Hydrocarbons (PAH). Naphthalene is the simplest PAH with just two benzene rings. Far more complex PAHs, often with more than 70 carbon atoms, are called asphaltenes. Asphaltenes are very complex high molecular weight (lots of carbon and hydrogen atoms in the molecule) hydrocarbons, with many benzene rings and molecular branches. Asphaltenes, because of their density, absorb light, giving crude oil its dark color and are found in large quantities in heavier crudes. Asphaltenes, along with long-chain paraffinic wax molecules, create challenges for pipeline operators in low temperature environments as they can clog pipelines.

The xylene molecule has three possible isomers, or possible molecular structures, each with the same number of atoms: para-xylene, ortho-xylene, and meta-xylene, all of which together are referred to by oil traders as mixed xylenes.

Fig. 4-4

Aromatic structure examples			
Benzene (C_6H_6) Simple Aromatic Ring (Note the alternating double bonds) 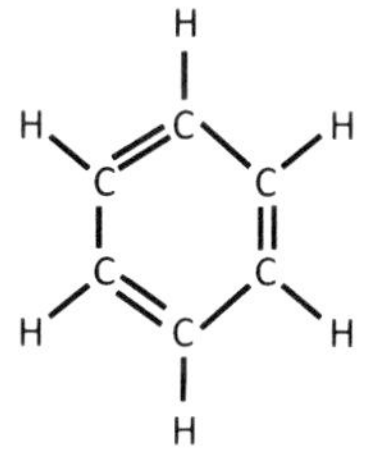	Toluene (C_7H_8) Branched Aromatic Ring 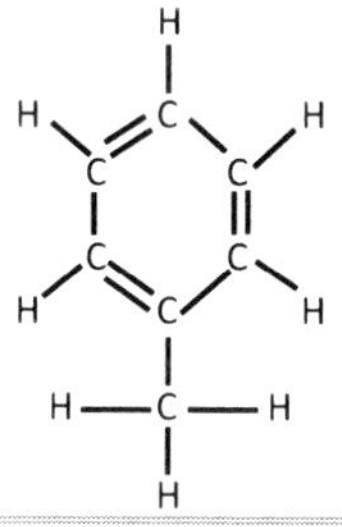	Para-Xylene (C_8H_{10}) Branched Aromatic Ring 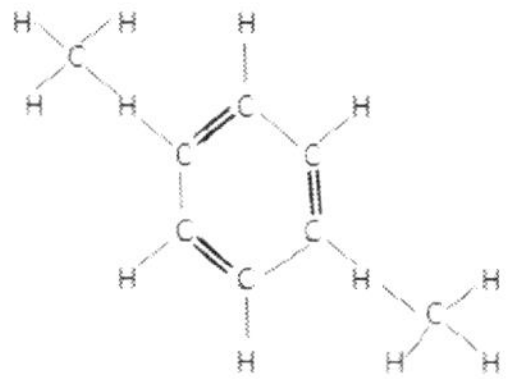	Naphthalene ($C_{10}H_8$) Two Fused Benzene Rings (Polycyclic Aromatic) 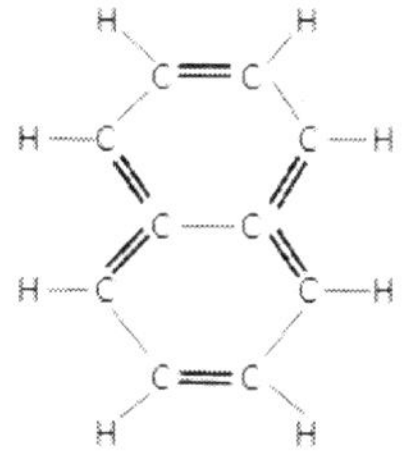

Some aromatic molecules have a detectable pleasant sweet smell, hence the term aromatic. When you are refueling your car with gasoline, if you detect an odor, you may be smelling the aromatic hydrocarbon molecules such as benzene. Benzene is a known carcinogen and therefore its content in gasoline is usually strictly limited by law.

Refinery-created hydrocarbon structures
In addition to the paraffins, naphthenes and aromatics which we have already discussed, the fourth group of hydrocarbon molecules, olefins, is rarely found naturally in crude oil. It most often results from refinery processes such as thermal cracking.

Olefins, or polyenes as they are known in modern chemical nomenclature, are unsaturated hydrocarbons with double or triple bonds, which result from the cracking of paraffinic molecules. The unsaturated nature of olefins makes them highly reactive, which is ideal for the petrochemicals industry, and the reason they rarely exist in a crude oil reservoir. Olefinic structures come in three forms: mono-olefins (alkenes), diolefins (dienes), and alkynes (acetylenes) – with modern chemical names in parentheses.

Mono-olefins (alkenes)

General formula: C_nH_{2n} (same general formula as naphthenes)
Molecular structures: 1. Chain (normal); 2. Branched chain (isomer); 3. Ring (cyclic); 4. Branched ring.
Bonds: Unsaturated
Examples: Ethylene (2 carbon atoms); Propylene (3 carbon atoms); Butylene (4 carbon atoms)
             (aka Ethene)   (aka Propene)   (aka Butene)

Mono-olefins are unsaturated, lacking sufficient hydrogen atoms to form single bonds, and contain only one double bond (hence the prefix mono-) between a pair of carbon atoms Olefinic molecules can be arranged in either open chain or ring structures.

As with paraffins, those olefins with four or more carbon atoms (the minimum number to create isomers of paraffins or olefins) can exist as structural isomers, with the same chemical formula and number of atoms as the normal version of the molecule, but a different molecular structure and therefore different physical properties.

The two most common olefins, ethylene and propylene, are used in the vast majority of consumer products made with plastics. Olefins are also, in a refinery process called alkylation, bonded with isobutane to form larger branched chain isoparaffin (isomers of paraffinic) molecules called alkylate, which are blended into the hydrocarbon pool making up high octane gasoline.

Fig. 4-5

Mono-olefin (alkene) structure examples		
Ethylene (C_2H_4) Simple mono-olefin 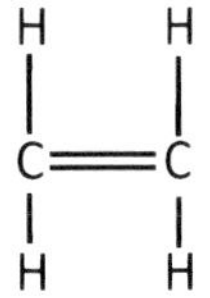	1-Butene (C_4H_8) Straight Chain Molecule (Note the single double bond of carbon) 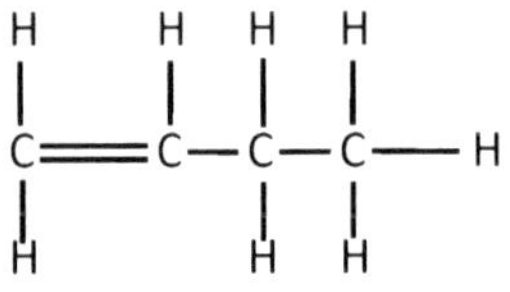	Iso-Butene (C_4H_8) Branched-Chain (Isomer) Molecule (Note the branch off the chain and the double bond of carbon) 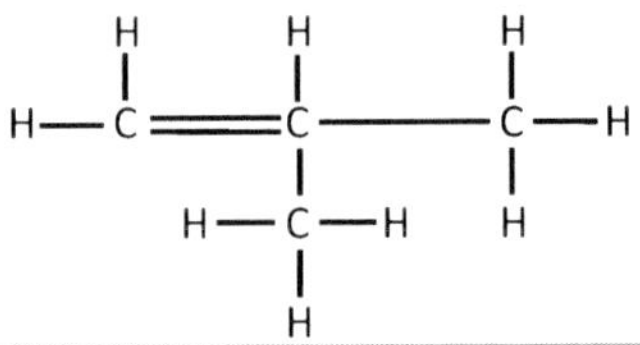

Diolefins (dienes)

General formula: C_nH_{2n-2}
Molecular structures: 1. Chain (normal); 2. Branched chain (isomer); 3. Ring (cyclic);
                       4. Branched Ring
Bonds: Unsaturated
Example: Butadiene

Diolefins have two double bonds, hence the prefix *di* in the name, between carbon atoms and, as with mono-olefins, are not naturally found in crude oil and instead result from the refinery molecule cracking process. Diolefins usually have fewer than 5 carbon atoms in their molecular structure. Diolefins are highly reactive and can easily combine with other atoms.

Fig. 4-6

Diolefin (diene) structure examples	
1,2 Butadiene (C_4H_6) (Note the two double bonds of carbon)	**1,3 Butadiene (C_4H_6)** (Note two double carbon bonds)
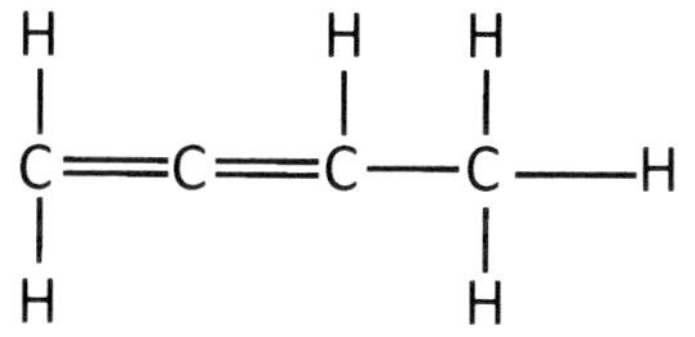	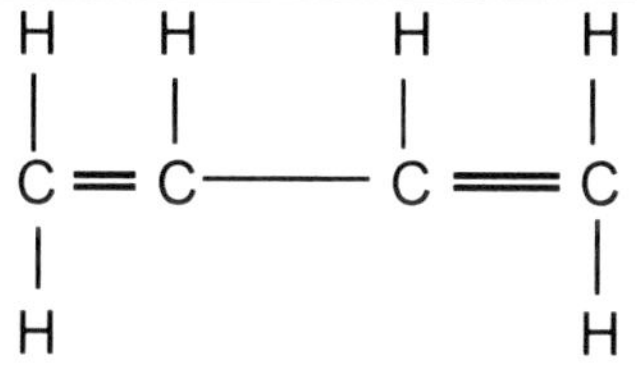

Alkynes (acetylenes)

General formula: C_nH_{2n-2}
Molecular structures: 1. Chain (normal); 2. Branched chain (Isomer); 3. Ring (cyclic);
                       4. Branched ring
Bonds: Unsaturated
Examples: Acetylene (aka ethyne)

Alkynes have a triple bond between carbon atoms and, as with mono-olefins and diolefins, alkynes do not naturally exist in crude oil and instead typically result from the refinery cracking process. Alkynes usually have fewer than 5 carbon atoms in their molecular structure. Alkynes are the most highly reactive of the three types of olefins.

Fig. 4-7

Alkynes (acetylene) structure example
Acetylene (C_2H_2) (Note the triple bond of carbon)

CHAPTER FIVE
INDUSTRY OVERVIEW

The oil industry, colloquially known as the oil patch, can be divided into three areas. Most companies operate in just one area. Vertically integrated organizations are those which have upstream, midstream and downstream operations.

Fig. 5-1

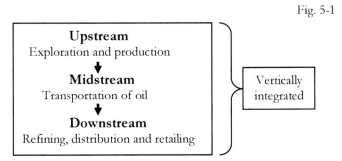

Upstream

Upstream, also called the exploration and production (E&P) sector, involves finding and extracting crude oil.

NOC crude producers

The majority of the world's crude oil production is controlled by government-owned National Oil Companies (NOCs). Most NOCs exist because they assumed ownership of oil within their national territory which private industry originally risked funds to discover. All OPEC production and around a third of non-OPEC production is controlled by NOCs. Most NOCs farm out some technical operations to private companies.

OPEC

From its inception in 1859, the oil industry has been able to produce more conventional petroleum than consumers demanded. In order to keep prices high enough to generate a required return on invested capital, some large oil producers had to withhold oil from the market. The surplus was managed, in turn, by Rockefeller's Standard Oil, its post-monopoly children, and then the TRC and its influence over the Majors and other International Oil Companies (IOCs). Finally, in the early 1970s, OPEC member nations took it upon themselves alone to deal with global spare crude oil production capacity.

Twelve OPEC cartel nations: Algeria, Angola, Ecuador, Iran, Iraq, Kuwait, Libya, Nigeria, Qatar, Saudi Arabia, United Arab Emirates (UAE) and Venezuela, have produced around 40% of the world's crude in recent years (Fig. 5-2 to 5.4). Sudan has expressed an interest in joining OPEC in 2009, which would increase OPEC's share of global production by 3%.

Fig. 5-2

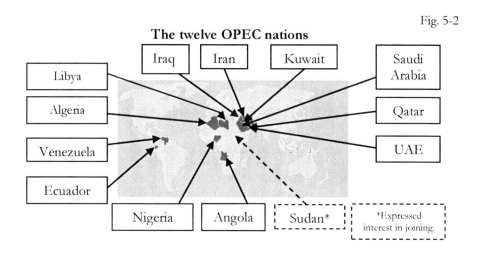

Fig. 5-3

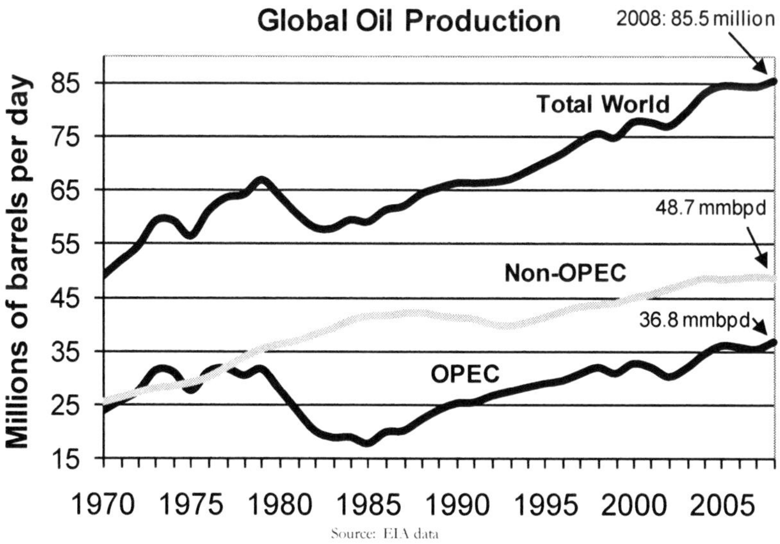

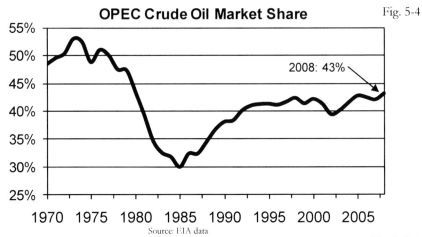

Fig. 5-4

OPEC Crude Oil Market Share

2008: 43%

Source: EIA data

Table 5-1

OPEC Member	National Oil Companies (NOCs)
Saudi Arabia	Saudi Aramco
Iran	National Iranian Oil Company (NIOC)
Venezuela	Petróleos de Venezuela, S.A. (PDVSA)
U.A.E.	Abu Dhabi National Oil Company (ADNOC)
Nigeria	Nigerian National Petroleum Corporation (NNPC)
Kuwait	Kuwait Petroleum Corporation (KPC)
Algeria	Sonatrach
Iraq	North Oil Company (NOC) and South Oil Company (SOC)
Libya	Libyan National Oil Company (NIOC)
Qatar	Qatar Petroleum
Ecuador	Petroecuador
Angola	Sociedade Nacional de Combustiveis de Angola (Sonangol)
Sudan	Sudan National Petroleum Corporation (Sudapet)

OPEC oil ministers meet regularly, often at their Vienna headquarters, to discuss the market and set production limiting quotas. Iraq, because of obvious political issues, has not participated in production quotas since 1998.

Price setting cartels such as OPEC are especially challenging economic organizations to manage. There is an incentive for individual members to cheat by secretly producing beyond quotas and relying on other members to take the pain. OPECs de facto leader is Saudi Arabia, although each member gets an opportunity to vote on and assume the organizations secretary general and presidency positions. Saudi Arabia as the largest producer has traditionally wielded the implied threat of flooding the market with oil and push prices to minimal levels to prevent egregious cheating occurring.

When most OPEC members had plenty of spare capacity, each member country sought a large quota and a method of allocating production had to be devised. OPEC members decided to apportion quotas on the basis of population sizes and reserves estimates. Linking quotas to reserve estimates led to artificial inflation of reserves from 1982 through 1988. Individual OPEC members simply increased their estimated reserve numbers overnight, often by double or more without any additional oil being discovered. Other members had to follow or risk having their quota reduced. Despite several decades of crude production (most oilfield reserves deplete by 2-8% per year) from these inflated estimated reserves, few OPEC countries have changed their estimates since 1988. This implies that OPEC oil production has been exactly offset by nonexistent discoveries over that time, which of course is nonsense. For this reason, OPEC reserves statements are unreliable. OPEC members published reserves estimates claim over two-thirds of the world's reserves of conventional crude oil. These are the numbers one sees published in oil industry statistical reviews.

In addition to reserves reports, OPEC statements on daily production rates are also inaccurate. OPEC, as an organization, never had a strict method of verifying compliance with quotas. To avoid accusations of cheating, individual OPEC member nations simply proffered numbers to the OPEC organization and the general public which were in compliance with quotas, regardless of true production levels. Several techniques have been developed by the oil market to determine the true levels of OPEC members' production. One method is tanker tracking, which involves surveying ship brokers and counting the size and number of tankers leaving OPEC ports and how low they sit in the water. UK based Oil Movements and Lloyd's Marine Intelligence Unit (LMIU) in addition to Geneva based Petrologistics are the three most commonly cited private tanker trackers.

In order to more closely manage oil prices, in March of 2000 OPEC created a mechanism known as the OPEC price band. OPEC collected the prices for a basket of seven crudes: Algeria's Saharan Blend, Indonesian Minas, Nigerian Bonny Light, Saudi Arabian Arab Light, UAE Dubai, Venezuelan Tia Juana and Mexican Isthmus, which is a non-OPEC oil. OPEC calculated the simple daily arithmetic average of the market prices of these crude oils. If the basket price was outside of a $22 to $28 per barrel range for a period of time then OPEC members were to increase or reduce supply in order to return prices to within this range.

OPEC only acted on this basket mechanism once, on October 31, 2000, when the Organization increased production by 500,000 barrels per day.

In June 2005, OPEC changed the basket to include eleven crude oils, with only the first four crudes of the original basket remaining. The new OPEC Basket comprises: Saudi Arabian Arab Light, Nigerian Bonny Light, Algerian Saharan Blend, Iranian Heavy, Iraqi Basra Light, Kuwait Export Blend, Libyan Es Sider, Qatar Marine, UAE Murban, and Venezuelan BCF 17. The new basket is heavier, at 32.7°API versus 34.6 °API, and more sour, at 1.77% versus 1.44% sulfur, compared with the old basket and initially was valued roughly $2 per barrel under the old basket.

The OPEC basket mechanism has been abandoned since the beginning of 2004 and the basket price is simply now used as a reference in OPEC discussions.

OPEC production quotas always only applied to conventional crude oil, excluding both condensates and natural gas liquids (NGLs). Condensates and NGLs are considered part of crude oil by most in the oil industry but were excluded as larger OPEC members sought to bring in extra revenue unrestricted by quotas. In addition, OPEC quotas excluded extra heavy oil below 10°API, although this is also considered part of crude oil production in the oil industry. Such heavy oil can only be handled by a few dedicated refineries and for OPEC to restrict production would have a minor affect on the general crude oil market.

The exclusion of NGLs, condensates, and oil below 10°API, resulted in a recent OPEC quota of 28 million barrels per day, for example. Condensates, NGLs, heavy oil and refinery gain at the time accounted for 6 million barrels per day (mmbpd) of OPEC production above quotas such that OPEC produced 34 mmbpd, or roughly 40%, of the world's 84 mmbpd.

When OPEC members cut production, they almost always just cut production of their heaviest grades of 10°API+ conventional crude oil because this oil sells for less than lighter grades of crude. Saudi Arabia, for example, cut production of Arab Heavy crude while continuing to sell their lighter grades as usual. For this reason OPEC cuts have a disproportionate effect on profit margins of refineries designed to handle heavy crude grades as the market price for these crudes would rise and fall with OPEC decisions. OPEC production cuts also had a disproportionate effect on residual fuel and other heavy oil product markets.

Another interesting feature of OPEC implementing quota cuts has been the tendency of Saudi Arabia, for strategic reasons, to cut supplies to Asia moreso than those to Europe or the US. Crude oil is to a certain extent relatively fungible and most producers could care less where the physical oil ends up. The Saudis, however, would put destination restrictions into crude oil sales contracts; penalizing those it found trying to circumvent the process. This led to what was called the Asian premium of up to a couple of dollars per barrel for crude oil in Asia over markets west of the Arab Gulf.

The year 2005 became a watershed for the oil market, as global demand temporarily caught up with global supply capacity. On September 19, 2005 OPEC gave its members full reign to produce as much oil as they were able beyond quotas. This announcement mirrors a similar announcement by the TRC in 1971 as US oil production peaked and began to fall. With no spare capacity, just as the TRC before it, OPEC lost the ability to control global prices. Oil prices soared to $148 per barrel. The oil supply crunch destabilized global economies. Global oil demand growth stalled for the first time since the early 1980s and oil prices collapsed. In late 2008 OPEC cut production to support prices. This created a small amount of spare capacity and OPEC quotas became relevant once more.

Relative to its large share of GDP in OPEC economies, oil production does not employ many people nor require the general population of a producing nation to be well educated. Additionally, oil revenue is easily controlled by very few people. As in any situation where revenue control is concentrated and education is not encouraged, checks and balances on power are easily removed allowing corruption and bad policy to flourish. This has resulted in what is referred to as the oil curse, such that OPEC members are controlled by mostly authoritarian, conservative, and often ideologically or religiously extremist and intolerant governments.

Another affliction caused by outsized oil production in OPEC countries is Dutch disease. Dutch disease is a term created in the 1970s to describe the demise of the Dutch manufacturing sector as North Sea natural gas revenue suddenly flowed into the non-OPEC Netherlands. Revenue from natural gas resulted in a rapid appreciation of the Dutch guilder, which made producing non-gas related goods and services there internationally uncompetitive. In addition, the sudden wealth generated by gas increased the price of domestic goods, such as real estate, resulting in higher wages across the economy and a further loss of competitiveness for non-natural gas Dutch industries.

Because of Dutch disease, OPEC nations which produce large amounts of revenue from oil relative to their economy size, find it difficult to maintain a diversified economy and often end up much worse off after natural resources have been depleted than if the resources had never been discovered.

One solution to avoiding Dutch disease, which is politically tough to implement, is to use oil revenue to create an offshore non-domestic currency stabilization fund into which income is placed for use by future generations. Abu Dhabi, Norway, Saudi Arabia, Kuwait, Russia, Qatar, Alaska, Libya, Brunei, Kazakhstan, Alberta and Iran, have such sovereign wealth funds which can be used to stabilize the economy when oil prices fall or production eventually ceases.

Contrary to popular perception, OPEC has never used oil as an economic weapon. The two oil price shocks of the 1970s and early 1980s were production cuts by individual members and groups within OPEC, but not by OPEC as an organization.

OPEC members have traditionally concentrated on producing crude and have not moved into the refinery business in a big way. The primary reason is that, except for a brief period between 2003 and 2008, refining had not been very profitable. A secondary reason is that those investing the billions of dollars building and running a refinery preferred it to be physically located in a non-OPEC nation where it is perceived to be a secure environment, particularly given NOC tendencies to nationalize oil assets.

Many OPEC members, such as Libya, Iran, Iraq, Nigeria and Venezuela have been especially difficult for investors to deal with for a variety of reasons, mainly political, and have not received much needed capital investments and technology to improve their production.

Underinvestment in oil fields can severely and permanently reduce the total quantity of oil ultimately recoverable. Water injection, for example, is usually beneficial to maximizing total oil output if carried out correctly. However, water injection is expensive to implement properly. A lack of capital may encourage badly managed water injection to be carried out for a short term increase in revenue without regard to the long-term total potential of a field. This appears to have occurred in Iraq during the 1990s.

A frequently mentioned cliché is that it costs some OPEC countries under $5 or $10 per barrel to find and develop oil fields. While this may have been true a long time ago, it is no longer the case.

Saudi Arabia, for example, allegedly has one of the lowest OPEC finding and developing (F&D) costs. Their F&D costs in the past were very low as most of its oil is located in a few gigantic dry land-based fields, including the massive Ghawar field, which has astonishingly been in production since 1948. However, no significant oil fields have been found in Saudi Arabia since the 1960s, and not for the lack of trying. The geology in that country still points to oil in some undeveloped fields, but not the massive dry-land-based oil discoveries of the past. The Saudis are now venturing into the expensive offshore arena simply to maintain current production rates. The F&D costs in Saudi Arabia for incremental new oilfields are well over $50 per barrel.

OPEC countries have very young populations. Even if OPEC producers' F&D costs on a pure oilfield cost basis are relatively low compared to non-OPEC, if one adds the expenditure of keeping young and restless populations content with government welfare and state-generated jobs, the true required market price for OPEC production climbs rapidly above that of many non-OPEC producers. In addition to social programs, several OPEC countries, including Saudi Arabia, have large government debts to service which require high oil prices in order to balance the state budget.

Non-OPEC NOCs
Around 60% of daily global crude production takes place outside of OPEC countries and roughly a third of this is produced by NOCs. Many of these are vertically integrated to some extent. The trend toward nationalization of privately owned oil production into NOCs has increased with oil prices since 2003.

Non-OPEC NOCs have tended to produce at maximum sustainable levels so long as it has been profitable or strategically necessary. Thus, unlike OPEC NOCs, non-OPEC NOCs have never had any spare crude oil production capacity.

The largest non-OPEC integrated NOCs are in Mexico, Brazil, Russia, China, Malaysia, Norway and India. Non-OPEC integrated NOCs include: Petróleos Mexicanos (PEMEX), Petróleo Brasileiro S.A. (Petrobras), Rosneft, Gazprom, PetroChina, Sinopec, CNOOC, Petronas, StatoilHydro, and ONGC, among others.

Private crude producers

Majors

The six Majors are ExxonMobil, Chevron, BP, Shell, ConocoPhillips and Total. The current Majors are a result of industry consolidation particularly during a period of particularly low oil prices and low refinery margins from the mid-1980s through to the late-1990s. Large private non-OPEC producers consolidated into larger firms in order to reduce overhead and better withstand such low prices. The steadily falling rate of discoveries since the 1960s and the rapidly increasing costs due to the world's remaining undiscovered oilfields being mostly offshore also forced private companies to merge in order to spread the now much higher risk of exploration across larger capital bases.

Second tier integrated

Integrated private companies without the same global reach or scale as the Majors are referred to as second tier integrated, and include Hess, Marathon, and Repsol YPF.

Facing a similar situation to the Majors, beginning in the late 1990s many of the largest privately owned oil companies merged in order to save on overhead costs and boost reserves. Consolidation of non-OPEC producers continues to this day as most producers are finding it less expensive and less risky to buy reserves from another company than finding oil themselves to offset depletion of their existing reserves.

Independent E&P

Independent E&P businesses are those which participate only in the exploration and production of crude oil. Independent E&Ps tend to be the highest risk and highest return part of the oil industry as such companies account for over 80% drilling of new field wildcats. Examples of independent E&P operators are Apache, Anadarko, Canadian Natural, Murphy, Devon, XTO Energy, Talisman, Nexen, EOG Resources, Woodside and EnCana.

Junior independents tend to be newly formed startup oil companies or smaller oil companies operating in the E&P sector.

Large upstream non-OPEC companies   Table 5-2

	Company Name	Home Country	Crude Oil mm bpd	Ticker (NYSE)
1	Pemex	Mexico	3.50	---
2	Exxon Mobil	US	2.70	XOM
3	BP	UK	2.45	BP
4	Petrochina (CNPC)	China	2.30	PTR
5	Lukoil	Russia	2.00	LUKOY.PK
6	Royal Dutch Shell	UK & Netherlands	1.90	RDS-A
7	Petrobras	Brazil	1.85	PBR
8	Rosneft	Russia	1.60	
9	Total	France	1.50	TOT
10	TNK-BP	Russia	1.40	
11	Chevron	US	1.30	CVX
12	Surgutneftegaz	Russia	1.30	
13	ENI	Italy	1.00	E
14	ConocoPhillips	US	0.90	COP
15	ONGC	India	0.90	---
16	Tatneft	Russia	0.90	
17	Sinopec (CPC)	China	0.80	---
18	Petronas	Malaysia	0.70	---
19	StatoilHydro	Norway	0.70	STO
20	Gazprom Neft	Russia	0.70	---
21	Repsol–YPF	Spain	0.50	REP
22	Occidental	US	0.50	OXY
23	CNOOC	China	0.38	CEO
24	Petro-Canada	Canada	0.32	
25	Hess	US	0.27	HES
26	Suncor Energy	Canada	0.22	SU
27	Devon	US	0.22	DVN
28	Apache	US	0.26	APA
29	Bashneft	Russia	0.24	
30	Anadarko Petroleum	US	0.24	APC
31	Marathon	US	0.20	MRO
32	BHP Billiton	Australia	0.16	BBL
33	EnCana	Canada	0.14	ECA
34	Others	---	16.44	
	Total non-OPEC Crude		50.50	

mmbpd = millions of barrels per day
Total/partial state owned companies in black.
Source: Financial Reports 2007.

Upstream breakdown by Country

Table 5-3

Producers	Total oil Production		Exporters	Net oil Exports	
	mmbpd	% of total		mmbpd	% of total
1. Saudi Arabia	10.1	12.0%	1. Saudi Arabia	8.7	20.3%
2. Russia	9.9	11.7%	2. Russia	7	16.4%
3. United States	8.5	10.1%	3. Norway	2.5	5.8%
4. Iran	3.9	4.6%	4. Iran	2.5	5.8%
5. Mexico	3.6	4.3%	5. United Arab Emirates	2.5	5.8%
6. China	3.7	4.4%	6. Venezuela	2.2	5.1%
7. Canada	3.3	3.9%	7. Kuwait	2.2	5.1%
8. Venezuela	2.8	3.3%	8. Nigeria	2.2	5.1%
9. United Arab Emirates	2.9	3.4%	9. Algeria	1.9	4.4%
10. Kuwait	2.5	3.0%	10. Mexico	1.7	4.0%
11. Nigeria	2.3	2.7%	11. Libya	1.5	3.5%
12. Norway	2.3	2.7%	12. Angola	1.4	3.3%
13. United Kingdom	1.8	2.1%	13. Iraq	1.4	3.3%
14. Algeria	2.2	2.6%	14. Kazakhstan	1.1	2.6%
15. Iraq	2	2.4%	15. Canada	1.1	2.6%
16. Brazil	1.8	2.1%	16. Qatar	1	2.3%
17. Angola	1.7	2.0%	17. Oman	0.7	1.6%
18. Libya	1.7	2.0%	18. Azerbaijan	0.5	1.2%
19. Kazakhstan	1.4	1.7%	19. Ecuador	0.4	0.9%
20. Indonesia	1	1.2%	20. Equatorial Guinea	0.3	0.7%
Subtotal :	69.4	82.1%	Subtotal :	42.8	100%
Other countries	15.1	17.9%			
Total	84.5	100%			

Consumers	Total oil Comsumption		Importers	Net oil Imports
	mmbpd	% of total		mmbpd
1. United States	20.9	24.4%	1. United States	12.4
2. China	7.7	9.0%	2. Japan	5.2
3. Japan	5.2	6.1%	3. China	3.7
4. Russia	2.9	3.4%	4. Germany	2.5
5. India	2.7	3.2%	5. South Korea	2.2
6. Germany	2.6	3.0%	6. France	1.9
7. Canada	2.3	2.7%	7. India	1.7
8. Brazil	2.3	2.7%	8. Italy	1.6
9. South Korea	2.2	2.6%	9. Spain	1.6
10. Mexico	2.1	2.5%	10. Taiwan	1
11. Saudi Arabia	2.1	2.5%	11. Netherlands	0.9
12. France	2	2.3%	12. Singapore	0.8
13. United Kingdom	1.8	2.1%	13. Thailand	0.6
14. Italy	1.7	2.0%	14. Turkey	0.6
15. Iran	1.6	1.9%	15. Belguim	0.6
16. Spain	1.6	1.9%	16. Poland	0.5
17. Indonesia	1.2	1.4%	17. Greece	0.4
18. Taiwan	1	1.2%	18. Ukraine	0.3
19. Australia	0.9	1.1%	19. Philippines	0.3
20. Thailand	0.9	1.0%	20. Portugal	0.3
Subtotal :	65.7	76.7%	Subtotal :	39.1
Other countries	20.0	23.3%		
Total	85.7	100%		

OPEC countries in black.
Source: EIA data 2007.

Upstream support

Drilling rig contractors exist as most upstream oil explorers and producers do not own rigs, rather they lease them from independent rig owners. Rig contractors usually provide crews to operate the equipment. The drilling business is highly cyclical, and is prone to long periods of low revenue, before short periods of very high rates for a few years until new rigs come online.

Oil equipment providers rent and sell drill bits, piping, pumps and all other tools required in the business of producing crude. The companies operating in this area tend to have more stable revenue streams over longer periods of time compared with drilling rig contractors.

Oil service providers are used by crude producers to find and develop crude oil deposits and to provide specialist support in niche areas of expertise. These oil services providers carry out, among many functions, seismic analysis, oil drilling and production services, oil field equipment maintenance, geophysical/reservoir services, and support services such as food, security and crew transportation. Oil services companies do not typically own any equity crude production and are hired on a contract basis.

The following are some of the larger upstream support corporations:

Table 5-4

Company name	Ticker (NYSE)	Specialization
Schlumberger Ltd.	SLB	Oil services
CGG Veritas	CGV	Oil services
Petroleum Geo-Services (PGS)	--	Oil services
Diamond Offshore Drilling, Inc.	DO	Drilling
Nabors Industries, Inc.	NBR	Drilling, equipment & services
Noble Corp.	NE	Drilling, equipment & services
Transocean Inc.	RIG	Drilling
Pride	PDE	Drilling
ENSCO Intl	ESV	Drilling
Rowan	RDC	Drilling
Todco	THE	Drilling
Baker Hughes, Inc.	BHI	Oil equipment & services
Halliburton Company	HAL	Oil equipment & services
National Oilwell Varco, Inc.	NOV	Oil equipment & services
Smith International, Inc.	SII	Oil equipment & services
Weatherford Int'l, Inc.	WFT	Oil equipment & services
Cooper Cameron Corp.	CAM	Oil equipment & services
BJ Services Company	BJS	Oil services
Keppel Corp (Singapore)	--	Rig builder
SembCorp (Singapore)	--	Rig builder
Daewoo Shipbuilding (Korea)	--	Rig builder
Samsung Heavy Industries (Korea)	--	Rig builder

Midstream

The midstream sector involves the transportation of crude oil and finished products by ship tanker, pipeline, railcar tanker and truck tanker. While several OPEC and non-OPEC producers own their own methods of transporting oil, most use independent transportation operators for some or all oil movements. Some of the larger transportation companies include:

Table 5-5

Company name	Ticker (NYSE)	Specialization
Teekay Shipping Corp.	TK	Ship tankers
Frontline Ltd. (USA)	FRO	Ship tankers
Overseas Shipholding Group	OSG	Ship tankers
A/S Steamship Company Torm	TRMD	Ship tankers
General Maritime Corp.	GMR	Ship tankers
OMI Corp.	OMM	Ship tankers
Ship Finance International Ltd.	SFL	Ship tankers
Seacor Holdings Inc.	CKH	Ship tankers
Kirby Corp.	KEX	Ship tankers
Colonial Pipeline Company	n/a - private	Pipelines
Enbridge Inc.	ENB	Pipelines
Kinder Morgan Energy Partners LP	KMP	Pipelines
Plains All American Pipeline LP	PAA	Pipelines
TEPPCO Partners LP	TPP	Pipelines
Buckeye Partners LP	BPL	Pipelines

Downstream

The downstream sector is sometimes referred to as the refining and marketing, or R and M, sector and also involves storage. As with the midstream sector, the downstream sector is fairly diverse with hundreds of organizations competing. Refineries without any attachment to upstream operators are known as independent refineries.

There were 720 refineries around the world in 120 countries with a total charge capacity of 85.4 million barrels per day. Because of net refinery processing gain, the production capacity from these refineries is approximately 88 million barrels per day. The US has 145 refineries with a combined charge capacity of 17.4 million barrels per day.

It costs over US$5 billion and takes between 3 and 5 years to build a new complex refinery which can process 500,000 barrels per day. Although OPEC nations produce around 40% of the world's crude, only 10% of refinery capacity is located in OPEC nations. From the 1970s to the early 2000s, oil refining was one of the least profitable parts of the oil industry.

Other reasons refineries tend to be located in non-OPEC countries are that investors usually wish to locate this expensive plant in a stable country. Thus, refineries tend to be located in oil consuming nations and not in oil producing nations. In addition, oil consuming countries want to have the ability to control as much of their finished oil product supply as possible for strategic reasons.

Over time refineries tend to become larger as economies of scale are sought. This is called capacity creep. Additionally, in what is referred to as complexity creep, existing refineries have been able to produce larger volumes of finished products by adding more expensive equipment which cracks large heavy crude oil molecules into lighter, more voluminous products.

Countries with the largest refinery capacity Table 5-6

		Refinery Charge (input) Capacity mmbpd	Capacity as % of Total Global	Oil consumption mmbpd
1	US	17.4	20.4%	20.9
2	China	6.3	7.4%	7.7
3	Russia	5.3	6.2%	2.9
4	Japan	4.7	5.5%	5.2
5	South Korea	2.6	3.0%	2.2
6	Germany	2.4	2.8%	2.6
7	Italy	2.3	2.7%	1.7
8	India	2.3	2.6%	2.7
9	Saudi Arabia*	2.1	2.5%	2.1
10	Canada	2.0	2.3%	2.6
11	France	2.0	2.3%	2.0
12	Brazil	1.9	2.2%	2.3
13	UK	1.9	2.2%	2.1
14	Mexico	1.7	2.0%	2.1
15	Iran*	1.5	1.7%	1.6
16	Singapore	1.3	1.6%	0.8
17	Taiwan	1.3	1.5%	1.0
18	Venezuela*	1.3	1.5%	0.6
19	Spain	1.3	1.5%	1.6
20	Netherlands	1.2	1.4%	0.9
	100 other nations	22.7	26.6%	20.1
Global Charge Capacity		85.4		85.7

*OPEC member nations. (mmbpd = millions of barrels per day.) Source: IEA 2007 data

Largest Refineries in the World
Table 5-7
(by charge capacity)

	Refinery Owner	Location	bpd
1	Reliance Industries I & II	Jamnagar, India	1,240,000
2	PDVSA	Cardón & Amuay, Venezuela	955,000
3	SK	Ulsan, South Korea	817,000
4	GS Caltex	Yosu, South Korea	650,000
5	ExxonMobil	Jurong Island, Singapore	605,000
6	ExxonMobil	Baytown, TX, US	562,500
7	Saudi Aramco	Ras Tanura, Saudi Arabia	550,000
8	Formosa	Mailiao, Taiwan	540,000
9	S-Oil	Onsan, South Korea	520,000
10	ExxonMobil	Baton Rouge, LA, US	503,000
11	Hovensa (PDVSA/Hess)	St. Croix, Virgin Islands	494,000
12	Shell	Pulau Bukom, Singapore	458,000
13	NIOC	Abadan, Iran	450,000
14	KNPC	Mina Al-Ahmadi, Kuwait	443,000
15	Citgo	Lake Charles, LA, US	429,500
16	BP	Texas City, TX, US	417,000
17	Shell	Pernis, Netherlands	416,000
18	BP	Whiting, IN, USA	410,000
19	Saudi Aramco/Exxon	Yanbu, Saudi Arabia	400,000
20	Saudi Aramco	Rabigh, Saudi Arabia	400,000
21	Rosneft	Angarsk, Russia	384,000
		Other refineries	74,996,000
		Total world capacity	85,400,000

Source: EIA 2008 data, Oil & Gas Journal and various news reports

In the US, there was a boom in domestic refinery construction during the 1970s due to the US government's Crude Oil Entitlement Program which ran from 1974 to 1981. The program encouraged the building of small inefficient refineries which quickly closed down when US price domestic crude oil price controls were eliminated in 1981.

Very few new refineries have been built over the past 25 years in the US due to the massive excess of global capacity added in the 1970s and it is only in the past few years that refiners have had decent profits. Due to capacity and complexity creep, US refinery throughput has not declined despite the reduction in the number of refineries.

Since 2000, new refinery capacity has mostly been built in the Middle East and Asia rather than in the US and Europe or South America.

Other uses of the upstream, midstream, and downstream terminology

The upstream, midstream, and downstream terminology, in addition to being used to describe the overall oil industry, is also used to describe processes within a refinery – upstream being initial distillation and downstream being cracking, reforming, coking, blending and other post-distillation processes.

Similarly in the petrochemical industry, upstream refers to the cracking of feedstock, such as naphtha, ethane, and propane, and downstream refers to further down the petchem production chain, such as the production of plastic resin for containers.

Additional oil industry participants

Energy Information Administration (EIA)

The US Department of Energy's (DOE) Energy Information Administration (EIA) was created in 1977 as the statistical wing of the DOE. The DOE formulates policy advice for the US Executive branch using EIA data. The EIA is most well known for its weekly US petroleum stock report, which is published on Wednesday at 10:30am Eastern Time or on Thursday if there was a holiday since the previous week's numbers. It is compulsory for those surveyed in the US to provide data. The EIA also puts together energy market reports, forecasts and policy independent research.

The EIA oil stock report each Wednesday shows US inventory levels as at 7am the previous Friday. The EIA's independence from Executive review is enshrined in law and is supposed to ensure that the Executive (President and executive agencies, such as the DOE, and the Department of the Treasury) branch of the US government cannot require changes to data in order to make them more politically favorable. The EIA has an excellent web site (www.eia.doe.gov) and appears to have the most reliable statistics in the oil industry globally.

International Energy Agency (IEA)

The Organization for Economic Co-operation and Development (OECD) is a forum for 30 of the most industrialized nations to discuss and agree on recommendations for policy to enhance economic growth. Following the Arab oil embargo of 1973-1974 the OECD decided to create an agency to collect and share information and to formulate and co-ordinate energy policies in order to prevent and mute economic shocks resulting from oil supply disruption.

The OECD created the International Energy Agency (IEA) in 1974, which is today comprised of 27 of the 30 OECD member nations. IEA Member countries commit to hold emergency oil stocks equivalent to at least 90 days of net oil imports and provides advice and recommendations on the coordinated release of such stocks. Such stocks can be held by government or industry.

A coordinated release of global strategic crude is hoped to lessen short term market panics. Since 1974, the IEA has required the coordinated release of global stocks only once. In September of 2005, following Hurricane Katrina damage to the US Gulf Coast refineries, the IEA made available 60 million barrels of oil to the market.

In January 1991, two days after the beginning of Operation Desert Storm, the IEA announced a contingency plan to make available to the market 2 million barrels per day. The US released oil from its Strategic Petroleum Reserve (SPR) alone and prices fell. The IEA did not activate its contingency plan to make available any oil in 1991.

When the IEA announces a stock release, it simply requires each member nation to allow the oil industry to buy from strategic reserves. The oil doesn't actually have to be sold if nobody bids for it.

The IEA collects data on global oil inventories which it publishes in its Monthly Oil Market Report (OMR). The IEA also produces forecasts of oil consumption, production and prices.

American Petroleum Institute
The American Petroleum Institute (API), established in 1919, is a private organization funded by the US oil industry to be its representative voice. The API also produces and maintains standards covering petroleum industry equipment and products, as well as lobbying the government to reduce taxation of the oil industry. The API produces weekly oil statistics, published on Tuesdays at 4:30pm Eastern Time. The oil market tends to place a low value on the API inventory numbers as several large oil market participants do not submit data to the API. The oil market tends to favor the weekly EIA numbers as all organizations have to submit data to the EIA.

Euroilstock
Euroilstock is a private organization, based in the Netherlands, which publishes statistics on European Union oil stocks, and supply and demand on a monthly basis. The stock data, while a decent guide, are heavily revised, and therefore some say its relevance is not as good as it could be. Also, unlike the mandatory US DOE EIA submissions, Euroilstock submissions are voluntary.

Minerals Management Service (MMS)
The mandate of the US Department of the Interior (DOI) is to protect and conserve the nation's natural and cultural resources. The DOI is most well known for running US national parks. The Minerals Management Service (MMS) is a bureau in the DOI which manages the nation's natural gas, oil and other mineral resources on the federally controlled outer continental shelf

(OCS) in addition to revenue from onshore mineral leases on Federal and Indian lands.

The OCS comprises submerged lands outside individual States' jurisdictions. Nationally, the OCS begins at 3 nautical miles from land, with exceptions of 9 nautical miles in Texas, and 3 imperial nautical miles in Louisiana. The outer edge of the OCS is defined by principles of international law and is typically several hundred miles from shore.

The MMS manages roughly $8 billion per year of income from federal leases of oil production areas on the OCS, mostly in the Gulf of Mexico.

The oil markets pay most attention to the MMS during hurricane season, which runs from June 1 to November 30. It is viewed to have the most accurate assessment of offshore oil and gas platform evacuations and production shut ins. The MMS monitors shut ins because when oil production stops, it doesn't receive lease royalty payments from offshore oil producers.

Commodity Futures Trading Commission (CFTC)

The US CFTC was created in 1974 to protect the public from fraud, manipulation and abusive practices in US futures and options exchanges. The CFTC oversees the New York Mercantile Exchange (NYMEX) on which three of the most liquid petroleum futures and options contracts globally are traded: West Texas Intermediate (WTI) crude, New York gasoline, and New York heating oil.

In addition to its regulatory function, the CFTC is most well known in the oil market for its weekly Commitments of Traders (COT) report which provides a valuable window into market psychology. The COT report, first published in 1962, is published every Friday at 3:30pm Eastern Time. The report provides a breakdown of the previous Tuesday's open interest on future and options exchanges. The report is broken down into futures only, and futures and options combined, numbers. The COT report also shows open interest in NYMEX cleared oil swaps.

Market positions are also subdivided into categories depending on what type of trader is holding the position: commercial; non-commercial; and non-reportable. Commercial traders are those, such as an oil producer or refinery, having futures and/or options position in order to hedge a real commercial exposure. The non-commercial positions are purely speculative in nature, such as those held by hedge funds or individual speculators.

The change in non-commercial position is sometimes viewed as being a leading indicator of market price direction.

Very large non-commercial positions are also seen as a contrarian indicator, as if traders are long, which means they all have positions which bet that the market will rise, then they will eventually have to sell those contracts to take profit or stop losses – thus a very large long position is actually bearish the larger it grows. The opposite is true if non-commercial traders have a large short position in that it indicates that the market may be approaching a bottom.

In December 2005 the CFTC launched a monthly report which lists open positions of U.S. and foreign banks active in futures and options markets. This Bank Participation report is published on the first Friday following the first Tuesday of each month.

National Oceanic and Atmospheric Administration (NOAA)
The NOAA's National Weather Service's (NWS) National Hurricane Center (NHC), a part of the US Department of Commerce, provides the oil market with detailed status and forecasts for weather in the high productivity Gulf of Mexico (GOM or GOMEX) especially during hurricane season which is from June 1 to November 30.

During winter heating demand season the NOAA's NWS forecasts for the US Midwest and Northeast are also looked at particularly closely.

US Geological Survey (USGS)
Created in 1879, the USGS is the natural science agency of the US Department of the Interior. In the oil market, the USGS is most well known for their occasional estimates of the global oil endowment which they last published in the year 2000. The USGS year 2000 assessment of global oil reserves puts remaining discovered and undiscovered global reserves at approximately 1,000 billion barrels (95% confidence) and 2,000 billion barrels (50% confidence.) To put this number in perspective, global consumption is currently just over 30 billion barrels (Gb) per year. The year 2000 assessment has some very vocal critics who point out that the USGS assessment has many basic flaws which render the report unreliable.

Japanese METI
Japan's Ministry of Economy, Trade, and Industry (METI) publishes monthly statistics on petroleum stocks, supply and demand in that nation which are viewed as being fairly reliable.

<u>IE Singapore</u>

Formerly known as the Singapore Trade Development Board (TDB), International Enterprise Singapore, or IE Singapore, is Singapore's vehicle to encourage foreign investment into the city state. Singapore is a major oil trading hub in Asia due to its geographic location, stable legal environment, and relatively large number of refineries, storage facilities and oil traders located there. IE Singapore releases oil stock numbers every Thursday at 4pm local Singapore time. Although it is one of the few honest storage data assessments in the region, with new storage and trading hubs appearing throughout Asia, particularly in mainland China, Malaysia, and Indonesia, the IE data is becoming less relevant as an indicator of regional supply/demand balances.

Investing in the oil industry

There are many choices for investors to gain exposure to the oil industry including shares and bonds of oil corporations, mutual funds which invest in oil company shares, exchange traded funds (ETFs), exchange traded notes (ETNs), mutual funds which track commodity indices, OTC financial instruments which track oil prices, capital guaranteed notes from banks, and specialized vehicles such as royalty trusts and master limited partnerships.

Specialized investment entities found in the oil industry

Royalty trusts and Master Limited Partnerships (MLPs) are two investment vehicles particular to the North American oil and natural gas industry.

<u>Royalty trusts</u> are a mechanism for the owner of an asset, an oil field for example, to raise cash secured solely on that asset. The original asset owner can then apply that cash to paying down debt or to other projects which they view as having higher returns.

Shares of royalty trusts, such as BP Prudhoe Bay (ticker: BPT), trade on equity exchanges similar to any other stock. Royalty trusts are essentially financing vehicles backed by assets which produce cash flow and have no operations or management. Royalty trusts pay out almost everything they earn in the year in which it is earned.

Royalty trusts are not taxed at the corporate level if they pass all income directly to their shareholders. Trust managers aren't allowed to issue fresh debt or equity which inhibits growth of the royalty's asset base. If the asset on which the royalty trust is based, such as an oil field, depletes the value of the royalty trust investment declines to zero.

<u>Master limited partnerships (MLPs)</u>, in the US, are publicly traded partnerships, the shares of which, similar to royalty trusts, trade on an exchange just like any other stock. MLPs were fashioned by the US federal government in the 1980s to encourage development in the energy industry. The MLP structure allows an energy company to spin off reliable cash flow generating production and transportation assets into an MLP structure.

Normally, corporations pay corporate tax and then shareholders pay income tax on dividends. Shareholders of MLPs are effectively partners and the MLP pays no federal corporation tax. Instead, the individual shareholders pay personal income tax on distributions, thus avoiding double taxation. Companies putting assets into MLPs do so to raise cash to pay down debt or invest in faster-growing businesses. MLP assets are typically oil or natural gas wells, oil or natural gas pipelines, or distribution companies for propane or home heating oil. MLPs are, because of the tax advantage, typically valued higher than the equivalent corporation. MLPs are exempt from corporate income taxes so long as it pays out almost all of its profits to investors.

MLPs differ from royalty trusts in that royalty trusts, apart from essential maintenance of the asset which produces revenue, do not reinvest any of their profits in expansion and have no management team apart from a trust manager. MLPs have a management team and can expand by raising additional debt and equity.

CHAPTER SIX
EXPLORATION AND PRODUCTION

Exploration for, and production of, crude oil, known as the E&P or upstream sector, is the most high risk and lucrative part of the oil industry. The costs involved in the earliest stage of E&P are finding and development, or F&D, costs.

Exploration
Obtaining rights to explore and drill
Before an oil company can begin to explore, it must obtain permission from two owners: the surface property owner where exploration and production facilities will be built, and the owner of the mineral rights to the petroleum contained beneath the surface.

In most countries any valuable minerals below the surface, such as oil, are owned by the government, even if the land is the property of private individuals or organizations.

The US is one of a rare few countries, along with Canada, in which private individuals and organizations can own mineral rights on privately held land. The US government only owns the rights to minerals on US federal lands such as in national parks and beyond individual State offshore boundaries.

In the US, a person called a landman often negotiates the purchase of exploration and production rights. Investing to find oil is called an oil play and a geological feature of interest is known as a lead. Once surface and mineral rights have been legally secured and the area ready for drilling, the oil play and lead become known as a prospect.

Oil underneath the surface is frequently governed by the rule of capture, a legal principle based originally on English common law rights to catch wildfowl. According to the rule of capture, if you have possession of oil, and you extracted it from land you own, then it is yours, even if the oil originally came from part of an oil reservoir under your neighbors land – just as you would have ownership of a wild pheasant that had flown from your neighbors land into your back yard whereupon you captured it.

One is typically not permitted to use slant drilling to physically put a well bore under a neighbor's land – which is referred to as subsurface trespass. Under the rule of capture, the oil must flow to your land and once it is under your surface you can extract it as your own. Subsurface trespass is not just an academic legal point. Iraq invaded Kuwait in 1990 under the pretext, among many others, of stopping Kuwait from allegedly meeting its OPEC quota entirely by slant drilling and producing oil from a field which lay underneath a border area between the two countries.

Those unfamiliar with the oil industry often think that oil is pulled across the reservoir by pumps. This is not the case. Oil moves across a reservoir to the bottom of a well because pressure in the reservoir is higher than the low pressure at the bottom of a well. Once the oil gets to the bottom of the well, if reservoir pressure is high enough, the pressure differential may be sufficient to additionally push the heavy column of oil in the well shaft up to the surface without any pump. Pumps are sometimes used to help to lift the column of oil up through the well shaft, but they generally cannot pull oil across a reservoir to the well.

Reservoir pressure is reduced every time one extracts oil, until pressure is so low that the remaining oil has to be simply left in the ground. Of the total original oil in place (OOIP), it is exceptional to have a recovery factor (RF) above 60% of the OOIP from a reservoir. The average worldwide RF is around 30%.

There is an optimal rate of production which maximizes the total amount of oil which can be removed from an oil field. This optimal rate makes the most efficient use of reservoir pressure until it is depleted. If one produces above the optimal rate, by increasing the number of bpd, the ultimate amount of oil produced will be reduced.

The rule of capture principle plays havoc with attempting to produce from a reservoir at an optimal rate. Individual competing producers with wells taking oil from the same reservoir will simply locate wells too close to one another and produce at maximum, not optimal, flow rates, thus rapidly depleting pressures, in a mad race to capture the oil before everyone else. The total amount of oil produced under the rule of capture is therefore usually much less than if a single producer, or a coordinated group, manages production rates.

In the early days of the oil industry, it would be common to find investors buying a few square feet of land above a reservoir to drill a well. An oil field would quickly become covered in hundreds, if not thousands of wells every few feet, with everyone trying to get as much oil out as fast as possible (Fig. 6-1).

Forest of oil wells at Spindletop in the early 1900s   Fig. 6-1

Source: USGS

These days the spacing around a single well is often kept to 40 acres, as there is a greater understanding of the benefit of controlling well flow rates to maximize total output over the production lifetime of a reservoir. The 40 acre well spacing is common, but not a hard rule. Drilling additional wells within the 40 acres around a producing well is referred to as infill drilling, or offset drilling, and outside the standard 40 acres is referred to as step-out drilling.

If an oil company can, and if the economics are favorable, it will buy mineral rights outright. Where mineral rights are owned by a party unwilling or unable to sell those rights outright, exploration and production of crude are typically carried out using one of five types of agreement between the exploring entity and the land/mineral rights owner: concessions, production sharing agreements, service contracts, joint ventures and production contracts.

Concession agreements are one of the oldest types of oil production agreement and were common up until the late 1960s between oil majors and OPEC governments.

Today, concession agreements are not used as much by OPEC nations and are instead used by the US federal government and other non-OPEC governments.

Concession agreements are also known as lease agreements, or direct equity agreements. They allow a private individual or organization (the lessor) to explore for and hopefully produce oil on behalf of the mineral right's owner (the lessee), which can be a government, individual, partnership, or corporation. Exploration must take place within a fixed primary term period, often 2-5 years from the date the concession is issued. If oil is discovered, the lease may be extended into a secondary term in which production takes place.

Most governments have regulations which divide onshore and offshore territory into sectors and smaller blocks. Concessions are often put out for tender, or bidding, in such blocks.

At the outset of exploration, a cash fee, referred to as a bonus, is paid to the lessee of the petroleum rights. A second cash fee called a rent is paid to the lessee to retain the concession if production is not taking place.

If production begins, a payment, called a royalty, is paid in cash or in physical crude oil. The share of physical production which the lessor receives can be called equity crude, equity production, or lease crude. The percentage share is called working interest. You may often hear a producer say "we are equity producers in that field", which means they have an active concession with ownership in part of the crude produced.

Under concessionary agreements the E&P individual or organization (the lessor) assumes all the risks and costs of finding and developing. The lessor must try to re-coup these costs from the revenue remaining after the royalty is paid.

In the US, the Minerals Management Service (MMS) manages concessions/leases on behalf of the US federal government. Most of the MMS managed oil exploration and production takes place on offshore underwater federal lands beyond State boundaries. Federal royalty crude is physical crude oil, in the form of in kind royalties, delivered to the MMS.

Production sharing agreements (PSAs), known also as buyback agreements, began to appear more often in the late 1960s. PSAs involve an oil company bearing the exploration and production costs and trying to recoup these costs over a fixed period of time. That period is typically at least 5 years, often with a maximum of 30 years, before transferring ownership of the entire venture back to the government/mineral rights owner – hence the term buyback. The E&P company has to recoup its costs out of a fixed percentage of production, if any, over the limited time window. This restricted production time window of PSAs is the major difference with concession agreements, which can remain in place until the oil field is abandoned.

PSAs are often used in circumstances where foreign direct equity ownership of oil exploration and production is limited by the producing nation, such as in Iran and Saudi Arabia.

Whether oil is discovered or not, the E&P organization has to pay a royalty to the mineral rights owner as consideration for entering into a PSA.

As there is a time limit on participation in a producing oilfield, most PSAs are designed to allow the E&P entity a decent chance to recoup its exploration and production costs upfront.

A certain limited portion of any initial production is set aside against which costs can be deducted before having to share revenue with the mineral rights owner. This portion of production is called cost oil, and is limited to ensure the oil company doesn't over inflate costs.

The other portion of oil produced is called profit oil and revenues from the sale of this oil are shared between the E&P entity and the mineral rights owner – hence the term production sharing.

Service contracts, or contractor agreements, involve an oil company carrying out exploration and production on behalf of the mineral rights owner for a flat fee. The flat fee is paid only if production takes place so this is a very risky contract for an oil company to enter into. The ownership of any oil produced remains totally with the mineral rights owner and never passes to the oil service contractor.

Exploring for oil is a high risk venture with a significant failure rate which is why oil companies rely on receiving big upside surprises and a share of large discoveries every once in a while, and is why service contracts are not very popular among IOCs.

Joint venture (JV) agreements are agreements to share the cost of exploration and production with a National Oil Company (NOC) or another oil producer. JVs are usually used for very high risk plays, such as ultra deep water exploration where the risks for a single producer going alone may be too high. JVs may also be required by governments in order to create jobs locally and facilitate the sharing of technical knowledge with NOCs.

JVs differ from concessions, PSAs and service contracts in that, with JVs, all parties to the agreement risk capital from the outset of exploration. Although there are several investors in a JV, one of the parties to the JV, known as the field operator, usually runs the oilfield processes on a day to day basis for a fee.

Production contracts are used to develop improved oil recovery techniques such as waterflooding or gas reinjection, both of which are discussed later in this chapter, to maintain or increase the rate of production from a field. The machinery to carry out such improvements is expensive to install and maintain. Using a production contract, the mineral rights owner will give the installation contractor a share of the revenue from any production rate increases generated by such methods. Such agreements are also called production contracts, or production new ventures agreements.

Now we have covered the common legal frameworks under which exploration and production can take place, we will move onto what everyone is looking for.

Essential oilfield geological features

Exploration involves searching for indicators of economically and technically producible oil. In particular, prospectors are looking for the three features necessary for the formation of a crude oil reservoir: a source rock, a reservoir rock, and a cap rock.

Source rock, also referred to as a mother rock, is rock laden with kerogen. Kerogen is a solid dark waxy rock.

In the normal cycle of life, most dead plant and animal material is consumed by living plant and animal material. Organic material can be taken out of this life-cycle under very specific conditions such as those present in dark, oxygen-poor areas at the bottom of some lakes, river deltas, seas, or oceans where decomposition by other plants or animals is not possible.

All living creatures are comprised of the elements carbon, hydrogen, oxygen, nitrogen, sulfur, phosphorus, along with minor amounts of other elements. Kerogen is comprised of these same elements in addition to minor amounts of inorganic materials picked up from the surrounding area.

The hydrogen and carbon within kerogen combine over time to produce very high molecular weight complex hydrocarbon molecules, often with 1,000 atoms or more in their structure. Kerogen is, therefore, dense and heavy.

Marine source organic material such as plankton and algae that has been deposited in sedimentary layers over many years in dark anaerobic dead zones on the ocean floor are believed to account for most kerogen which goes on to form oil.

Terrestrial source material tends to form kerogen which becomes solid hydrocarbon fuels such as peat or coal rather than oil.

There are three types of kerogen which produce hydrocarbons:

Table 6-1

Kerogen type:	Type I (Algal)	Type II (Liptinic)	Type III (Humic)
Kerogen source:	Marine algae and plankton	Marine and terrestrial plant and animal material	Terrestrial plant material
Hydrogen to Carbon ratio	> 1.25	< 1.25	< 1.0
Oxygen to Carbon ratio	< 0.15	0.03 – 0.18	0.03 – 0.3
Hydrocarbon formation propensity	Rich in hydrogen and poor in oxygen and thus tends to produce liquid oil.	Produces liquid oil and/or gas.	Low in hydrogen and high in carbon so produces mainly coal.

As type I and II kerogen are formed mostly by sedimentation of marine material, oil tends to be found in sedimentary rocks produced when ancient rivers slowed down and suspended organic material fell to the river bed. Geological rock formations that have common features are referred to as provinces. Examples of oil-bearing provinces are the Permian sedimentary basin running through Texas, and the Appalachian sedimentary basin, where the modern oil industry began in Western Pennsylvania.

Sedimentary rocks containing oil are predominantly either sandstone or carbonate. In turn, carbonate rocks holding oil are most often either limestone or dolomite.

Geologists use the time period the rocks were created to classify a rock province (Table 6-2). Just over half of the oil discovered since 1859 was formed during the Cenozoic Era. In other words, if you are searching for oil, your chances of discovery in newer rock formations are higher. Petroleum was also formed in earlier eras. However, such oil has had much more time to escape to the surface where it evaporates, is metabolized by bacteria, or is oxidized. A summary familiarization with geological time scale is useful as geologists will frequently refer to the scale when discussing an oil play or prospect.

Geological time scale

Table 6-2

Era	Period	Epoch	Approximate duration (millions of years)	millions of years ago
Cenozoic (Most oil still existing today)	Quaternary	Holocene	10,000 years to the present	(most recent ice age ends)
		Pleistocene	2	0.01 (first humans)
	Tertiary	Pliocene	11	2
		Miocene	12	13
		Oligocene	11	25
		Eocene	22	36 (Himalayas form)
		Paleocene	71	58 (dinosaurs have become extinct)
Mesozoic	Cretaceous		71	65 (first birds)
	Jurassic		54	136
	Triassic		35	190 (dinosaurs and mammals)
Paleozoic	Permian		55	225
	Carboniferous - Pennsylvanian - Mississippian		65	280 (coal forming swamps)
	Devonian		60	345 (amphibians)
	Silurian		20	405 (land plants)
	Ordovician		75	425 (vertebrates appear as fish)
	Cambrian		100	500 (shelled animals)

The biological theory of petroleum's origin is supported by the presence of organic debris in petroleum in addition that oil is almost always found in rock formed by the sedimentation of organic material. Connecting the dots of the oil formation chemical and physical processes is relatively straightforward. Versions of the same theory are commonly accepted for the formation of other hydrocarbon fuels, such as coal and natural gas. This is why hydrocarbon fuels are called fossil fuels.

An alternative to the biological, biogenic, theory is the abiogenic theory, also known as the abiotic theory. The abiogenic theory holds that since the most basic hydrocarbon, methane is known to exist on planets with no known life, petroleum and natural gas may originate deep within the earth independent of biological activity. In other words, crude oil close to the surface may be a result of the combination of methane molecules from deep within the earth. Abiogenic theory suggests that the existence of biological debris in petroleum and natural gas deposits is due to the ability of bacteria to live in extreme conditions deep underground. Recent discoveries of bacteria living in and around super hot boiling sulfur springs under immense pressure in the absence of sunlight and oxygen at the bottom of the ocean have lent some support to the abiogenic argument. Abiogenic theory also garners support from unusual and rare hydrocarbon deposits such as extremely deep oil deposits which are below what geologist call the oil formation window.

Although abiogenic theory is quite interesting and should be considered and tested further, the current overwhelming and scientifically convincing hard data supports biological sources for fossil fuels such as crude oil.

So, now that we know that a kerogen laden rock is the starting point in oil formation, we move onto how the transition to oil occurs.

If a kerogen deposit is buried deep enough; the heavy complex hydrocarbon molecules of kerogen will, over time, be cracked by heat into smaller, lighter hydrocarbon molecules. These smaller hydrocarbon molecules are what comprise liquid oil and natural gas.

Cracking requires heat to occur. Just like a stove-top pressure cooker, increasing pressure reduces the amount of heat required for chemical changes to occur.

As one digs or drills beneath the earth's surface, the temperature one observes increases closer to the planet's hot core. The temperature increase, known as the geothermal gradient, varies slightly around the world, but is approximately +15°F per 1,000 feet, or +25°C per kilometer (Fig. 6-2 and 6-3). Some people heat their houses using this heat, which is known as geothermal heating.

Pressure also gradually increases with depth underground. This increase is known as the pressure gradient. Pressures increase with depth due to the weight of rock being pulled toward the earth's core by gravity. The additional pressure reduces the temperature required to crack molecules.

As kerogen is buried, the heat and pressure breaks down heavy hydrocarbon molecules into light molecules which make up liquid oil. The deeper the kerogen is buried, the higher the temperature and pressure and the lighter the resulting oil tends to be. If the remaining kerogen and any oil created are buried deeper still, the hydrocarbon molecules are all broken down into methane, the simplest and lightest hydrocarbon.

The range of depths at which kerogen is heated sufficiently to form liquid oil is known as the oil window. The deeper level at which any remaining kerogen and the crude oil formed at shallower depths are cracked into methane is known as the gas window.

Fig. 6-2

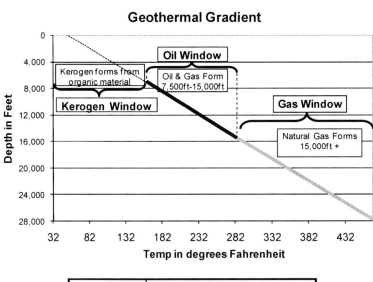

	Depth	Temp. F
Oil Window	7,500ft - 15,000ft	165 - 285
Gas Window	15,000ft +	285 +

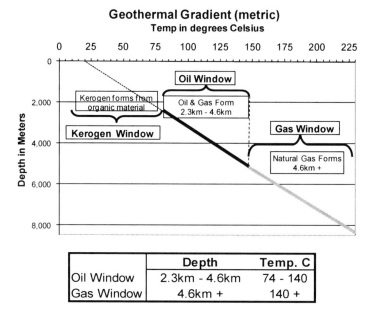

Fig. 6-3

	Depth	Temp. C
Oil Window	2.3km - 4.6km	74 - 140
Gas Window	4.6km +	140 +

Maturation is the process by which crude oil is formed from kerogen laden source rock. The area in which the maturation occurs is referred to as the kitchen, as the waxy kerogen is effectively cooked into oil and/or gas.

Conditions in the petroleum formation kitchen are crucial. There is little point looking for oil in a rock formation which has been buried so deep to have been heated above 200°C, as that temperature would have been outside the oil formation window of temperatures.

When a well is drilled, the temperature measured at the bottom of the borehole only indicates current status, not temperatures present when the oil formed. Geologists analyze rock samples and carry out tests to estimate paleotemperatures, which are temperatures rock experienced in the past.

Once the kerogen is cracked into less dense crude oil and natural gas hydrocarbon molecules, it migrates upward. Just as a less dense hot air in a balloon moves higher away from the more dense air at the earth's surface; the oil rises from the dense kitchen area until it collects in a reservoir rock that is capped by an impermeable trap rock. If there is no suitable reservoir rock capped by an equally suitable cap rock, the oil and gas migrates all the way to the surface where it evaporates over time, is metabolized by bacteria, or is oxidized by contact with oxygen into the atmosphere. Most oil formed in the past has succumbed to this fate.

Crude oil formed in the oil window tends to become denser during migration toward the surface, as lighter molecules are degraded first. For this reason, crude which has migrated to shallow depths is often heavy and called old oil. Crude found at deeper levels is said to be young oil as it is lighter and yields higher value products.

Reservoir rock must have communication with the source rock, sufficient porosity to hold oil and sufficient permeability to allow the oil to pass through it to an oil well (Fig. 6-4). Porosity and permeability are a function of what is referred to as the lithology of a rock, which is a description a rock's mineral content, texture and grain size.

Although kerogen laden shale is a good source rock, it is not permeable enough to be a good reservoir rock. Migration of the oil into a reservoir rock has to occur.

In effect the perfect reservoir rock is a solid sponge-type matrix material.

Oil in pores within rock matrix

Fig. 6-4

Oil within a pore which is connected to other pores.

There are three general rock classifications depending on how the rock was created. Igneous rocks, such as granite or basalt, are formed by heat. Sedimentary rocks, such as sandstone, limestone and dolomite, are formed by layers of sediment. Metamorphic rocks, are formed by physical or chemical changes in an igneous or sedimentary rock due to pressure and heat. Metamorphic rocks include slate formed from shale and marble formed from limestone.

Most hydrocarbon reservoirs are found in sedimentary rock basins, so geologists rule out looking for oil in igneous and metamorphic rocks (Fig. 6-5). Sedimentary rocks make good reservoir rocks as they are more porous and permeable than igneous and metamorphic rock and can hold oil and allow it to travel through the rock.

Major Sedimentary
Basins of the World

Source: USGS

Sedimentation occurs as moving water slows and suspended debris drops to the bottom; hence the term sedimentary basins. Sedimentary rocks commonly found holding oil are sandstonnes, carbonates, conglomerates and unconsolidated sands. Carbonates such as calcium carbonate and calcium-magnesium carbonate are more commonly known as limestone and dolomite, respectively. Conglomerates are coarse large grain sedimentary rocks held together with dissolved minerals. Unconsolidated sands are fine grained sedimentary sands.

Most oil is found in sandstone and carbonates. Conglomerates and unconsolidated sands account for a very small minority of conventional reservoir rocks.

Cap rocks prevent migration of oil. If a suitable cap rock does not trap the oil and gas then they will rise to the surface over time, where first the light hydrocarbons will evaporate, be metabolized or oxidized, and leave denser heavy hydrocarbon molecules. This is why surface oil seeps are tar-like in consistency. Eventually, even the heavy hydrocarbon molecules will succumb to the same fate and the seep will no longer exist.

The impermeable cap rock, or trap, can be shale. Although a quite porous sedimentary rock, shale is relatively impermeable. The cap rock can also be salt (sodium chloride) or anhydrite (calcium sulfate.) Salt and anhydrite caps are far less porous and permeable than shale and can cap oil columns over 200 feet.

In order to trap oil, the cap rock has to be in a shape that can prevent the oil from moving upward. There are three categories of trap depending on how the structure of the cap rock was formed: structural, stratigraphic and combination traps.

<u>Structural traps</u> result from rock movement such as folds, faults and salt domes (Fig. 6-6).

Fold traps, also known as anticline traps, are rock formations which have been bent into an undulating pattern by tectonic processes. The oil traps created are finger-like projections which appear every few miles at the top of an anticline, which is a convex up fold. Over 80% of the world's oil is found in anticline traps. Geologists are always on the lookout for surface indications of anticlines in provinces which have suitable source, reservoir and trap rocks.

Fault traps are caused by the movement of rock along a tectonic fault line to form a trap. Oil is trapped next to the fault, and this is where drilling occurs.

Salt dome traps are also called plug traps, and are formed by underground salt movement. Salt deposited by ancient oceans is usually less dense than most other forms of rock. Therefore, it will slowly rise in vertical underground formations, known as salt domes, toward the surface. As the salt rises to the surface over many years, it penetrates layers of rock and creates impermeable pockets in which crude oil can collect and be trapped. Salt domes can result in donut shaped reservoirs, or the salt can trap the hydrocarbons within the dome itself.

Structural traps   Fig. 6-6

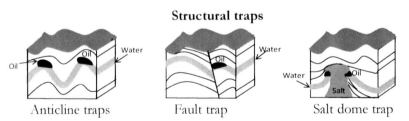

Anticline traps   Fault trap   Salt dome trap

<u>Stratigraphic traps</u> result from sedimentation of different rock strata. The three types of stratigraphic traps are unconformity, pinchout, and lens traps (Fig. 6-7).

Unconformity traps, also called a truncated traps, form when rock formations are thrust upward. They are then worn away by surface elements like the wind and sea. After the weathering, a stratum of impervious rock is laid down over time to form the cap rock. Because unconformity traps are formed by both movement and sedimentation they can be considered a hybrid of structural and stratigraphic traps.

Pinchouts are traps formed where the reservoir is defined by the meeting of two impervious strata.

Lens traps appear a little like the lenses of spectacles and are formed when oil bearing reservoir rock is penetrated by bands of nonporous rock such as shale or clay. The result is that the oil reservoir is broken into small reservoirs underneath each lens. Lens caps are often formed by clastic deposits. Clastic deposits occur when large rocks have been broken into smaller rocks, such as silt being converted to clay. The clay may then settle as a result of sedimentation into lens type formations at, for example, a bend in an ancient river.

Reef traps are a specific type of lens trap in which the cap rock was originally created by shell secretions from organisms such as corals.

Stratigraphic traps   Fig. 6-7

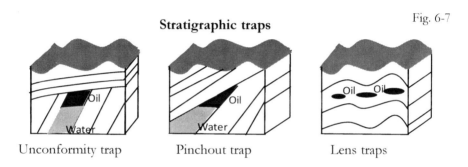

Unconformity trap   Pinchout trap   Lens traps

Combination traps are combinations of various types of traps. Most crude oil reservoirs are a combination of different types of traps.

So now we know that geologists are looking for a source rock containing kerogen which has in the past been buried at oil window depth for a period of time and is in communication with a suitable reservoir rock which is most likely a sandstone or carbonate. Finally, geologists look for signs, such as anticline bumps or fault lines on the surface, which may indicate that a cap rock is in the right shape to trap rising oil.

As you can already tell, there are a lot of factors which have to align for oil to form (Fig. 6-8). If there was no kerogen laden sedimentary source rock, or if it is too young, or if it was never buried deep enough, or buried too deep, then there is no point looking for oil in that area.

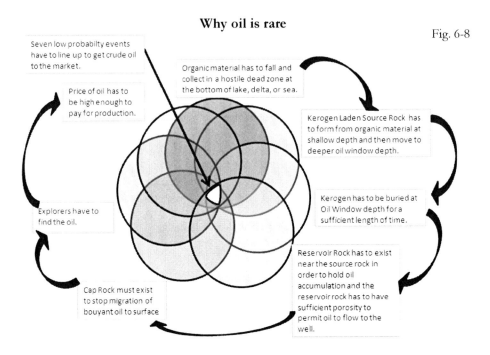

Why oil is rare

Fig. 6-8

Seven low probabilty events have to line up to get crude oil to the market.

Organic material has to fall and collect in a hostile dead zone at the bottom of lake, delta, or sea.

Price of oil has to be high enough to pay for production.

Kerogen Laden Source Rock has to form from organic material at shallow depth and then move to deeper oil window depth.

Explorers have to find the oil.

Kerogen has to be buried at Oil Window depth for a sufficient length of time.

Reservoir Rock has to exist near the source rock in order to hold oil accumulation and the reservoir rock has to have sufficient porosity to permit oil to flow to the well.

Cap Rock must exist to stop migration of bouyant oil to surface

Now that we know what oil explorers are looking for, we can move on how one goes about finding the three essential indicators of an oil field.

Tools used to identify oilfield geological features

Two types of professionals tasked with finding reservoirs are geologists and geophysicists. Geologists study the physical structure of the earth using a variety of sciences including chemistry, biology and astronomy. Geophysicists study the physical nature of the earth based on the interaction of energy and matter, such as sound waves and gravity with rocks and fluids.

Geological surveys involve geologists collecting and analyzing rock samples in order to find areas with sedimentary rocks. Today, the world's rocks have been picked over such that it is almost a certainty that there is no new Saudi Arabian Ghawar field, or other similar massive land-based undiscovered oil province out there. Any large elephant oil fields will have to come from deep offshore areas or polar regions where few geological surveys have taken place so far.

Remote imaging analysis involves explorers poring over photography and radar images taken from aircraft and satellites. They are looking for signs of anticlines or fault lines, the cap rock formations most commonly found to have trapped rising oil.

Gravimetric surveys take advantage of the fact that the force of the earth's gravity varies slightly with changes in density of subsurface rock. Minute variations in the earth's gravitational field can be used to indicate the type of rock, and any fluids, underneath the surface.

Magnetic field surveys involve looking for variations in the earth's magnetic field. Most oil is contained in nonmagnetic sedimentary rocks. Igneous and metamorphic rocks, which contain no oil, are magnetic.

Seismic surveys are used to create an image of subsurface rock. Seismic testing involves bouncing sound waves against subsurface material and then measuring the time it takes for sound reflections to be picked up using devices called geophones. Geophones are sometimes referred to as jugs, or if used over water, hydrophones (Fig. 6-9). Subsurface terrain is mapped with isochronic (iso = same, chrono = time) lines which link points of the same sonic distance.

Large vibrator trucks, sometimes called thumper trucks, are used on land to create the sound source. In the past, dynamite was used to produce sound waves. Over water, an array of air guns is used to create sound waves. Seismic surveying is actually much easier at sea than on land because it is not necessary to have a large group of people involved placing and moving the geophones and one usually doesn't require permission of individual land owners before conducting the tests.

Fig. 6-9

Land-based seismic survey

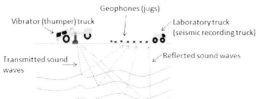

Marine-based seismic survey

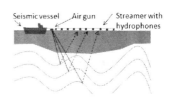

Primitive seismic analysis began to be used to find oil in the 1920s. Even up until the 1980s, the results were sometimes printed out on in long sheets of paper placed on the floor of very large rooms with seismologists poring over the charts. They are looking for what is known as a bright spot, which is a relatively clear change in returned seismic energy at the transition point between a liquid, such as oil and water, and a gas, such as natural gas. The gas may be located as a gas cap over an oil reservoir or maybe a gas field in its own right. Either way, it can indicate the presence of a hydrocarbon deposit.

Up until the 1990s, 2-dimensional seismic analysis was used almost exclusively. 2-D seismic, which shows images of individual slices of a reservoir, is laborious to analyze and the results are somewhat difficult to conceptualize (Fig. 6-10).

The development of 3-D seismic surveys, which requires increased computing power, has revolutionized the process of exploration and production since the early 1990s. In addition to finding and proving the size of a reservoir, 3-D imaging can make it easier to pinpoint the optimal point for drilling. In 3-D seismic analysis, 3-dimensional isochronic depth, width, and height data are together used to form a virtual image of what a reservoir looks like.

Fig. 6-10

2-D image of an oil reservoir

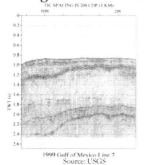

1999 Gulf of Mexico Line 7
Source: USGS

3-D image of an oil reservoir

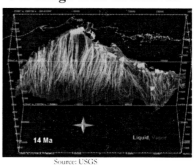

Source: USGS

Since the 1950s, the oil industry has always been one of the world's largest users of supercomputers. Many pioneering computer companies such as Texas Instruments began as providers of technology to the oil industry. These days, large arrays of connected computers are most often used in 3-D surveys.

4-D seismic analysis, which is a time-lapse of 3-D analyses, is used once production has begun to see how the production flows are changing the reservoir.

Seismic analysis is not without challenges and one cannot know the true state of a reservoir until one actually drills into it. For example, seismic analysis cannot tell if reservoir pressure is sufficiently high for commercial production, or how permeable the reservoir rock is.

We now know the indicators (source, reservoir, and cap rocks) of a hydrocarbon reservoir and the people (geologists and geophysicists) and tools (photographic, gravimetric, magnetic, radar and seismic analyses), used to find them.

All of the tests conducted so far can only generate a good idea of where to drill. However, nothing beats drilling a test well.

Drilling for oil

There are many situations in which drilling takes place. Exploratory, appraisal and step-out wells are test wells, used to discover and define a reservoir. Development and infill wells are production wells, used to extract oil from a reservoir.

Exploratory wells, referred to as wildcat wells, are drilled in an unproven area to discover oil. A wildcat well is considered dry if it contains no commercially viable oil. The ratio of dry to successful wildcat wells globally tends to be about 4:1. If a wildcat is drilled in an area which no well has ever been drilled before, it is known as a rank wildcat. Wildcatters usually try to carry out 'tight hole operations', which means keeping the geological results of exploratory wells confidential, to prevent competitors bidding up nearby drilling rights.

Appraisal wells, also known as definition wells, are wells drilled at locations around a well which is currently in production to determine and outline the size and quality of the reservoir. In recent years, as drilling is expensive, the number of appraisal wells has declined. Instead of drilling appraisal wells, oil producers increasingly rely on 3-D seismic analysis for reservoir definition in an attempt to lower production costs.

Step-out wells are drilled outside the proven limits of an oil field. Whereas an appraisal well is used to merely confirm a reservoirs size, a step-out well is specifically drilled to test whether production can be expanded beyond a producing area.

Development wells are used to begin production after a reservoir has been discovered and defined. Development drilling usually takes place over the area with the largest pay thickness in the reservoir.

Infill drilling involves placing additional wells in a proven area already producing oil. Infill drilling is used to recover oil from deposits previously inaccessible from the original well, speed up the extraction of oil from a reservoir, and reduce the amount of water or gas produced with oil. The use of 4-D seismic analysis, which shows the changes of fluid location over time, provides information to use when deciding locations to apply infill drilling.

Drill bits

Drill bits have evolved from the cable tool and fishtail drills used in soft rocks, to roller cone and fixed cutter bits used in more dense rock.

Cable tool drilling, or percussion drilling, is relatively simple. It involves dropping a drill bit shaped like a chisel over and over from a height into the hole being drilled. The bit is turned a little each time it is dropped into the hole. Water must be used to flush out broken rock and other debris. This method was predominantly used up until 1909, but is rarely used nowadays and only for shallow wells in soft beds of rock.

Fishtail drilling bits were also used in the early oil industry, and have a profile similar to a fish's tail. Fishtail drilling is rarely used today as it is very slow and can only be used in soft rock.

Rotary cone drilling bits, sometimes known as roller cone drill bits, are the most common bits in use today. The drill bit acts as a series of chisels mounted on up to three turning cones (Fig. 6-11). The roller cones' chisels, or teeth, are either milled from the same piece of metals as the cones (known as a mill tooth bits, or steel tooth bits), or inserted into the cones later (known as a tungsten carbide insert, or TCI, bits). The bi-cone rotary drill was invented by Howard Hughes Sr., father of 'The Aviator' Howard Hughes, in 1909. It was a revolutionary technology which sped up drilling immensely. The tri-cone drill bit was invented by the Sharp Hughes Tool Co. in 1933 and replaced the bi-cone as the dominant drill bit.

Rotary drills are powered by a steel pipe known as drill-string, which turns the drill bit cones. The drill string is turned on the surface by a device connected to an engine. The drill string is hollow and, in addition to turning the cones of the drill bit, carries specially formulated mud.

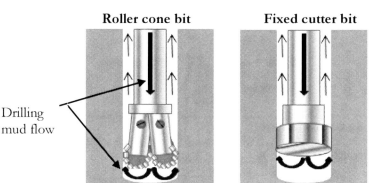

Roller cone bit   **Fixed cutter bit**   Fig. 6-11

Drilling
mud flow

Fixed cutter drills are used when drilling through very hard rock (Fig. 6-11). Fixed cutter drilling is slower than rotary cone drilling and produces much finer rock cuttings, called flour. Fixed cutter bits, unlike rotary drill bits, do not have any moving parts such as bearings, or rolling cones. Instead, the entire drill bit itself turns. A commonly used fixed cutter bit is called a polycrystalline diamond compact, or PDC, bit, which has embedded inserts, typically made of tungsten carbide steel topped by natural or synthetic diamonds.

The drilling process

Drilling begins with setting up a drilling rig with its associated equipment.

Few E&P companies own rigs, preferring instead to contract out the drilling. Most drilling rigs, whether land-based or on the sea, are owned by independent companies specializing in drilling. E&P companies hire the drilling rigs with their associated labor crews and pay the going rig dayrates. In addition to dayrates, drilling contractors may also enter into footage contracts. These are more risky than daywork contracts for the rig owner, as they get paid by the foot drilled. Another even more risky contract for the rig owner is a turnkey contract, in which the rig owner agrees to a total fee up front and must attempt to meet the customer's specifications under this fixed job rate in order to make a profit. Rig hire rates fluctuate with drilling supply and demand and can be quite volatile.

The following diagram shows the basic setup for a rotary drilling mechanism:

Fig. 6-12

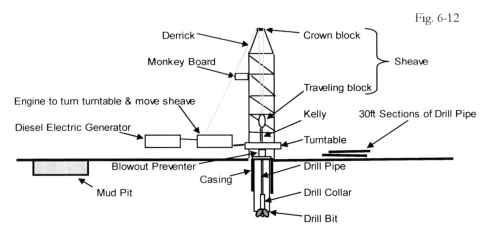

An oil derrick, or mast, is a tall metal structure which rises from the ground to hold cables and pipe sections used to lower the rotary drill bit. The derrick can be custom-assembled on the drilling site, or a special truck called a jackknife rotary rig with a pre-assembled derrick can be wheeled in place. The initial stage of drilling is called spudding. A sheave is a pulley system comprising of a stationary crown block at the top of the derrick, and a moving traveling block. The sheave is used to lift the heavy drill string into and out of the well bore. As the bit is lowered, it is powered by a large diesel engine on the surface. The Kelly is a four or six-sided pipe that is turned by the rotary table, which in turn is powered by the diesel engine. The Kelly is connected to the topmost drill pipe of a series of turning drill pipes, known together as the drill string. The drill pipes turn the drill bit. Since the pipes are hollow, they are also used to convey drilling fluids to the drill bit. The drill pipes are connected to a heavy drill collar, which puts weight on the drill bit, which is in turn connected to the bit. The drill pipes are stored individually in a rat hole on the rig floor just before it is attached to the next section of drill pipe.

Drilling fluids, or drilling mud, are comprised of water, clay, weighting material and chemicals. Pressurized drilling fluids are pushed through the hollow drill pipe and drill collar to the drill bit (Fig. 6-13). The mud is re-circulated. This mud mixture cools the drill bit, forces rock cuttings to the surface, and coats the drilled hole to prevent cave-ins. The drilling fluid is prepared by the mud engineer, known as a hoghead. The hoghead has to carefully calculate the density of the mud being used as downhole pressure changes at each new depth and also has to bear in mind the geothermal and pressure gradients as heat and pressure changes affect the stress the mud places on the well bore wall. The viscosity of the mud affects the flow, and thus the cooling ability, of the mud. Rock cuttings are analyzed when they reach the surface by a geologist known as a mud logger.

Drilling mud circulation system

Fig. 6-13

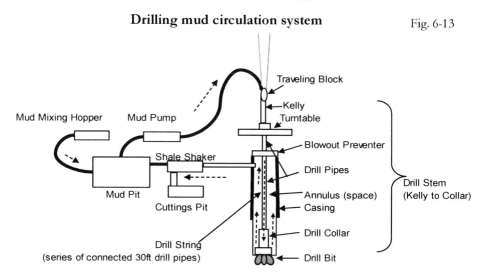

As the bit sinks into the ground, the Kelly attached to the drill pipe will also move lower to the ground. Every thirty feet, after a period of drilling, the Kelly is disconnected from the drill pipe, and a new section of standard 30-foot pipe is attached to the one in the ground. The Kelly is then reattached to the top of this new pipe section. This messy and potentially dangerous process on the drilling rig floor is carried out by workers known as roughnecks. Roughnecks are most often seen operating the large mechanical tubing tongs which are used to handle the drill string components. Sections of drill pipe are screwed together by roughnecks using either spinning chains or hydraulically powered drill pipe spinners. Roughnecks are assisted by less skilled roustabouts to move pipes and other equipment on the drill floor. Roustabouts are also known as worms. A roughneck working from a platform called a monkey board suspended from the top of the derrick is known as a derrick man. A floor man, or slip-puller, performs tasks on the derrick floor.

Drilling rigs usually operate 24 hours a day. Each of three eight hour drilling crew shifts includes a drilling foreman in charge of the roughnecks and roustabouts.

A drilling engineer with an onsite office is responsible for the overall correct technical operation of the well-drilling process. The drilling engineer is also sometimes called a tool pusher.

A Company Man is an onsite representative of the E&P company contracting the drilling.

Drilling speeds usually average less than 100 feet per hour. Oil wells can reach as deep as 25,000 feet underground, although most oil is found within 15,000 feet of the surface.

It is very difficult to drill down vertically in a precise straight line. As the drill bit meets different rock types, which are harder, softer or slanted, it may deflect, or dogleg, slightly.

If the drill bit becomes dull, if the drilling crew encounters substantially harder or softer rock formations, or the drill string snags, the drill crew have to raise the drill bit out of the hole. This involves pulling the entire drill string out of the hole, in a process known as tripping, and replacing the bit or making other adjustments. If the drill bit or drill pipe breaks in the hole then the crew has to try to fish it out or use other means to drill around it. Any objects in the way of a well bore, such as broken sections of drill pipe or a broken bit, are known as fish.

Pressure differentials between the oil in the reservoir and the pressure at bottom of the well can result in oil shooting out of the well as a gusher, or blowout, when the pay zone containing oil is finally reached. Such an explosive situation is very dangerous for the drilling crew and also is wasteful of underground reservoir pressure which can be used to remove oil. Another dangerous situation occurs if the drill bit penetrates a gas cap above an oil reservoir. Due to pressure differentials between an underground gas cap and ground level atmosphere, a bubble of gas can rapidly rise through the well bore, expanding as it heads toward the surface. This bubble of gas is known as a kick. Deep sea divers know that as bubbles of air rise from the sea floor they expand rapidly as the water around the air bubble becomes less dense – the inverse relationship between pressure and volume known as Boyle's Law. The same thing happens with a bubble of gas in a well bore.

In order to prevent uncontrolled rushing of oil or gas to the surface while drilling, a blowout preventer, or BOP, stack is used. A BOP stack is a series of valves at the base of the derrick, or on the sea floor for underwater drilling. It seals the high-pressure drill lines in the event of blowout, referred to as killing the kick, and allows pressure to be released in a controlled manner when safely possible.

Most well bore holes are lined with carbon steel casing which encases the open hole (Fig. 6-14). Casing may occur periodically during the drilling process at sections of the well bore hole or over the entire well bore hole. The series of steel casing pipes screwed together are used to line the walls of the drilled hole, or specifically at a problematic area such as weak rock, or an area where water is entering into the drilled hole. Casing prevents contamination of groundwater from oil production, prevents water encroachment into the well, and also enhances the structural integrity of the hole, preventing collapse as it passes through weak rock or rock containing fluids. The process of installing casing is referred to as running pipe. The sections of steel casing, typically 40 feet long, are screwed together and cemented to the wall of the hole. Once the cement has set over the cased portion of the well hole, the drill crew bores through the set cement at the bottom of the well to resume drilling. Casing pipes get more and more narrow as the bore goes deeper because new casing has to be passed down through any casing previously installed, resulting in the need for smaller and smaller bits to pass through the casing and punch through the cement and the rock. Oil wells can start out as wider than 20 inches on the surface. It gradually narrows to around 6 inches at the pay zone.

Casing is given various names depending on the depth and purpose.

A steel plate on the surface, known as a casing head, is attached to the first section of wide-bore casing known as conductor casing. This casing section is also known as the short string.

The next section of casing, known as surface casing, prevents contamination of fresh groundwater by any oil met by the drill head.

Intermediate casing is used next to prevent collapse of the well borehole. The intermediate string is typically the longest section of casing in the well.

Finally, production casing, also called the long string, or the oil string, is used through the pay zone. Once the desired depth in the pay section of the reservoir is reached, the production casing is perforated with customized explosives.

Holes which are cased all the way to the pay zone are referred to as cased-hole completions. To save on costs, the last and deepest section of the well bore is sometimes not cemented, and the pipe is referred to as lining rather than casing. Liner, sometimes called hanger, is simply hung on the intermediate casing above and not cemented to the well bore. Also, often the last and deepest section of well bore through the pay zone is sometimes left open to the reservoir without any casing or liner, known as an open hole completion.

Surface, intermediate and production casing is a feature of onshore wells. In offshore drilling conductor casing is typically the only one used.

The space between the drill pipe casing is known as the annulus.

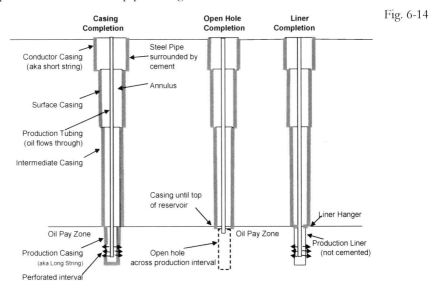

Fig. 6-14

Completions

Once the desired depth has been reached, the drill crew removes the drill string and a decision on well completion is taken. The term completion is used to refer to the final installation of equipment to produce crude or not. Completions of a well can take several forms: dry hole, production well, or injection well completions. Installing completions requires workover or well-servicing rigs after the drilling rig has been removed.

Dry hole completions are used when encountering dry holes, often referred to as dusters. Even if a well produces oil, there may not be sufficient quantities to justify construction of production facilities. A well where insufficient oil or gas is available for economic production is completed by plugging and abandonment, referred to as P&A-ing. A cement plug is poured down to the bottom of the borehole. It is important that cement is poured down to the very bottom of the hole. If the drilled hole has poked through the cap rock, oil may now escape from the reservoir over time and rise though the well bore to contaminate groundwater at shallow depths. Borehole P&A-ing should not be confused with shutting in a well, which is a temporary closure of the well for maintenance, or during violent storms, such as hurricanes.

Injection well completions are used as a method of enhanced recovery in which a gas or liquid is pumped at high pressure into a reservoir through wells drilled separately from the production well. Injection is used to maintain reservoirs at high pressure.

Production well completions involve building the surface installations required to extract crude, store it, and transport it to market.

The drilling rig, including the tall derrick, can be removed and replaced with a wellhead system, often only a few feet tall. The wellhead is the mechanical interface to the casing. Attached to a producing wellhead is a system often referred to as a Christmas tree, which connects with the producing flowline tubing (Fig. 6-15). The tree is an arrangement on top of the wellhead with a series of safety valves, nozzles, pressure gages, chokes, sampling points, thermometers, and chemical injection points.

A tree on an oil well

Fig. 6-15

Source: EIA

Pipelines may have to be constructed if production is large enough. As part of a production completion, a battery of storage tanks are constructed to enable production to continue if delivery systems shut down. A flare bleeder is occasionally lit a distance from the wellhead to burn off any unused gases in a safe manner.

As part of the downhole production completion, the drill string is removed. Production flowline tubing, which is a small diameter pipe, often between 2 and 7 inches in diameter is run down the well. In order to ensure that the crude oil and any gas only come to the surface through the production tubing and not through the annulus between the tubing and the casing, packing is added at the bottom of the well bore hole. The production tubing is connected to the tree which in turn is attached to the wellhead structure.

A subsurface safety valve (SSSV) is frequently attached to the downhole production tubing. The SSSV automatically shuts down the production flow if the tree or wellhead structure is destroyed or damaged. Apparently Kuwait did not use such safety valves on its fields, which had catastrophic results when the Iraqi army destroyed well heads in 1991.

The final stage of a production completion process is to perforate the production casing and the cement surrounding it at the bottom of the well in the oil pay zone. The perforations open the well bore to the reservoir and allow communication with the pay (Fig. 6-16). Perforations use custom-built high precision explosive charges to create many pencil thin channels in the casing, cement and many feet into the reservoir rock.

In conjunction with perforation, acidic fluids are often forced into the new perforations at a pressure exceeding the reservoir rock formation's ability to accept fluids. This results in fractures extending well beyond the fractures which would occur from the explosive charges by themselves. This process, with explosives and acid, is referred to as extreme overbalanced perforation, or acid fracing.

Perforations are sometimes made at various intervals along the well bore in order to make improved contact with the pay. Underperforming wells are also often reperforated in order to increase production flow rates when existing perforations become clogged. Reperforating a well may also provide access to structurally isolated portions of the reservoir above the original perforations.

Well perforations and packer

Fig. 6-16

Tests performed while drilling

Cutting analysis involves scrutinizing cuttings brought to the surface by the drill fluids/mud to see what type of rock the drill bit is meeting. The primary advantage of this process, a part of mud logging, is that it can be carried out without stopping drilling.

Well-logging takes place at intervals in drilling with a series of downhole tests in order to define the pay zone. Well-logging, developed by French brothers Conrad and Marcel Schlumberger in the 1920s, involves lowering electrical resistance (oil and gas do not conduct electricity as well as water), in addition to seismic and radiation testing equipment down the borehole. The instruments are used to determine the rock formation type, the deviation from true horizontal of the rock formation, known as the dip, and the rock porosity. Well-logging is carried out by an engineer known as a wireline logger. In the past, drilling usually had to stop in order to carry out well-logging. Recent advancements in technology allows sensors to be attached to the downhole drilling equipment which allow real-time feedback on drilling. These are known as measurement while drilling (MWD) and logging while drilling (LWD) technologies.

Core sampling occurs once a drill bit reaches depths of interest previously defined on a seismic analysis. The entire drill string is pulled, or tripped, out of the hole and the bit is replaced by a hollow drill bit, called a coring tool, which is used to extract a cylinder of several meters of rock for analysis. This process can confirm the rock type, fluid content, dip, porosity and permeability.

Flow testing is used once oil or gas is found. A meter called a choke is attached to perform the flow tests, which measures downhole pressure. Measuring the capacity of oil to flow within a reservoir gives a good indication as to the ultimate productivity of a reservoir. Flow rates also indicate rock permeability. While flow testing can be carried out at any time after a well is producing, drill stem tests (DSTs) are flow tests carried out on the well while drilling is taking place and are used to obtain fluid samples and early measurements of bottomhole pressure. For mature wells, flow tests carried out over time help to determine the rate of decline in production capacity of the well.

People involved in drilling and completion

In addition to the roustabouts, roughnecks, toolpushers, hogheads, mud loggers, wireline loggers, geologists and geophysicists mentioned already, there are many others involved in exploration and production.

Once oil has been discovered and a decision has been made to produce, a petroleum architect designs the overall system by which the reservoir will be developed. The system will include the number and location of wells, surface installations, in addition to pipelines or other mechanisms by which the crude will be transported.

Reservoir engineers analyze reservoir characteristics and decide on the optimum rate to extract the most profitable amount of oil out of the reservoir. Reservoir engineers are responsible for reservoir pressure management and maximizing reservoir output. This includes attempting to extend primary production, minimize damage to the reservoir and maximizing the return from secondary or enhanced production techniques.

The production engineer's role is to maximize crude oil output to the optimal level set by the reservoir engineer, and also to manage the process of temporarily shutting in production during a storm or a trade union strike.

Process engineers are responsible for managing individual processes such as separation of crude oil and gas and the removal of water and sand.

Drilling directions

Linear drilling is used in vertical or slanted wells (Fig. 6-17). Vertical wells simply involve drilling straight down. Despite best efforts to drill vertically, most wells are very slightly deviated. Slant drilling is a slight variation of vertical drilling, and is also known as deflected drilling, or deviated drilling. Slant drilling may be used to drill under a lake while basing the drilling rig on land.

Advanced drilling techniques, also known as kick-off drilling, involves drilling in whatever non-linear direction one can conceive of and have become more popular since the early 1990s (Fig. 6-17).

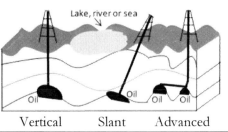

Fig. 6-17

Vertical   Slant   Advanced

Advanced drilling has been made possible by three developments: downhole sensors attached to the drill bit, mud motor drill bits and coiled tubing. Downhole sensors allow a driller to determine the bit position in three dimensional space and to geosteer the bit in real time. Mud motor driven drill bits do not rely on the typical rotating rigid drill string to be turned. Instead the mud motor located just above the drill bit uses the hydraulic pressure of the circulating drilling fluid to turn the drill bit. As the drill bit is turned solely by the downhole mud motor and not the drill string, the bit can be offset at an angle without having to attempt to angle the entire drill string. Advanced drilling also uses a technique called coiled-tubing drilling (CTD). Usually drill strings are sectional drill pipes connected every thirty feet by roughnecks. CTD differs in that a continuous steel pipe is fed from a large coil on the surface.

The yield from an advanced drilling technique can be enhanced compared with conventional vertical well bore, as it is possible to steer the drill across a pay zone. Advanced drilling techniques can also reach pockets of attic and basement oil. Attic oil is oil which may be structurally isolated or located up-dip from the production well – such oil deposits are said to have been bypassed. Basement oil is oil which is structurally isolated below the production well.

Additionally, advanced drilling techniques save on surface completion costs. A single platform can be used to access an entire reservoir. This is particularly useful when drilling offshore where platforms are very expensive or in environmentally sensitive areas, where many surface platforms may be unacceptable.

Horizontal drilling is where one drills down vertically to a desired depth, then the drill bit is turned horizontally, or sideways, using a downhole mud motor driven cutting head (Fig. 6-18). A steel wedge, known as a whipstock, is sometimes used to deflect the drill bit from the vertical bore hole. Horizontal drilling is particularly useful when the pay zone is not very thick but extends over a wide area, such that drilling a conventional vertical well would be uneconomical.

Fig. 6-18

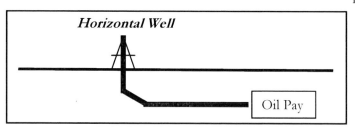

Maximum Reservoir Contact (MRC) wells, also known as multilateral wells or bottle-brush wells, are where multiple new well bores are drilled, but they all share the part of the well that reaches the surface (Fig. 6-19). Multilateral well bores are frequently drilled in forked or fishbone patterns, hence the term maximum reservoir contact. In addition to increased communication with the pay, having multiple wells use a single surface completion cuts development costs.

Often the packing and tubing in MRC wells can be made intelligent by attaching sensors which shutdown flows from an individual leg of the well if water levels are too high relative to oil from that leg.

Fig. 6-19

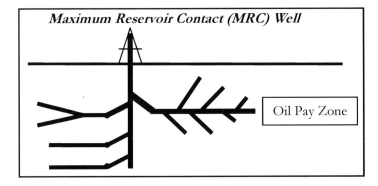

Designer wells are also known as directional drilling wells (Fig. 6-20). Using 3-D seismic analysis, reservoir engineers can plot the location of small areas containing oil, and drill into those isolated pockets using advanced directional equipment. To drill for these small oil traps individually would have been uneconomical with conventional vertical wells. Advanced equipment allows the well bore to take turns of close to 300° to tap separate oil pockets which would have been economically inaccessible otherwise.

Fig. 6-20

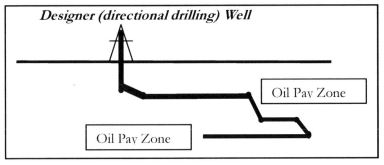

Sidetrack drilling
Sidetracks are holes drilled off the side of an old well bore used to produce from a previously closed well (Fig. 6-21). Sidetrack drilling uses coiled tube drilling combined with an innovation called through-tubing rotary drilling (TTRD), in which a small and slender drillstring is used drill using existing well tubing, in other words through the existing casing.

Fig. 6-21

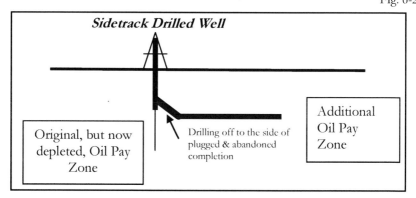

Sidetrack Drilled Well

Original, but now depleted, Oil Pay Zone

Drilling off to the side of plugged & abandoned completion

Additional Oil Pay Zone

Offshore drilling and production

Until the late 1930s, almost all drilling was done on land. Drilling from fixed piers jutting out into lakes and seas or from shallow fixed wooden pile platforms had taken place in a very small way since 1897. The first mobile (where the rig could move from one place to another) offshore drilling was carried out by Texaco (now part of Chevron) in 1933 on Lake Pelro, Louisiana. Mobile offshore drilling in shallow water was carried out by attaching derricks to barges which were floated into place and then weighed down so they rested on the sea floor. The first bottom-supported platform out of sight of land was constructed by Kerr McGee (now part of Anadarko) in 1947. It was as a depth of 30 feet of water, still very much on the continental shelf, one mile from the Louisiana coast in the Gulf of Mexico. In Europe, following discoveries in the late 1960s, the North Sea became another major offshore production area. The Brazilian Campos Basin, offshore West Africa (Nigeria and Angola especially), the Caspian Sea, and the Mexican Cantarell field in the Bay of Campeche later also became other major offshore producing areas.

The discovery of oil on the continental slope, the declining seabed which extends many miles offshore beyond the continental shelf, led to the need to drill in greater depths of water. In 1979 the first rig exceeded a depth of 1,000 feet of water on the continental slope.

Drilling on the continental shelf in less than 1,000 feet of water is known as shallow water drilling. Drilling on the continental slope between 1,000 and

2,500 feet of water is known as deep water drilling, and between 2,500 to 12,000 feet of water is known as ultra deep water drilling.

Beyond 12,000 feet, in what is known as the abyssal zone, which can be as deep as 30,000 feet, organic sedimentary deposits have been so dispersed by the time they reach the ocean floor as to be unlikely to form any economic oil deposits.

In the offshore oil world, there are drilling rigs and production platforms. Very often drilling rigs and production platforms can operate from the same structure. On an offshore drilling rig or production platform, the floor of the rig is called the deck and the structures built on the deck are referred to as topsides. Topside facilities are contained in modules, including processing (dewatering, gas separation etc.), drilling, wellhead, mud, power, and crew quarter modules. Topside structures will also typically include a helideck, flareboom, piperack, and lifeboat station.

Even the most stable drilling rig will move slightly as a result of water movement and wind. A subsea drilling template is cemented to the sea floor and has a hole which acts as a guide for the drill stem. This enables accurate drilling to take place while allowing for rig motion on the ocean surface. The drilling rig is attached to the template by cables. A floating rig or platform will be secured to the sea floor using mooring lines and massive anchors, often put in place before the floating platform arrives.

Drilling and production take place inside flexible steel pipelines known as conductors, or marine risers, which are used to connect the sea floor bore hole to the drilling rig and, later, the production platform. These risers can sometimes be disconnected from floating drilling rigs or production platforms ahead of storms. Remotely Operated Vehicles (ROVs) can be used to retrieve and reconnect heavy underwater lines. Alternatively, for more delicate work, highly trained divers are used to depths of up to 1,000 feet. In comparison, most recreational divers only go to 100 feet. Extreme depths require replacing nitrogen with helium or other gases in divers' compressed air (to prevent narcosis), in addition to lengthy decompression periods.

Instead of going straight down, risers are often hung in catenary, or chain shaped, curves, in which case they are referred to as steel catenary risers (SCRs). The catenary shape allows a riser, because it is not taut, to absorb movement of the platform without causing undue stress at the point where it connects to the platform or the location where the riser touches the sea floor, known as the touchdown zone. A particular type of riser called an export riser is used to send fluids, such as methanol used as antifreeze, down to the sea floor. Risers which contain electrical power lines to heat the crude or cables connected to monitoring equipment on the sea floor are known as umbilicals.

A challenge with producing oil offshore is ensuring that crude, which is heated underground due to the geothermal gradient, does not form paraffin waxes or gas hydrates as it travels up the production flowline through frigid ocean water. The crude is kept warm by insulating and heating the production tubing. This requires large amounts of energy. Antifreeze chemicals are also sometimes added to the crude to enable it to flow more freely at lower temperatures.

When drilling offshore, the viscosity of drilling mud must be controlled even more precisely than on land because of the additional pressure of sea water. Too much pressure can rupture the rock surrounding the drill string, and potentially causes blowouts of oil or gas.

Depending on where the well head is located, offshore production completions are divided into two categories: dry tree and wet tree.

Dry tree completions are where the wellhead is located on a platform out of the water.

Wet tree completions, also known as subsea systems (SS), are where the wellhead is located underwater on the sea floor. Wet tree systems can be used to reduce costs by connecting several wellheads on the seafloor by pipeline to use a single strategically placed surface production platform, instead of building a separate production platform for each wellhead. Where a wet tree shares a production platform with other wells, the flowline connecting the subsea system to a platform is called a tieback. Wet tree completions are particularly useful for producing from small satellite fields around a main field.

When constructing an offshore drilling rig or production platform, a major concern is that they must often remain moored at a fixed location throughout a storm.

Production on platforms in the Gulf of Mexico can be shut in temporarily. Personnel are evacuated for safety ahead of particularly strong storms during hurricane season, which runs from June 1 through November 30. Similar temporary shut ins can occur as a result of severe storms on the North Sea and on the North West Shelf (NWS) of Australia.

Wave heights during storms typically reach up to 25 feet (7 meters) and under rare conditions up to 50 feet (15 meters.) Freak waves of up to 100 feet (30 meters), once considered a once in 1,000 year phenomenon, were proven to be quite common, as 10 were detected globally during a three week survey period in a analysis carried out by the European Space Agency using satellite radar imaging in its 2001 MaxWave study.

With greater appreciation of ocean waves, the air gap, which is the distance between the height of the rig or platform deck and the waterline, has been steadily increasing since the 1960s from an average of 35 feet to 55 feet today.

Mars Tension Leg Platform in the GOMEX
Fig. 6-22

Platform with a 114 foot air gap
Source: NOAA

After 2005 Hurricane Katrina
Source: MMS

In addition to the damage waves can inflict above the waterline, large waves can cause underwater mudslides due to the sudden increase in ocean floor pressures. These can damage production pipelines and other infrastructure on the ocean floor.

There are several types of offshore oil drilling rigs and production platforms varying mostly on the depth of water in which they can operate:

Submersibles are mobile structures used in relatively shallow water less than 80 feet deep. When towed into place, pontoons are filled with water and the structure sits on the sea bed. The pontoons are sometimes referred to as bottles or caissons. A version of the submersible called a drilling barge is used on inland waterways.

Jackups are mobile, just like submersible rigs and platforms, in that they can be moved relatively easily from one location to another (Fig. 3-23). Jackups are used to depths of up to 500 feet. Once the floating jackup is towed into place, the legs (typically three) are lowered to the sea floor which raises the deck out of the water to a desired height. Some jackups can move from place to place using their own power. The vast majority of jackups are Independent Leg Cantilevers (ILC) which have legs situated such that they can be extended out and away from the derrick, allowing the rig to be placed adjacent to a production platform. ILCs capable or operating in 300 feet or more of water are the most sought after and valuable jackups. Less common jackups are Independent Leg Slot (ILS) rigs, which have slots in the hull of the rig for the legs and cannot be placed over any other structure. Mat Cantilever (MC) and Mat Slot (MS) jackups are similar to independents except the bottoms of the legs are attached to a horizontal platform which makes them more suitable for use on soft sea beds.

Jackup   Fig. 6-23

Source: MMS

Fixed platforms are large immobile structures with the topsides sitting out of the water on frames, called jackets, built from concrete and steel, which rest on the sea bed. There are three types of fixed platform: gravity platforms, fixed steel platforms and compliant towers.

Fixed concrete gravity platforms are used in depths up to 1,000 feet, and the structure is made from steel reinforced concrete and is held in place by gravity alone. The gravity platform merely rests on the sea floor and does not connect to steel pilings anchored deep on the sea floor as other fixed platforms do.

Fixed steel platforms (FP) are used for drilling and production up to water depths of 1,500 feet (Fig. 6-24). The deck sits on a support structure jacket consisting of tubular steel connected to steel piles driven up to 400 feet into the sea bed.

Fixed Steel Platform   Fig. 6-24

Source: MMS

Compliant towers (CT) are used for drilling and production in water depths between 1,000 and 3,000 feet. A compliant tower is constructed with flex legs, comprising of steel axial tubes, fixed to the sea bed. The flexible jacket structure allows the CT to withstand much stronger winds, currents and waves than other fixed platforms. The narrow footprint of a compliant tower on the sea floor is much less than that of a fixed steel platform. The savings in steel make the CT economical.

Semisubmersible is a term created in 1961 by Shell Oil engineer Bruce Collipp in order to avoid calling the vessel a ship when applying for a license from the US Coast Guard. Describing the vessel as a ship would have brought it under the influence of maritime trade unions. Semisubmersibles are used very often for deep and ultra deep production where a platform resting on stilts (an expensive steel jacket) would not be cost effective. Semisubmersibles have one, or more, pontoons which are filled with water, or more dense material, to partially submerge the vessel. This gives it much more stability, although it is still floating. The pontoons are referred to as the hull. Although waves can move the semisubmersible structure, it is generally fairly stable and is attached to the sea bed using mooring cables and anchors. While they are often designed for a specific location, semisubmersibles can be moved from one location to another under tow from barges. Sometimes floating hotels, or flotels, which are converted semisubmersibles, are used to provide extra crew accommodation. The flotel is attached to the production semisubmersible via a gangway.

There are four types of semisubmersibles: tension leg platforms, monocolumn TLPs, and spar platforms

Tension leg platforms (TLP) are floating structures which are held in place by vertical steel tensioned tendons secured to piles on the ocean floor directly below the platform. The tendons pull the vessel into the water a little deeper than it would float without the tendons, thus keeping tension on the mooring lines and stabilizing the vessel. TLPs are used in depths up to approximately 5,000 feet. TLPs tend to have dry trees although they can also be used with subsea systems in very deep water.

Monocolumn TLPs, also known by the trademarked name SeaStar® (SStar), are small tension leg platforms used for production in depths of 600 feet to 3,500 feet (Fig. 6-25). Smaller, and thus less expensive than a regular TLP, the hull consists of a central column with three pontoons at the bottom of the column extending horizontally away from the central column in a star pattern. The three pontoons are attached to the sea floor using tubular steel tendons attached to pilings on the sea bed.

Monocolumn TLP

Fig. 6-25

Source: MMS

Spar platforms, also referred to as Deep Draft Caisson Vessels (DDCV), have a single central column pontoon on which the deck sits (Fig. 6-26). The hull is moored to the sea floor. Spar platforms are used in up to 10,000 feet of water. The column is weighted at the bottom with a material denser than water to keep it vertical. Although the platform has to be anchored with mooring lines to prevent sideways movement, because of the massive column, the spar doesn't move up or down much at all relative to other floating platforms. Spars also have large fins called strakes running in a spiral pattern along the outside which reduce vibrations from underwater vortices. As with TLPs, spar platforms tend to have dry trees although they can also be used with subsea systems in very deep water. Very rarely, spar platforms are used as offshore oil storage facilities, such as the Brent spar which was used in the North Sea.

Spar Platform

Fig. 6-26

Source: MMS

Floating production systems (FPSs) can be used at any depth up to and over 10,000 feet (Fig. 6-27). FPSs are semisubmersibles with several pontoons, or can have the profile of a ship. A variation of the FPS is the Floating Production, Storage and Offloading (FPSO) vessel which is used to gather oil from wet tree wellheads on the ocean floor and then to store oil until a tanker arrives. A variation of an FPSO is known as a Floating Production and Storage Unit (FPSU). Shuttle tankers arrive periodically to haul away the FPSO or

FPSU stored crude oil. FPSO and FPSUs can sometimes be disconnected from risers and moved out of the way of potentially severe storms, although most tend to stay moored in the one spot throughout even the worst storms.

FPSs are almost always used in conjunction with wet trees and subsea systems (SS).

Fig. 6-27

FPS   **FPSO**

Source: MMS   Source: MMS

Dynamically positioned (DP) drill ships are mobile drilling rigs which are typically self-propelled and thus do not require barges to be towed into place (Fig. 6-28). Drill ships use GPS satellite navigation and dynamically-positioning thrusters to keep the vessel in position while drilling. Drilling occurs through a large hole in the center of the hull called a moon pool. Once the drill ship has finished, it will simply move on the next job and leave the shut in well head on the sea floor (a wet tree) ready for attachment to a subsea system or a production riser when a production platform is ready.

A drill ship

Fig. 6-28

Source: MMS

Tender rigs are semi-submersible or self-supporting rigs used to provide additional deck capacity for drilling rigs.

Fig. 6-29

Offshore Systems Summary

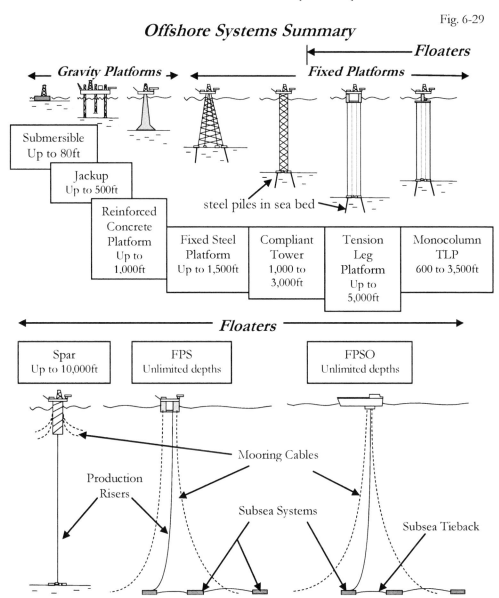

In order to visualize large depths, the Empire State building in New York is 1,250 feet high and the Eiffel Tower in Paris is 986 feet high.

All gravity platforms, fixed platforms, and spar platforms can be used for drilling in addition to production. FPSs are used solely for production, with the exception of a vessel called a Floating, Drilling, Production, Storage and

Offloading (FDPSO) vessel. Drill ships are solely used for drilling and not for production.

Except in the case of FPSOs and FPSUs, oil is not stored on production platforms and is instead loaded onto a tanker or connects to a subsea pipeline as it is produced.

Inside a reservoir

Reservoir rock formations can be homogenous or heterogeneous. A homogenous reservoir rock formation is in a consistent form throughout the reservoir. If a reservoir rock formation is heterogeneous, or complex, it means that the rock formation changes significantly through the reservoir. In a heterogeneous formation the behavior of fluids in the reservoir are much more difficult to predict than homogenous formations.

Sedimentary rock is found in layers, or strata, which can result in reservoirs being vertical, horizontal or slanted. A slanted reservoir is also referred to as dipping.

The reservoir, or petroleum trap, often consists of a water leg on the bottom in an aquifer, then an oil leg/column, and then a gas leg on top of that (Fig. 6-30). Not all reservoirs have water underneath or gas above the oil. The water, oil and gas, are contained within pores in the rock. Hydrocarbon gases found with oil are known as associated gas, or casinghead gas.

Fig. 6-30

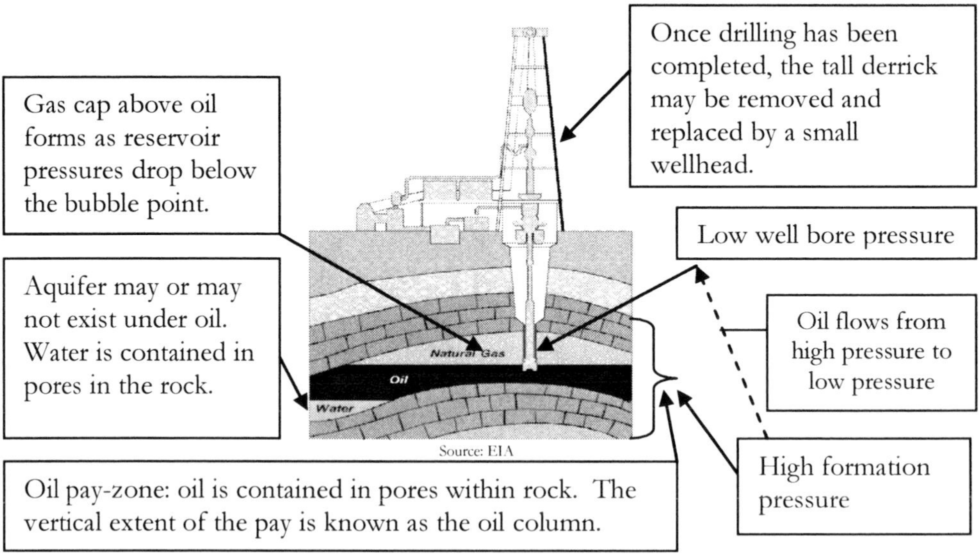

Once drilling has been completed, the tall derrick may be removed and replaced by a small wellhead.

Gas cap above oil forms as reservoir pressures drop below the bubble point.

Aquifer may or may not exist under oil. Water is contained in pores in the rock.

Low well bore pressure

Oil flows from high pressure to low pressure

High formation pressure

Oil pay-zone: oil is contained in pores within rock. The vertical extent of the pay is known as the oil column.

Source: EIA

Due to pressure variations, hydrocarbons within a reservoir can be found in three phases: solid, liquid, gas. Reservoirs are described by the number of phases they contain. Single-phase reservoirs contain just liquids, or just gases. If reservoir pressure is high enough, above what is called bubble point pressure, the lightest hydrocarbons, methane, ethane, propane and butane remain as a solution in the crude oil within the reservoir, and the reservoir will exist in a single-phase. Two-phase reservoirs contain oil below and gas above. A two-phase reservoir occurs if pressure has dropped below the bubble point pressure level. Below the bubble point pressure level, gases leave the oil solution and form a gas cap above the oil.

Key reservoir statistics
A well data sheet shows parameters used to describe qualities of a reservoir:

Well Data Sheet Summary Example   Table 6-3

```
        Date:
        Completed by:
        Installation (new, workover):
        Phase:
        Well location:
        Primary power supply:
        Formation type (sandstone, limestone, other):

Well Data
        API Casing:        x" O.D.        X Pipe length in FT      x FT   TO  x FT
        Liner (if any):    x" O.D.        X Pipe length in FT      x FT   TO  x FT
        Open hole:                                                 x FT   TO  x FT
        Total depth:                                                     x FT
        Perforation intervals:                                    x FT   TO  x FT
                                                                   x FT   TO  x FT
                                                                   x FT   TO  x FT
        API Tubing:        x" O.D.    x Threads
        Flowline Data (Size/Length):
        Artificial Lift (if any: Rotary, Jack, Plunger, other)?
        Special problems (sand, scale, corrosion, paraffin, H2S, power supply, other)?

Reservoir Data
        Flowrate (at present)
                Gas volume:  X
                Oil volume: X
        Flowrate (normal)
                Gas volume:  X
                Oil volume: X
        Oil fluid level above producing interval (Y/N)?
        If gas produced then presures: Flowing  X    Shut in Tubing  X    Shut in Casing  X
        Producing GOR:  X
        Oil API Gravity:  X
        Oil Viscosity:  X
        PVT Data:  X
        Bubble Point Pressure:  X
        Bottom hole formation temperature and pressure:  X
        Water Cut:  X
        Water sensitive formation (Y/N)?
```

Porosity is measured by the percentage of void space within reservoir rock versus solid rock. Reservoir rock is like a sponge, full of tiny pores contained within a solid rock matrix. Increased porosity means more space where oil is potentially trapped. Porosity of an oil reservoir typically ranges from 5% to 25%. Porosity usually declines the deeper the rock is buried as pores collapse under increased pressure. Sandstone normally requires a porosity of at least 18% to form an economic oil reservoir, or 12% for a gas reservoir. Limestone and dolomite reservoirs can be economical at lower porosity levels than sandstone. This is because limestone and dolomite are often fractured with cracks along which oil can easily flow.

Permeability, sometimes merely referred to as K, measures how connected the pores within the rock are. It indicates how easily oil will flow from within the reservoir to the bottom of the well. A unit called a Darcy is used to measure the ability of a fluid to move through a porous rock. Permeability of oil reservoir rock is measured in millidarcies (mD) which is 1/1,000 of a Darcy. Rocks within a reservoir typically have a permeability of between 5 and 1,000 millidarcies (Table 6-4). The larger the number the more permeable the rock is and the easier it is to remove the oil. A reservoir with permeability of 5 mD or less is referred to as a tight formation and production may be difficult as the oil simply may not move.

Permeability (K)   Table 6-4

tight	<5	mD
fair	5-10	mD
good	10-100	mD
very good	100-1000	mD

While permeability is typically measured across a rock matrix, flow can sometimes be enhanced by fractures within the rock. Areas of very high permeability within a reservoir may be due to fluids traveling across such fractures. These are referred to as super-conducting, or super-K, zones.

Viscosity of oil at reservoir conditions indicates how easily oil will flow through a reservoir to the bottom of the well. Viscosity of oil in a reservoir is measured in centipoises (cP). Viscosity is usually measured at reservoir conditions, which, because of geothermal heat, is much hotter than at the surface and therefore viscosity of crude at reservoir conditions can often be less than 1 cP which is the viscosity of water at 68.4°F (20.2°C).

Oil density, also known as the oil's gravity and measured in degrees API, gives an indication of the products, such as gasoline, diesel etc., which crude oil in a reservoir can easily be refined into.

Reservoir vertical extent is also referred to as net reservoir thickness, net pay, or the oil zone. This is the thickness, in feet, of the productive portion of a reservoir in which oil is trapped.

Reservoir areal extent is measured in acres. Along with porosity, permeability and reservoir vertical extent, this allows us to arrive at an estimation of the amount of the total oil contained within a reservoir.

Gross pay is the average thickness of the entire reservoir, not just the area currently in production.

Depth of oil-water contact (OWC) is the point where oil makes contact with water from an aquifer. If the OWC dips, which can occur if the water is moving, the OWC depth can be a range. Not all oil deposits make contact with an aquifer.

Reservoir pressure (in psi) is the average pressure across the reservoir.

Bubble point pressure (in psi) is the lowered pressure at which hydrocarbon gases leave the liquid oil solution and begin to form a gas cap. Bubble point pressure is important as additional facilities to handle high levels of gas production often have to be built once pressure drops below the bubble point.

Drawdown is the pressure differential between average reservoir pressure and pressure at the bottom of the well bore. Drawdown is the engine which propels the oil out of the reservoir.

Productivity index (PI) is an index which enables the relative productivity of different reservoirs to be compared. The PI measures the average numbers of barrels of oil produced per day divided by the psi of the drawdown.

Recovery efficiency, also known as the recovery factor (RF), is the maximum percentage of oil in place in a reservoir which is technically recoverable.

Water saturation (Sw) is the percentage of pore space in a reservoir filled with water. In general, in an economic deposit, 40% or more of pore volume should contain hydrocarbons and less than 60% water. If water saturation is greater than 60% then the reservoir is referred to as being wet.

Formation volume factor is the volume of space one stock tank barrel (STB) of fluid on the surface occupies in a reservoir, known as a reservoir barrel (RB). A barrel of oil in a reservoir will occupy more space because it includes rock surrounding the pores.

Solution Gas-Oil Ratio (GOR) measures how much gas is in produced oil. GOR is the standard cubic feet of gas released as a oil stock barrel is produced.

Production: how oil is removed from a reservoir

Fluids move from high pressure to low pressure. A key factor underpinning all conventional crude production is that the pressure on the oil in a reservoir has to be above the pressure at the bottom of the well hole. If there is no pressure differential, or drawdown, then the crude will not flow across the reservoir to the bottom of the well and will be left in place.

Once oil reaches the bottom of a well, ideally the natural drive pressure differential continues to push the heavy column of oil up the well bore toward the surface, which is called natural lift. The well is known as a flow well. The natural lift stage of production is referred to as flush production.

If force needs to be applied by a manmade pump to lift the column of oil through the well bore to the surface, it is called artificial lift, and the recovery stage is referred to as settled production.

Any methods used to extract oil other than natural and artificial lift are known as improved oil recovery (IOR) or well stimulation techniques.

Primary recovery: flush production

There are six natural drive mechanisms which keep reservoir pressures high relative to the pressure at the bottom of a well hole: solution gas drive; gas cap drive; natural water drive; gravity drainage drive; compaction drive; and combination drive.

Solution gas drive is where there is no active gas cap drive and no active water drive. As crude oil is removed from a reservoir the pressure drops on the remaining oil in the reservoir. The pressure drop causes the hydrocarbon gases in the crude oil solution to expand, providing the force to move oil across the reservoir to the low pressure well bore.

The factor critical to managing reservoir pressures in a solution gas drive reservoir is the bubble point. All reservoirs in production pass through this point eventually, and producers try to keep pressures above this level by water flooding and gas reinjection, among other means. Once pressures fall below this points, production continues, but challenging new obstacles appear.

When reservoir gases are in solution above the bubble point pressure, they flow at the same rate as crude oil across the reservoir rock. When the reservoir pressure is below the bubble point pressure, gases leave the oil and flow across the reservoir to the well more quickly than liquid crude. Suddenly a much larger amount of gas will be produced with the crude oil, as increasing amounts of free gas reaches the well before liquid crude. To handle the increased amount of produced gas, some method of handling the gas will have to be built. An additional gas-oil separation plant (GOSP) may have to be built, which is expensive. The separated gas may then be flared if there is no pipeline available to bring the gas to market. Flaring gas is wasteful and often illegal. Alternatively, it could be reinjected into the reservoir to slow the decrease in pressure.

Solution gas drive reservoirs are also known as depletion drive reservoirs. As there is no gas cap drive or active water drive, pressure falls rapidly as the gas comes out of solution and there is no further mechanism for natural recovery of the remaining crude. Solution gas drive may result in a gas cap as the gas comes out of solution, but this gas cap will not be large enough or have sufficient pressure to produce oil. Also, a gas cap above the oil may result in the producing well becoming surrounded by gas instead of oil. A new deeper well may have to be drilled.

As a mechanism for generating drive, solution gas drive is very short lived and provides a very low recovery factor. Solution gas drive typically produces around 5-25% original oil in place (OOIP) and approximately 60-80% gas initially in place (GIIP.)

Gas cap drive is where a reservoir has an original gas cap in place above the crude oil. Gas which is out of solution and in a gas cap is referred to as free gas. As oil is removed from the reservoir, pressure drops and the gas cap expands to displace the produced oil. In order to generate sufficient pressure, the gas cap must be several times the size of the crude oil pay zone.

The bottom of the production well is placed as low as possible in the oil pay in order to prevent gas coning (Fig. 6-31). Coning literally means that the gas is pulled in a cone shaped pattern toward the producing well and displaces the flow of crude oil to the well. Coning can sometimes be reduced by allowing the well to rest for a period of time by shutting in production or reducing production rates. Unfortunately, coning often stays in place for months and even years after the well has been allowed to rest, in which case a new well may have to be drilled.

Fig. 6-31

Oil reservoir with gas cap drive

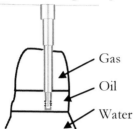

Gas

Oil

Water

Gas coning in an oil reservoir

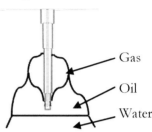

Gas

Oil

Water

As with solution gas drive, pressure drops continuously with gas cap drive production, although gas cap drive pressure falls more gradually compared with solution drive. With solution gas drive, the drive gas is produced with crude from the outset, whereas with gas cap drive very little drive gas is produced with the crude and therefore pressures remain higher within the reservoir for longer.

Gas cap drive can be used to produce approximately 20-40% of the OOIP and often as high as 60% depending on the size of the gas cap relative to the amount of oil. Once crude production has ended, up to 80% of the GIIP can potentially be removed from the reservoir.

As coning is avoided, neither gas drive mechanism is rate sensitive. If production is sped up or slowed down, the ultimate amount of oil recoverable will not change.

Natural water drive is possible when water below from an aquifer is in communication with the oil in the reservoir. Although a combination of drives is most common, water drive is the dominant and most productive of the individual drive mechanisms. The oil-water contact can be under part of a reservoir, in which case it is referred to as edge water drive, or underlying the entire reservoir, where it is referred to as bottom water drive (Fig. 6-32).

Edge water drive

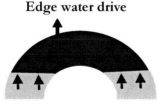

Bottom water drive

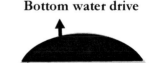

Fig. 6-32

An aquifer is rock containing water. The water contained in an aquifer is called groundwater and can be found under large parts of the earth's surface, even under deserts in Saudi Arabia, Iraq or Libya. Aquifer water is usually contained in pores rather than in large lakes, just like the oil in the reservoir rock. The upper level of an aquifer is known as the water table. Aquifers are recharged by rain, rivers, lakes or oceans. For an aquifer to be sustainable, it is important that the rate of withdrawal of fluids does not exceed the rate of recharge. Aquifers are often massive. A single aquifer can underlie millions of acres. Aquifers can take thousands of years to be filled, and may contain what is referred to as fossil water, which was deposited thousands of years ago. Not all aquifers contain fresh potable water, as the water may have high levels of dissolved minerals.

As crude oil is removed from a reservoir, pressure is maintained by water inflow. The water in the aquifer must be many times the size of the oil in the reservoir to generate sufficient drive pressure. If there is no gas cap, the driller will try to locate the producing well as high as possible in the oil pay zone away from the water. If there is a gas cap above the oil pay then the driller will try to locate the production well somewhere between the gas and the water at the thickest part of the pay.

Water drive is the highest yielding type of drive and can produce between 35-75% OOIP and 60-80% GIIP. If the water in an aquifer is replenished, reservoir pressures can remain high for very long periods of time. Water drive is one of the reasons the famous Ghawar field in Saudi Arabia has been in production for the past 60 years, which is an unusually long time for an oil field.

The primary challenges with a well using water drive are preventing water breakthrough and preventing a gas cap from increasing in size.

Water breakthrough, where water encroaches on the production well, results in higher processing costs for a producer and is therefore avoided if at all possible. Often water breakthrough is so severe that a new well has to be drilled as water levels become too high.

Water coning is a common breakthrough scenario (Fig. 6-33). Water coning, like gas coning, can be alleviated by allowing the producing well to rest for a period of time. In a horizontal completion well, where the well bore is turned sideways, water coning is called water channeling.

Fig. 6-33

Oil reservoir with water drive   **Water coning in an oil reservoir**

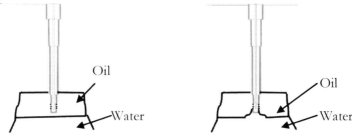

If water breakthrough has occurred, the amount of water produced with oil is measured as a water-oil ratio (WOR), or water cut. Water in a reservoir is known as formation water. When it is produced with crude oil it is known as produced water. The cost of dewatering the oil on the surface ranges between US$0.10 and US$2 per barrel of water. Handling water is expensive as it requires energy to lift the water out of the well, to separate water from the oil, to treat the water, and then to dispose of the water. The produced water is often pumped back down into the aquifer.

Water encroachment and breakthrough is very common. Globally, three barrels of water are currently produced for every barrel of oil. In the US, for example, the average ratio of water to oil is over 9:1. It is as high as 12:1 in parts of Texas, and as low as 3:1 in Prudhoe Bay, Alaska. Recently, much has recently been made of the water cut in Saudi Arabia. However, at less than 1 barrel of water per barrel of oil produced, the Saudis are well under the global average. The Saudis seem to maintain their low water cut by simply shutting down production and drilling new wells away from the zone where water breakthrough occurs. Other techniques include drilling horizontally and multilaterally, as well as the use of intelligent completions which shut off a well if high water levels are detected. What is worrying, of course, is the recent rapid increase in the Saudi's water cut, which indicates they are producing above their optimal sustainable level.

Water production along with oil is not a new phenomenon. It is simply a cost of doing business when producing from water drive wells. The cliché often mentioned is that crude oil could be considered a byproduct of water production. High water-oil ratios can be tolerated so long as the price of crude is sufficient to profitably compensate for dewatering and reservoir pressures are being maintained.

Production using water drive is rate sensitive. Rate sensitivity means that if crude is produced at a more rapid rate than optimal, the amount of oil ultimately recoverable may be significantly diminished. Two factors causing water drive reservoirs to be rate sensitive are aquifer replenishment speed and the ability of water to move across some reservoir formations faster than oil.

Water in an aquifer underneath oil has to be allowed to be replenished so that it can occupy the spaces left by crude oil produced. If oil is produced faster than the rate at which aquifer water is replenished then reservoir pressure will drop.

Additionally, the more quickly one removes oil from a reservoir, the more likely water, which can often move more quickly than oil across reservoir rock, can head toward the producing well to displace oil production. Reservoir pressures are depleted and wasted producing increasing amounts of water rather than crude oil. If one maintains a slower rate of production then water is overcome by gravity. Water is typically denser than crude oil at reservoir conditions, and it sinks below the oil.

The haphazard management of reservoir pressure is referred to as damaging a reservoir, and most often involves rapid increases in production above the optimal rate in order to maximize short-term production goals.

Gravity drainage drive occurs in all reservoirs to some degree, but is used as a primary drive mechanism in reservoirs with tilts in excess of 30° from horizontal. The producing well is located at the lowest point in the reservoir. Oil simply moves to the bottom of the reservoir due to the force of gravity, and gas moves in the opposite direction.

Gravity drainage is rate sensitive because as the oil moves down to the bottom of the dip, gas leaves the oil and moves updip to replace the volume left unoccupied by the oil. If production is sped up then this gas will be produced with the oil and nothing will be available to replace the updip space in the reservoir left by the oil. The oil will not flow unless there is something to replace it.

Compaction drive occurs when the pores surrounding the oil collapse as fluids are removed and reservoir pressure falls. Often compaction drive unexpectedly occurs to provide a boost for reservoir pressure and increase the total amount of oil which can be produced.

One of the most well known cases of compaction drive occurred in the North Sea Ekofisk field. Once a large amount of oil had been removed, the weak chalk rock matrix collapsed, increasing reservoir pressure on the remaining oil.

<u>Combination drive</u> involves a combination of several different types of drive, to varying degrees, and is the case in the majority of reservoirs. For example, a reservoir may have gas cap drive, water oil drive and gravity drainage.

Primary recovery: settled production

Once reservoir pressures have dropped and flush production has ended, pumping, also known as artificial lift, may be required to continue production. Artificial lift only lifts the oil column in the well bore and does not do anything to increase pressures in the reservoir. Switching from natural lift to artificial lift, or any change in which a producing well is changed, is known as working over a well.

Six types of pump are commonly used in oil production. These pumps listed in order of the volumes they can assist production are: sucker-rod pumps; plunger lift pumps; hydraulic lift pumps; progressing (or progressive) cavity pumps (PCPs); gas lift pumps and electric submersible pumps (ESPs).

<u>Sucker rod pumps</u> are referred to by a variety of names, including beam pumps, walking beams, nodding donkeys, horse heads, rod pumps, grasshoppers, thirsty birds, and reciprocating pumps (Fig. 6-34).

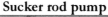

Sucker rod pump   Fig. 6-34

Source: EIA

Sucker-rod pumping is the most popular type of pumping for very low volume applications such as from wells producing less than 10 barrels per day, which are known as marginal, or stripper, wells. Sucker-rod pumping involves a sucker rod within the production tubing creating a small vacuum. The pump is powered by an engine called a prime mover, or power train, which can be fueled by diesel, gasoline, natural gas or electricity. As the sucker-rod drops through the well bore, it pushes a ball down into a seat, closing a valve (Fig. 6-35). Then as the pump rises it lifts the column of oil closer to the surface and the ball in the valve rises to let more oil into the column. The process repeats over and over, in the rhythmic nodding associated in the public eye with oil production.

Fig. 6-35

Ball and seat valve

There are several methods used to produce the reciprocating motion of the sucker-rod. The most popular method of driving a sucker-rod has tended to be the beam pump as it is relatively low maintenance, economical and a proven simple technology. Depending on the pump size and reservoir conditions, a beam pump produces between 1 and 10 gallons of fluids with each stroke (there are 42 gallons in a barrel.)

A beam pump without its own individual power source is referred to as a pump jack, or jack pump. Several pump jacks share power from a single source.

Often in locations where the sight of an oil pump is not acceptable, such as a residential neighborhood, low profile sucker-rod pumps driven by horizontal belts are used.

Of the over 500,000 producing oil wells in the US, the vast majority (80%) are marginal wells producing 10 or less barrels per day.

Even though stripper wells produce less than 10 barrels per day, and often less than one barrel per day, they collectively account for about 20% of US domestic crude production, or over one million barrels per day. US stripper well production rivals Alaskan production or US imports from Venezuela. Marginal wells are often drilled on oil fields which have been sold or leased from a larger oil company because of the low volume production rate. There is usually very little underground pressure remaining to provide natural lift. Due to the lack of economies of scale, these mom and pop operations are relatively high cost, often as high as US$20 per barrel, hence the term marginal.

Stripper wells are most often be seen in a pasture or even in front lawns in California, Texas, and Louisiana. A producer will pump the oil into a separation tank to remove water and then into small stock tanks. A local oil broker arrives with a tanker truck to haul the oil away every once in a while.

As marginal wells frequently operate over a large reservoir which may be tapped into by many independent producers, production from each well is frequently limited by a local organization to conserve overall reservoir pressure and ensure that the optimal rate of production is not exceeded. The limited allocations of production to each well are known in the US as allowables.

Plunger lift pumps involve a plunger with an open valve falling with the aid of gravity to the bottom of the well bore. As the plunger falls, it will pass through some crude oil in the well tubing. When the plunger reaches the bottom of the well, the valve in the plunger closes and effectively shuts-in the well. Pressure differentials build between the reservoir formation pressure and the pressure on top of a plunger. After a period of time, an electronic controller at the well head opens a choke at the surface attached to the flow line to allow the plunger and the column of crude above it to move to the surface. Once all of the liquids have been discharged at the surface, the plunger valve is opened and the plunger descends once again. As the electronic controller can be solar powered, plunger lift requires no external power source and is therefore quite efficient in low volume applications.

Hydraulic lift pumping involves pumping a liquid downhole under pressure. Often crude oil is used, which is called power oil when used for this purpose. The pressurized power crude operates either a downhole reciprocating pump or a jet pump. A downhole reciprocating pump is similar to the sucker-rod described previously. A jet pump uses a venturi tube, a cone shaped structure, which increase flow rates by creating low pressure below that of the reservoir formation.

Progressing (or progressive) cavity pumps (PCPs) are also known as a mono pumps or helical pumps. A PCP is a low-maintenance single thread steel screw which turns within a rubber-lined steel pump body, known as the stator (Fig. 6-36). The pump body has a helical hollow space, called an aperture. As the rotor sits within the stator, a continuous cavity is formed within the pump. As the rotor turns this cavity moves from the bottom of the pump to the top. The oil moves through the pump cavity. PC pumps are often used for highly viscous, or heavy, crude oil applications.

Progressing cavity pump

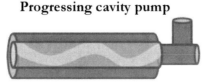

Fig. 6-36

Gas lift pumps have a relatively small footprint. They are often used on offshore applications where deck real estate is extremely valuable, and in multilateral completions where high volumes are produced. Gas lift can be used to enhance production rates on wells producing from as few as 40 to as many as 20,000 barrels per day. Gas lift pumps involve pumping compressed methane under pressure down the well bore. The compressed gas is released by gas lift valves at various carefully chosen depths in the well bore. The gas reduces the density of the column of fluid in the production tubing. The lowered density of the column creates lower bottomhole pressure and therefore causes an increase in the flow of oil from the reservoir to the bottom of the well.

Electric submersible pumps (ESPs) are used in high volume applications, up to over 90,000 bbls per day. An ESP is an electrical pump attached to production tubing and lowered downhole. As little surface installation is required, ESPs are often used as artificial lift devices on offshore wells where deck space is quite limited.

Secondary Recovery
All six of the pumping mechanisms described above are pull-based in that they seek to reduce the pressure at the bottom of the well by pulling liquids to the surface which allows high pressure oil from the reservoir to flow into the well bore.

Secondary production, involves push mechanisms, which attempt to maintain or increase reservoir pressure so that oil flows toward the low pressure well.

Although they are used when primary recovery production rates fall, secondary recovery techniques, particularly water flooding, are also often used to maintain pressure from early stages of primary recovery such as in deep offshore US gulf coast fields and in Saudi Arabia's Ghawar field.

Water flooding is the most commonly used method of secondary production. Water injection, although costly, is a highly successful method of maintaining reservoir pressures.

Injection wells drilled around the production well are used to pump water into the aquifer. Water flooding creates water drive by maintaining or raising reservoir pressure. Fresh water, or brine, which has been separated from produced crude oil, is pumped under high pressure into the aquifer. Water is often treated before injection to remove contaminants such as bacteria or other living organisms which might produce waste material which can clog production piping. The depth at which water is injected is important as one wants to increase reservoir pressures without having the water encroach on the oil pay.

Using natural aquifer pressures or water injection to enhance reservoir pressures is known as a water sweep. The area in which water has displaced the oil is known as the swept zone.

A challenge when using water flooding in a reservoir with a gas cap is that one does not want to increase pressure too much, or the gas cap may contract. This will pull oil into the gas cap zone. If oil is pulled into the gas cap zone, then it may stay there as attic oil. Attic oil is out of the reach of the flow of oil to the producing well.

Many templates are used for locating water flooding wells around producing well, including four spots, five spots, seven spots, and nine spots. The five spot pattern is most common. Four injection wells are placed in corners of a square around a production well in the center. A contour pattern around known edges of a reservoir, not necessarily in a pattern equidistant around the producing well, is also sometimes used. A method known as line drive can also be used where the injection wells are placed in a row, particularly when the reservoir dips.

Reservoir pores in the oil pay contain water also. Water contained in a reservoir can be either interstitial or free. Interstitial water is that which forms a film attached to the rock surface and which does not move. The interstitial water acts as a sort of lubricant for the oil and allows the hydrocarbons to move through the reservoir with less friction. Free water is water which is able to flow freely through reservoir rock.

Reservoir rocks tend to exhibit a preference for either water or oil to form the film around them (Fig. 6-37). In the past it was believed that most reservoir rocks' surfaces preferred water adhesion. Water usually adheres to the walls of the rock pore, and the hydrocarbons are contained by the water. Recently it has been discovered that oil-wet and intermediate reservoirs are more common than had been believed. Preferential wettability is a reference to the preference of rock pore surfaces for: water (water-wet reservoirs), oil (oil-wet reservoirs), or both water and oil (intermediate wettability).

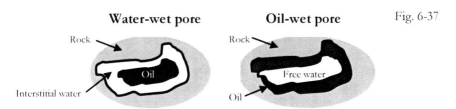

Fig. 6-37

The preferential wettability of a reservoir rock is especially important for water injection. If water is injected into a water-wet reservoir then it will displace oil toward the well. However, if water is injected into an intermediate or oil-wet reservoir then the injected water may simply become free water and move toward the well itself, bypassing the oil which is adhering to the pore surfaces.

Wettability, therefore, is a key determinant in the efficiency of waterflooding. Waterflooding is much more efficient in water-wet reservoirs than in oil-wet reservoirs.

Gas reinjection, the other method of secondary production, involves separating the hydrocarbon gases extracted along with crude oil, and injecting them back into the reservoir via a separate well. Again, the goal is to try to maintain or raise reservoir pressure.

Associated gas is usually a wet gas. Wet gas is methane mixed with ethane, propane, butane and natural gasoline. Dry gas, is simply methane by itself.

In a surface facility known as a separator, or gas-oil separation plant (GOSP), the wet gas is separated from crude oil. After the GOSP, the wet gas can then be sent to a gas fractionating plant, known as a gas stripper, which separates the methane from heavier gases including ethane, propane and butane (NGLs).

Enhanced oil recovery (EOR)

Once water flooding and gas reinjection methods have been exhausted, a final stage of extracting oil called tertiary production, or enhanced oil recovery, is the only method that remains to be used. Tertiary methods attempt to enhance reservoir rock permeability by creating fractures within the matrix of rock formations; displacing oil by pumping inert gases into the reservoir; and lowering the viscosity of crude in order to allow it to flow more easily across the reservoir formation to the well.

Hydraulic fracturing is used to enhance a reservoir's permeability. It involves pumping water and some low viscosity fluids down the well bore for a period of time at very high pressure. The fluid under pressure creates a large fracture, perhaps several hundred feet long, beginning at the bottom of the well in the reservoir rock, and may connect some pre-existing fractures. Sand and guar gum are sometimes pumped down with the fluid as propping agents to hold the new fractures open once the fluid pressure is lowered.

Acid fracturing, also known as acid fracing, involves pumping acids, such as hydrochloric acid (HCl) into the reservoir at a pressure higher than the rock can withstand. Deep fractures are created which extend far into the formation. Flow channels are created through which oil can move toward the well.

Miscible flooding, also known as a 'huff and puff' technique is commonly associated with heavy oil and involves injecting a material which is capable of mixing with the oil it comes in contact with.

A miscible material is one which will mix with another material. Water and oil, for example, are not miscible. Materials used in miscible flooding include inert gases such as carbon dioxide and nitrogen, as well as light hydrocarbons under high pressure such as methane, ethane, propane and butane. The material is allowed to soak for a short period of time before pumping begins. The miscible materials improve flow of the crude it is mixed with and also displaces the hydrocarbons in the reservoir.

The process of miscible flooding, or huff and puff, involves three stages: the huff stage, where miscible materials such as carbon dioxide and nitrogen are pumped into reservoir; the soak stage where the well is shut in for a period of time to allow materials to mix; and the puff stage where oil, having been displaced by the gases, is produced as the well is opened

The use of carbon dioxide that has been stripped from power plant exhaust, has been becoming more popular. CO_2 is a major greenhouse gas. The use of CO_2 injection thus has dual benefits. In addition to aiding oil recovery, it is a method of carbon sequesterization as it removes carbon dioxide from the atmosphere.

Thermal recovery, or steamflooding, involves pumping very hot water or steam containing surfactants, which are surface active agents such as detergents or emulsifiers, into the well at extremely high pressure. This loosens the oil and reduces its viscosity so that it can flow through the reservoir to the production well. Thermal recovery is most often used with heavy, low API gravity, crude oils.

Chemical flooding involves using three types of chemicals floods to reduce the viscosity of the crude oil in the reservoir: polymer floods, micellar/polymer floods, and micellar/alkaline floods.

In-situ combustion involves introducing air into the reservoir via an injection well. The air and oil in the reservoir are then ignited and the resulting hot gases, at high pressure, move toward the production well, which is at a lower pressure. The fire is controlled as a limited quantity of oxygen is introduced to the reservoir. Dry combustion involves simply air and the oil in the reservoir. Wet combustion involves injecting water along with the air into the well to enhance the viscosity reduction in the reservoir.

Microbial enhanced oil recovery entails the use of high heat-tolerant, anaerobic bacteria (which can survive without oxygen). Underground oil wells are typically 70°C – 150°C. The bacteria are injected into the well at high pressure. Once there, they multiply and produce gases such as carbon dioxide and methane, which create the gaseous pressures necessary to send the oil through the reservoir to the producing well.

Following tertiary production, as much as 25-75% of the remaining oil may left in the ground and is not recoverable in high volumes. Production ceases or occasionally continues as a marginal stripper well.

Depletion terminology
Reservoir depletion happens to every oil field. As oil is removed there is obviously less oil remaining.

Production rate declines eventually occur in every reservoir. Many reservoirs, especially ones which use water drive, can have long plateaus at a given production rate. Then, as pressures within a reservoir eventually declines, the daily production rate from a reservoir declines.

A production total decline happens when a well is over-produced, or pushed too hard too quickly. In this case the total amount of oil produced from a well can be reduced dramatically. This is due to rate sensitivity, in that the total amount of oil recoverable from a reservoir is dependent on the rate at which the oil is removed. If one removes oil more slowly, more crude oil in total may be removed.

In a high interest rate environment, oil companies may be acting financially rationally by over-producing a well and reducing the total amount of physical oil which can be removed. If interest rates are high then the present value of cash flows from oil production in the future is low. Thus, oil companies, assuming constant oil prices, may have a rational financial incentive to maximize short term cash flows as the value of having the cash today may more than offset the value of having less oil produced in the future.

Site remediation

Once a reservoir has been exhausted or it is no longer economical to produce, the entire production surface completion, including wellheads, piping, associated machinery, equipment and buildings are usually dismantled, the well is plugged, and the site is returned to its original state in a process called site remediation. This also involves removing all storage tanks.

Before production began, the oil producer would usually have performed an environmental impact analysis (EIA). This analysis outlines the effect on the environment of exploration and production and how the producer will endeavor to minimize and correct any damage.

In 1980, the US Congress created a trust fund, known as the Superfund, to enable cleanup of sites in the US which have been contaminated by hazardous waste. A Superfund site is one which is designated by the EPA to require a federally sponsored cleanup and is used only if a responsible party is unable to pay for cleanup. If a production site is properly plugged and abandoned, it should not become a Superfund site. The Superfund is financed by a special tax on chemical and petroleum industries. Legislation underlying the Superfund is the Comprehensive Environmental Response, Compensation, and Liability Act (CERCLA) of 1980 as amended by the Superfund Amendments and Reauthorization Act (SARA) of 1986.

CHAPTER SEVEN
REFINING

Feedstock

Feedstocks are inputs to refinery processes. End results, such as gasoline and kerosene, are finished products. In the oil market one may hear of an oil company taking a cargo of crude oil into their system. This refers to the company using the crude oil as a feedstock as opposed to selling it to another trader. It is common for grades of crude oil to be blended with other crude oils, or sometimes even finished products, to obtain feedstock which a refinery is configured to use.

The mix of crude oils used as a feedstock is called a refinery's crude slate. Refinery operators speak of the *avails*, or availability, of certain types of crude or product, and will try to purchase those grades of crude which will produce a desired product slate, which is a portfolio of finished products. Avails can be tight, or the opposite, which is plenty of avails, for a crude oil feedstock or finished products.

In a refinery, a barrel of crude oil, which contains 42 gallons, can produce, depending on refinery equipment, configuration and the type of crude oil, several additional gallons of finished products due to refinery gain, or processing gain. The *average* crude oil refined in the *average* US and global refinery produces the following (Table 7-1):

Table 7-1

Input	Refinery	Output Product	US Refineries Gallons	% Yield of output	Global Refineries Gallons	% Yield of output
		Motor Gasoline	19.66	44%	11.12	26%
		Distillate Fuel Oil (diesel & heating oil) ⎤ Middle	10.04	22%	11.59	27%
		Jet Fuel.................................... ⎦ Distillates	4.07	9%	2.46	6%
		Petroleum Coke	2.18	5%	1.72	4%
Crude Oil		Still Gas (aka refinery gas)	1.85	4%	1.90	4%
1 barrel = 42 gallons		Residual Fuel Oil (aka #6 Fuel Oil)	1.72	4%	5.99	14%
		Liquefied Refinery Gas (methane, ethane, propane, butane)	1.68	4%	2.11	5%
		Bitumen and Road Oil	1.34	3%	1.72	4%
		Naphtha for Feedstocks (petrochemicals, plastics etc.)	0.67	1.50%	1.29	3%
		Other Oils for Feedstocks (petrochemicals; plastics etc.)	0.55	1.22%	1.08	3%
		Lubricants	0.46	1.03%	0.86	2%
		Special Naphthas (paint thinners, cleaners, solvents)	0.13	0.28%	0.13	0.3%
		Kerosene (heating, cooking & lighting)	0.17	0.38%	1.03	2%
		Aviation Gasoline (used by small light aircraft)	0.04	0.09%	0.03	0.08%
"Refinery Gain"		Waxes (used for candles, coating fruit)	0.04	0.09%	0.03	0.08%
		Miscellaneous Products	0.17	0.38%	0.02	0.04%
		Total	44.77 gallons		43.10 gallons	

Source: EIA data

143

If you follow weekly US DOE petroleum statistics you will notice that the actual refinery output yield in the US of finished motor gasoline is over 55%, and not 44% as shown in the above table. The increase from 44% to over 55% is because molecules other than those from the average barrel of crude are added, such as oxygenates (MTBE and ethanol), NGLs and other hydrocarbons. You may also observe that US refineries have larger refinery gain than the average refinery globally. This is because US refineries are more complex with coking and cracking capabilities.

Refinery economics is the process of deciding which grades of crude oil to purchase for a feedstock slate, and which refinery processes to use on feedstock. Refinery economics is influenced by prices of crude oils, the cost of individual refinery processes and the prevailing prices for each of the finished products

The difference between crude oil input prices and finished product output prices is referred to as the refinery margin, the crack spread, or simply the crack.

Every oil refinery is configured differently in its ability to produce desired finished products from a given slate of crude oils. Some refineries are complex, with the ability to produce a lot of gasoline from heavy crude oil. Other refineries are simple and will not have the ability to produce as much gasoline from the same barrel of crude oil.

Stages in refinery process
There are four steps in refining: separation, conversion, treatment and blending.

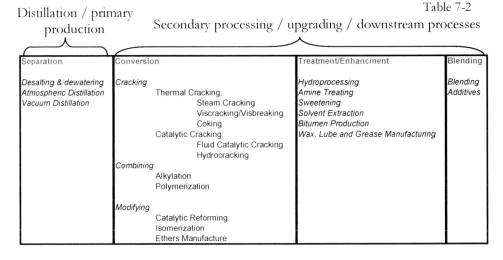

Distillation / primary production

Secondary processing / upgrading / downstream processes

Table 7-2

Separation	Conversion		Treatment/Enhancment	Blending
Desalting & dewatering	Cracking		Hydroprocessing	Blending
Atmospheric Distillation		Thermal Cracking:	Amine Treating	Additives
Vacuum Distillation		Steam Cracking	Sweetening	
		Viscracking/Visbreaking	Solvent Extraction	
		Coking	Bitumen Production	
		Catalytic Cracking:	Wax, Lube and Grease Manufacturing	
		Fluid Catalytic Cracking		
		Hydrocracking		
	Combining			
		Alkylation		
		Polymerization		
	Modifying			
		Catalytic Reforming		
		Isomerization		
		Ethers Manufacture		

Separation

Desalting and dewatering

Desalting is the first process crude oil is put through at a refinery. Salt must be removed as it can be corrosive to refinery piping. The desalting process removes salt and other contaminants before the crude is sent to settling tanks. There, water is removed before it is sent to the atmospheric distillation unit using a charge pump.

Atmospheric distillation

A furnace warms a continuous feed of crude with pipes containing super-heated steam up to temperatures of around 400°C. Some refineries power these furnaces with gases (methane, ethane, propane and butane) and heavy residual fuel oil actually produced in refinery processes. Gases are used as fuel because they are difficult to transport and store. Residual fuel is used because it sells for very little compared with other products. When used in this way the gases and residual fuel are known as refinery fuels.

The tallest silver column one sees in refineries is usually the atmospheric distillation unit (ADU), also known as the crude distillation unit (CDU), crude unit, crude fractionator, primary distillation unit (PDU), or simply the tower. Heated crude from the furnace enters the ADU in a continuous feed. Distillation involves heating crude until various components, referred to as product fractions, or cuts, boil into a gas and then condense separately. There is always at least one, but there may be several of these towers in a refinery. They are called atmospheric because they operate at normal atmospheric pressures.

Inside the ADU are horizontal plates called trays with mushroom shaped bubble caps to encourage the condensing of various products at different temperatures (Fig. 7-1). The lightest products, such as methane, ethane, propane, butane, naphtha and gasoline, rise through the trays to the top of the tower before condensing. The medium weight products such as kerosene, jet fuel, diesel and heating oil, condense in the middle. Heavier products, such as residual fuel oils, condense lower down. Finally the heaviest, tar-like products, such as bitumen settle to the bottom of the distillation tower, referred to as the bottom of the barrel. Separating light tops from heavy bottoms in the ADU is often referred to as topping the crude.

The boiling points of products overlap slightly as they contain some of the same hydrocarbon molecules. The boiling range cross-over is called the tail end of a cut. For example, diesel will contain some molecules with a similar boiling point as kerosene or jet fuel.

Products which go through separation in a refinery tower and are processed no further are known as straight-run products. Although it is possible for straight-run products to be sold from the refinery to an end user, in practice very few products are sold in this manner. Instead, they are subjected to further treatment before sale. In a very basic refinery, straight-run products are sometimes sold to other refineries that have more sophisticated processing units.

Fig. 7-1

Bubble caps force vapors to bubble through liquid standing on trays. Vapor cools as it bubbles through liquid, condensation occurs, and liquid is drawn off.

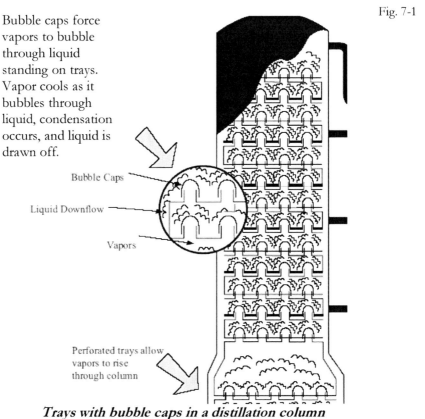

Bubble Caps

Liquid Downflow

Vapors

Perforated trays allow vapors to rise through column

Trays with bubble caps in a distillation column
Source: EIA

Vacuum distillation (vacuum flashing)

A second stage of separation occurs in vacuum distillation units (VDU). The process is similar to the one described in the ADU, except this time distillation occurs at less than normal atmospheric pressure. The reduced pressure acts to reduce the boiling point of the various fractions/products. This means that they can vaporize at lower boiling points without thermal cracking occurring. Thermal cracking is a process by which large hydrocarbon molecules are broken into smaller ones as a result of heat. A refiner does not want thermal cracking to occur in the ADU or VDU because it is difficult to control how much cracking occurs in these processes. Refiners prefer cracking to occur at a later stage in other devices in which they can control more precisely the amount of each product produced.

Vacuum distillation is usually only applied to heavy residuals from the atmospheric distillation unit. The main products of the VDU are vacuum gasoil (VGO), residual fuel oils and heavy residuals referred to as flasher bottoms.

In atmospheric distillation and vacuum distillation, the refiner is carefully separating groups of hydrocarbon molecules in order to further treat them in more specific processes to produce individual finished products.

Although the ADU and VDU are the largest distillation columns in a refinery, many subsequent downstream processes separate their output by distillation. Smaller downstream distillation columns are referred to as fractionators.

Conversion

With simple separation processes such as distillation, refineries are at the mercy of the type of crude they run as they cannot change the yields of various finished products. For example, if a crude oil contains 20% gasoline boiling range molecules, a refinery will only be able to distil 20% of the crude into gasoline using an ADU and VDU.

Gasoline and the group of products known as middle distillates (jet fuel, kerosene, diesel, and heating oil) are the finished products with the highest demand and the largest profit margin. There is, therefore, an incentive to maximize production of just these two product groups by converting as many hydrocarbon molecules as possible contained in crude oil into gasoline and middle distillates.

Most straight run gasoline produced in an ADU and VDU has a fairly low octane rating. Conversion processes are used to generate high octane blendstocks for gasoline. High octane gasoline blendstocks are among the most lucrative products of conversion processes.

Conversion is any process which cracks (break apart), combines, or modifies non-gasoline or non-middle distillate molecules to turn them into gasoline or middle distillate molecules using heat, pressure and catalysts. Conversion can also be used to produce petrochemicals and to reduce the viscosity of heavy residual fuel oil. Conversion is sometimes referred to as upgrading, as a refinery is taking lesser valued products and chemically altering them to produce higher value products.

Conversion takes place in secondary processing units, which are any units downstream from the ADU and VDU. Downstream processing units in a refinery are very expensive to build and operate and are sometimes referred to as pieces of refinery kit.

Although the refinery golden rule is to maximize gasoline or maximize middle distillates, refineries with conversion capabilities do produce less valuable products such as residual fuel oil, bitumen and coke. Less valuable products are produced when it is uneconomical to convert the molecules within those products to gasoline or middle distillates.

Refineries with secondary processing units have typically two modes of operation: max gasoline mode, in place just prior to and during summer driving season; and max distillate mode, implemented during winter heating season (Fig. 7-2). In the US, the swing in production is about a 2% yield change from gasoline to distillates.

Fig. 7-2

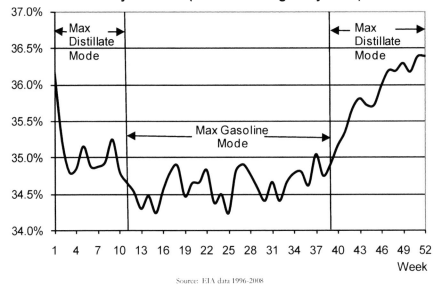

US Refinery Distillate (diesel + heating oil + jet fuel) Yield

Source: EIA data 1996-2008

The three types of conversion are: cracking, which involves breaking large hydrocarbon molecules into smaller ones; combining, in which smaller hydrocarbon molecules are combined into larger ones; and modifying, which is the re-arranging of atoms in hydrocarbon molecules.

Cracking

The rapid increase in demand for gasoline and middle distillates for transportation led to the commercial development around 1913 of cracking. Cracking involves using heat, pressure and sometimes catalysts to break large heavy hydrocarbon molecules into smaller lighter gasoline or middle distillate range molecules. Cracking processes are also employed to produce petrochemicals and to reduce viscosity of heavy fuel oil.

Cracking units in a refinery are tall, thick-walled, rocket-shaped structures accompanied by furnaces and heat exchangers. The temperature to which the cracker feed is heated is referred to as the severity of the process. The resulting stream of hydrocarbons produced in the cracking process usually has to pass through a distillation column to separate it into various product fractions.

There are two types of cracking: thermal cracking and catalytic cracking. Thermal cracking involves breaking apart heavy molecules of less valuable petroleum products with heat and pressure alone. Catalytic cracking also uses heat and pressure but takes place in the presence of a catalyst. Thermal cracking processes include steam cracking, viscracking, and coking. Catalytic cracking processes include fluid catalytic cracking (FCC) and hydrocracking.

Thermal cracking

Steam cracking is used to produce petrochemicals. Steam cracking involves using very high temperature steam to break apart the molecules of ethane, propane, light naphtha, heavy naphtha and light fuel oil/gasoil. Hydrogen (H) in the steam (H_2O) bonds with the cracked molecules to form new hydrocarbon molecules.

The main goal of steam cracking is to produce ethylene and propylene. Byproducts include butanes, pygas (pyrolysis gasoline - used to make benzene and as a gasoline blendstock), light fuel oil, hydrogen and methane. Steam cracking is also referred to as ethylene cracking, because of its primary use in producing ethylene for petrochemicals. Ethylene, propylene, and benzene are the principal petrochemicals used in products such as consumer plastics.

In the US and the Middle East, the NGLs ethane and propane are the most common feedstocks of a steam cracker/ethylene cracker. Ethane and propane gases can be inexpensively transported via pipeline to a petrochemical plant. In Asia, the feedstock to steam cracking units is commonly naphtha which is transported in a tanker, as NGLs are not readily available via pipeline. Steam crackers using naphtha as feedstock are referred to as naphtha crackers. The outputs are similar regardless of the feedstock used, although the ratio of outputs (ethylene, propylene etc.) differs with the feedstock used.

Viscracking, also known as visbreaking, is a cracking process designed to reduce the viscosity of residual fuel oil. The reduced viscosity fuel oil is called visbroken residual fuel oil. As the viscosity of heavy residual fuel oil is reduced, fewer expensive diluents have to be added to residual fuel oil before it can be used by a consumer such as power generation utility or a ship owner.

The viscracking process uses heat to break the heavy molecules in a soaker, or reaction chamber, and is followed by rapid cooling, or quenching, to stop the cracking and minimize coke formation. The products of a viscracker are then separated in a vacuum distillation tower.

Atmospheric tower residues, which are heavy products produced in the atmospheric distillation unit, and vacuum tower residues can be used as a feedstock to the visbreaker unit. In addition to reducing viscosity of resid, the visbreaker also produces some blendstock for gasoline, naphtha, middle distillates and bitumen.

Coking uses moderate pressure and heat to crack the residual end of the barrel into lighter products including gases, naphtha, blendstock for gasoline, middle distillates, and coke, a low value product which is used as a fuel in heavy industry. Producing coke is not the goal of coking, coke is merely a byproduct in the process of cracking residuals into higher value products.

The coking process takes place in tall barrel-shaped drums called cokers. Coke is a coal-like substance and usually has to be physically cut out of a coking drum using high pressure water drills. Refineries with cokers may have several coking drums so that one is always being filled while the others are having the coke removed, which is termed batch processing. The feed to the coker is continuous, so that it is just the collection drums which are batch processed. There are three types of coking: delayed-coking, flexi-coking and fluid-coking. Delayed coking is the most common and is called delayed because it takes around 12 hours for cracking to occur within the coker, with each batch taking approximately 22 hours. Each of the three coking processes produces differing amounts of coke and other products. Flexi and fluid coking also differ from delayed coking in that they are continuous and not batch processes.

Coking allows refineries to use lower priced heavy crude oils which without coking would yield a lot of residuals and very little high value products. The refinery yields shown on the first page of this chapter show a low yield of residual fuel in the US compared with a high yield of resid globally. The US has a large amount of coking refineries which crack the resid fraction into gasoline and middle distillates. Globally, refineries do not have the same level of coking as the US because building and operating a coker is very expensive.

Coking also results in a large amount of refinery gain in the US, which why the US has a higher level of gain than refineries globally.

Catalytic cracking
Catalytic cracking uses heat and pressure just like thermal cracking, but also uses a catalyst. Catalysts quicken cracking processes or enable them to occur at lower temperatures. Catalysts are used in a powder, solid bead, or pellet form. Although the catalysts are separated from the oil with a centrifuge and reused again and again, fresh catalysts need to be added to the process every once in a while as the centrifuge cannot capture all of the catalyst.

When using a catalyst, carbon molecules from oil eventually coat the surface of the catalyst rendering it ineffective. The carbon is removed in a piece of kit called a catalyst regenerator, in which the carbon is oxidized with super-heated air. The catalyst is then separated using another centrifuge to be fed back into the cat cracking unit.

Catalytic cracking is very flexible and the outputs from the process can be tailored to adapt to changes in individual product demands due to seasonality.

The output from cat cracking is put through a dedicated distillation column also known as a fractionator. The outputs are prefixed with the term cat cracked, and include: cat cracked-gas, gasoline, light gas oil, heavy gas oil. The residues, or bottoms, from a cat cracked fractionator are called cycle oil and are sent back to the cracker to be cracked into the higher value products. Cat cracker bottoms are effectively cycled out of existence.

The two common types of catalytic cracking units (CCUs) are fluid catalytic crackers and hydrocrackers.

Fluid Catalytic Cracking (FCC) is a process which uses catalysts, low pressure, and heat. When applied to gasoil, the FCC unit creates molecules which can be used in gasoline, middle distillates and other valuable products. The fluid in FCC refers to fluidizing of the solid catalyst. Fluidizing the catalyst allows it to react more quickly with the oil. Similar cracking processes to FCC are called moving-bed catalytic cracking and thermofor catalytic cracking (TCC), both of which are less common. An FCC is often one of the largest units in a refinery and is technically very complex. Most FCCs have custom one-of-a-kind construction and upgrades and are therefore unique in the world of refinery engineering. The small amount of heavy oils produced by the FCC are known as cat cracker bottoms, cat slurry oil, heavy cycle oil, or decant oil, and are used as feed for a coker. Slurry oil is also used for blending into residual fuel oil or producing another finished product called carbon black.

Hydrocracking is a process similar to fluid catalytic cracking but uses hydrogen and lower temperatures than an FCC, different catalysts, and higher pressure. Hydrocracking is not as common as FCC, as it is usually a more expensive process. Hydrocracking requires a large amount of energy. It also consumes hydrogen, which is difficult and costly to produce. The primary advantage of a hydrocracker is that it can convert the entire feedstock to a desired range of molecules because the process adds hydrogen. The primary hydrocracking feedstock is heavy vacuum gasoil (HVGO) and the process produces gasoline or middle distillate molecules, among other desired products.

Many oil traders associate hydrocrackers with jet fuel production, as these units are very good at producing the narrow range of molecules required in jet fuel.

The output from a hydrocracker used as a gasoline blendstock is sometimes referred to as hydrocrackate.

Carbon rejection and hydrogen addition
Refinery cracking processes can be classified as either hydrogen addition processes or carbon rejection processes (Table 7-3).

Hydrogen addition processes, such as steam cracking and hydrocracking, are those which add hydrogen to the cracked hydrocarbon molecules.

Carbon rejection processes such as visbreaking, coking, and FCC involve replacing carbon-carbon bonds within hydrocarbon molecules with carbon-hydrogen bonds, thus making the molecule lighter. However, as no additional hydrogen is added, the processes result in some very heavy, carbon laden products such as coke.

	Hydrogen addition processes	Carbon rejection processes
	Hydrocarbons from crude oil ↓	Hydrocarbons from crude oil ↓
Carbon rejection stage	Cracking makes all molecules ready to bond with hydrogen (hydrocarbon molecules are made to reject carbon-carbon bonds)	Cracking makes all molecules ready to bond with hydrogen (hydrocarbon molecules are made to reject carbon-carbon bonds)
	↓ Additional hydrogen is added	↓ NO additional hydrogen added
	↓ All molecules can become light hydrocarbons as additional hydrogen is available for all of the carbon to bond to.	↓ Some molecules take most of the hydrogen that was in the feed and become very light hydrocarbons. The remaining molecules are left with very little hydrogen and become heavy molecules (forming products such as decant oil and coke.)
Refinery processes	Steam cracking, Hydrocracking	Visbreaking, Coking, FCC

Combining

Combining is the opposite of cracking. Whereas cracking breaks the molecules of large hydrocarbons into smaller more valuable ones, combining joins smaller hydrocarbon molecules together to produce larger more valuable ones. Once again, these refinery processes are used to maximize the sweet spot for profits which surround molecules adding to gasoline and middle distillates. Because one is taking smaller molecules and adding them to make larger molecules, there is a volume but not a weight reduction in the outputs from these processes, i.e. refinery losses. Two common methods of combining are alkylation and polymerization.

Alkylation is a process by which unsaturated gas byproducts of cracking, in particular the olefins propylene and butylene, are combined with an isoparaffin, usually isobutane, in the presence of catalysts sulfuric acid or hydrofluoric acid to produce a high-octane blendstock for gasoline. This product in as an isoparaffin called alkylate. In addition to having a high octane rating, alkylate has a low vapor pressure and contains almost no sulfur, olefins or benzene – all of which make it an excellent gasoline component. The alky plant also produces small amounts of propane and normal butane as byproducts. The feed to an alkylation unit has to be chilled because of the heat generated by the chemical reactions taking place. The acids used as catalysts in alky plants are highly toxic and corrosive and strict safety controls are put in place to ensure they do not escape into the atmosphere.

A butane isomerization unit, sometimes referred to as a C4 isom plant or a butane isomerization (BI) plant, takes normal butane and to produce isobutane as a feedstock for the alky plant.

Polymerization, as with alkylation, involves linking a series of molecules. Polymerization, either with heat alone or in the presence of a catalyst at lower heat, combines light olefin gases, especially propylene and butylene, into higher octane gasoline blending stock called polymerate. Similar to alkylation, the gas feedstock for the polymerization unit can come from either cracking processes or the refinery light-ends unit/gas plant. Polymerization is a slightly older technology, and has generally been replaced by alkylation in refineries.

A particular version of polymerization called dimerization, as it involves linking just two molecules, has recently generated increased interest in the US, since MTBE has been phased out of US gasoline. Dimerization is used to manufacture isooctane. Isooctane contains a mixture of the 18 isomers of the octane molecule. Isooctane can be manufactured by combining two isobutylene molecules with hydrogen using the same equipment which has been used to manufacture MTBE (which cuts down on new plant construction costs), with the addition of a hydrogenation unit. Isooctane can be used as an octane raising gasoline blendstock to replace MTBE.

Modifying
Modifying involves the re-arranging, or alteration, of hydrocarbon molecules. There are three main refinery modification processes: cat reforming, isomerization and ethers manufacture.

Catalytic reforming modifies naphtha using heat, pressure and a catalyst, most commonly platinum group metals, into high octane gasoline blending components called reformates. In addition to being used as gasoline blending component, reformate has a high aromatics content and is the primary source of individual aromatics, benzene, toluene and mixed xylenes, known together as BTX, which are often separated out from reformate for use in petrochemicals or as gasoline blendstocks themselves.

Catalytic reforming takes place in a reforming unit, or reformer, of a refinery and can be recognized by its series of spherical reactors, one of which is spare to use when one of the others is taken offline in order to regenerate the catalyst when carbon from the oil eventually coats it.

The naphtha boiling range hydrocarbon molecules used as feedstock for a reformer can be from the ADU, VDU or downstream processing units. Apart from a small amount used as commercial solvent, straight run naphtha boiling range molecules have almost no direct end uses. Almost all straight run naphtha is modified into gasoline or petrochemicals.

It is very important for the reformer naphtha feedstock to be sulfur free, or less than 1ppm sulfur by weight, as sulfur poisons platinum and palladium catalysts. This is the same reason sulfur is removed from finished gasoline and diesel, so as not to damage the catalytic converter in a vehicle.

Hydrogen is a significant byproduct of reforming. It is separated for use in other refinery processes such as the hydrogen addition processes described earlier, as well as sulfur removal processes. The naphtha cat reformer is typically the primary source of hydrogen in a refinery.

Isomerization is a process which rearranges the molecular structure of normal paraffins: n-butane (nC4), n-pentane (nC5) and n-hexane (nC6), into higher octane rating isoparaffins: isobutane, isopentane and isohexane, respectively.

Normal pentane and hexane have very low octane ratings. Rearranging the molecules to form isopentane and isohexane increases octane rating dramatically. Isopentane and isohexane are then used as gasoline blending components which together are referred to as isomerate.

Normal butane converted to isobutane is used as a feedstock for alkylation.

Ethers manufacture began to be used in the 1970s and early 1980s. Refiners had to find an alternative gasoline octane enhancer to lead, which was being phased out as an additive. MTBE, an ether made from isobutylene and methanol, has a high octane rating and relatively low RVP, which makes it a good gasoline octane rating enhancer.

Methanol is created from natural gas in a petrochemicals plant. Isobutylene is a byproduct of cracking processes, but can also be created in a dehydrogenation plant from isobutane or sometimes normal butane. The process of dehydrogenation uses catalysts to remove hydrogen from the isobutane feed which converts it to isobutylene.

Treatment/Enhancement

Hydroprocessing is the primary method by which hydrogen is used to remove unwanted elements like sulfur, nitrogen, nickel, and vanadium.

Sulfur and nitrogen are usually bonded to carbon atoms contained within products such as gasoline or diesel. Hydroprocessing involves reacting sulfur and nitrogen compounds with hydrogen molecules in the presence of a catalyst at high pressure and temperature to create hydrogen sulfide and a small amount of ammonia. Hydrogen sulfide and ammonia can easily be removed from the product. Hydrogen sulfide is removed using an amine wash.

Hydroprocessing requires a readily available supply of hydrogen to bond with sulfur and nitrogen molecules. Hydrogen can be obtained from the reforming unit where it is produced as a byproduct. If a refinery has a hydrocracker, which is a large consumer of hydrogen, then it may be short of hydrogen for hydroprocessing. Additional hydrogen can be produced by reacting methane with steam and oxygen to produce carbon dioxide and hydrogen. This process of creating additional hydrogen takes place in a steam methane reformer (SMR.)

Hydroprocessing is relatively expensive. Producing hydrogen is generally the highest component of cost.

In the past few years, governments around the world have begun imposing sulfur content restrictions on gasoline and middle distillates. Refineries have had little choice but to build hydroprocessing facilities. In addition to government regulations, refineries have been facing a crude oil feedstock supply containing higher levels of sulfur. An increased amount of hydroprocessing has to be carried out in order to continue to meet product specifications.

Hydroprocessing is also referred to as catalytic hydrotreating or catalytic hydrodesulfurization. A hydrotreater (HDT) unit and a hydrodesulfurizer (HDS) unit are very similar. A hydrotreater is typically used on kerosene and lighter products. A hydrodesulfurizer is used for heavier middle distillates, such as diesel and heating oil. Hydrotreatment of kerosene and jet fuel can, in addition to removing sulfur, create a smoke point improvement (SPI) which ensures a more thorough burning of the fuel. HDTs or HDSs are also referred to as sulfur recovery units (SRU.)

Amine treating uses amine solvents such as MonoEthanolAmine (MEA) or DiEthanolAmine (DEA) to remove the highly toxic gas, hydrogen sulfide, produced as a result of bonding sulfur in oil to hydrogen during hydroprocessing. The amine treating process removes hydrogen sulfide and other unwanted components. Amine treating is also known as amine washing.

Solvent extraction, also known as solvent recovery, is a process which can be used at any stage in the refinery system to selectively remove components from product streams by dissolving them into a solvent individually. The components which one wishes to separate dissolve in the solvent and can easily be removed by separating the solvent solution from the oil product. Amine treating, aromatics/BTX recovery, solvent deasphalting and solvent de-waxing are all refinery processes in which solvents are used to separate and extract various components from a hydrocarbon stream.

Sweetening involves oxidizing a particularly foul smelling sulfur compound called mercaptan in gasoline and other intermediate and finished products. In the refinery sweetening process, mercaptan is oxidized into odorless disulfides, which may remain in the fuel.

If the mercaptan was not oxidized, storing and burning gasoline and other products would result in a very strong rotten egg smell. Because of its powerful smell, mercaptan is added to natural gas in trace quantities to give it an odor so that leaks can be quickly detected.

Bitumen production takes place by extracting asphaltene molecules from very heavy aromatic hydrocarbon molecules found in crude oil to produce bitumen. There are three processes for producing bitumen: straight run, solvent deasphalting, and bitumen blowing.

Straight-run bitumen is produced by vacuum distillation of atmospheric distillation unit residuals. Solvent deasphalting is a method of removing asphaltic molecules from the residual fraction of the ADU by using propane or butane as solvents. Bitumen blowing involves blowing hot air through a residual fraction feed to produce bitumen which is harder than straight-run bitumen. The bitumen produced in a blowing unit is sometimes called brown bitumen to distinguish it from straight-run bitumen.

Wax, lubricant and grease manufacturing
Solvent de-waxing is a process which removes wax from distillation residuals by using a solvent. The de-waxed oil is sent to lube oil manufacturing as a base stock.

Lube oil manufacture involves treating and blending de-waxed oil with other refinery products in order to meet the desired motor oil/industrial lubricant specifications. The components of lube oil are called lube oil base stock and are blended to produce finished lubricant formulations.

Grease manufacture, also called grease compounding, entails blending lubricating oils with metallic soaps and other additives at a moderately high temperature.

Blending/Finishing

Finished products are the result of blending a range of hydrocarbon molecules from many different processes within a refinery (Table 7-4).

For example, as straight run gasoline does not have a high enough octane rating for modern engines, blending higher octane rating components into gasoline is one of the most important blending processes in refineries.

The quest for low cost, high octane rating, low volatility gasoline blendstocks has driven most innovation in refinery technology. Individual gasoline blendstocks are often traded amongst refineries as blenders search for all important octane rating.

Each gasoline blendstock has a different octane rating, vapor pressure, and sulfur content, among other characteristics. Blending gasoline is challenging. Octane rating and vapor pressure of a blend are not simple linear weighted averages of the octane rating and vapor pressure of the blendstock components. The octane rating and volatility of each blendstock is affected by interaction with the other blendstocks used.

Additionally, octane rating and volatility of individual blendstocks differ. They contain a range of molecules which vary depending on the exact processes they have gone through and the original crude oil. For example, there are different versions of reformate, depending on the reformer technology used.

In descending order of magnitude, FCC gasoline (~33%), reformate (~33%), naphtha (~10%), alkylate (~10%), isomerate (~5%), and normal butane (~5%), together account for approximately 96% of the blendstocks used to make gasoline globally.

Gasoline blendstock components Table 7-4

FCC gasoline: aka catalytic cracked spirit, cat cracker gasoline, or FCC Naphtha
Reformate: aka catalytic reformate, which is heavy straight run naphtha (HSRN) having been processed in a reformer.
Naphtha: aka light straight-run naphtha (LSRN), straight-run gasoline, or natural gasoline.
Alkylate
Isomerate: isopentane and isohexane
Normal butane (nC4)
Hydrocrackate
Isobutane (iC4)
Butylene
Polymerate
Dimerate: Isooctane
Benzene
Toluene
Mixed (meta-, para-, ortho-) xylenes
Visbreaker gasoline/naphtha
Coker gasoline/naphtha
Raffinate
Pyrolysis gasoline
Ethers: MTBE
Ethers: TAME*
Ethers: ETBE*
Alcohols: Ethanol
Alcohols: Methanol*
Alcohols: TBA*
Lead: TEL (tetra ethyl lead)*

*globally only minor amounts of these are used directly in motor gasoline for environmental or cost reasons.

Some blendstocks are not added at the refinery. For example, ethanol is added much later as the distribution point such as at a tanker truck depot, known as the truck rack. Mixing octane rating enhancers at the last possible stage in distribution allows refineries the flexibility to sell their products to other customers which may not wish to have the additives. Another reason the additive ethanol cannot be blended in until the last minute is that it draws water into gasoline during pipeline transportation which is not desirable.

Refiners also often mix in additives to improve engine performance and storage longevity. Also, several oil companies have their own special formula or group of additives which they say will enhance the performance of the fuel.

There are hundreds of additives in use today. Most additives fall into the following categories:

Additives to finished refinery products   Table 7-5

Oxidation inhibitors (increases storage life of the fuel.)
Anti-static agents; Anti-wear agents; Anti-foam agents.
Deposit reducers (keep the cylinders clean and stops spark-plug fouling.)
Dyes (to color fuels so that they can be more easily identified.)
Corrosion/rust inhibitors (prevents fuel from corroding tanks.)
Anti-icing agents
Oiliness agents
Pour point depressants
Demulsifiers
Emulsifiers
Detergents
Dispersants
Extreme Pressure (EP) additives
Tackiness agents
Viscosity improvers

Types of refineries

Refineries can be classified into five categories: basic/topping; hydroskimming; cracking; coking; and full conversion/complex.

Basic topping refineries are the least sophisticated type of refinery. A topping refinery simply separates the components of crude oil into its various products by distillation. A topping refinery contains only an atmospheric distillation tower and sometimes a vacuum distillation tower. The types of product produced is directly determined by the type of crude being run as the refinery has little in the way of downstream units to change product yields.

As their capacity to produce market specification products is very limited, there are very few of these basic 'tin can' or 'teapot' refineries around the world, apart from in China, where there are over 100. Topping refineries often rely on selling their half-finished intermediate products to other refineries with excess downstream capacity. These straight-run products usually have to be further processed to bring them within even the most lax product specs.

Hysdroskimming refineries, in addition to the basic topping units, hydroskimmers have naphtha reformers and because reformers generate hydrogen, this hydrogen can be skimmed off for use in desulfurization units such as hydrotreaters. Thus a refiner can begin to increase the octane of gasoline with reformate, and lower the sulfur of light distillate production.

Cracking refineries have catalytic cracking units such as an FCC or hydrocracker in addition to a combining unit such as an alkylation plant which can increase the yield of gasoline and other high value products and minimize production of lower value products. Cracking refineries also tend to have gas plants which can separate light end gases.

Coking refineries have the same capabilities as cracking refineries, and have cokers capable of turning heavy vacuum residue into higher value products.

Full conversion/complex refineries are coking refineries with a steam cracker, which is used to produce the petrochemical ethylene, and its major byproducts, such as propylene. Ethylene and propylene are the primary starting points for most plastics. A full conversion refinery is essentially an oil refinery in addition to a petrochemicals plant.

The following table shows the approximate yields from an intermediate API gravity medium-sour crude such as Saudi Arab Light in three different refineries (Table 7-6). Obviously these are rough numbers as refineries are configured in many different ways and even a single refinery can be running to maximize distillate instead of gasoline or vice versa:

Table 7-6

Approximate percentage yields (by volume) in each type of refinery

	Hydroskimming	Cracking	Coking
Gasoline	25	45	50
Kerosene/Jet fuel	10	10	10
Diesel/Heating oil	20	25	30
Residual Fuel	35	10	-
Propane	-	3	4
Coke	-	-	3
Other products	5	5	5
Refinery fuel	8	12	13
Processing gain	(3)	(10)	(15)

Source: modified from Leffler, W.L. 2000

Processing gains shown as negative as yields are by volume. The sum of weights do not change.

Nelson complexity factor

The terms topping, hydroskimming, cracking, coking and full conversion are rough classifications for refineries. In order to compare refineries more precisely Wilbur Nelson created an index (Table 7-7) to measure refinery complexity. Each additional piece of refinery kit adds to the index based on the cost of running the unit in addition to its capacity.

Approximate Nelson complexity factors   Table 7-7

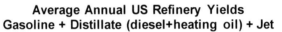

Basic topping =	1.0
Hydroskimming =	2.0
Cracking =	9.0
Full Conversion =	15.0

Complexity creep

Refineries in the US have steadily become more complex as can be seen from the increasing yield of gasoline plus middle distillates since the 1980s (Fig. 7-3). Part of the increase in gasoline yield is due to the addition of blending components such as ethanol.

Average Annual US Refinery Yields
Gasoline + Distillate (diesel+heating oil) + Jet   Fig. 7-3

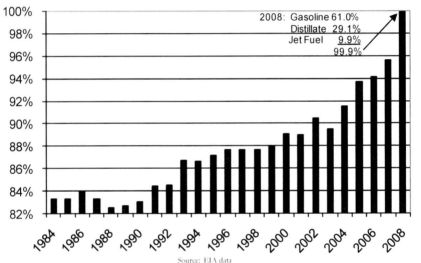

2008: Gasoline 61.0%
Distillate 29.1%
Jet Fuel 9.9%
99.9%

Source: EIA data

Defining refineries by size

In addition to complexity level, refineries are also described with reference to their processing capacity. Refineries are defined by either the *charge capacity* of feedstocks, which is the term used for inputs, or the *production capacity* of outputs. The volume of inputs in a refinery may not equal the volume of outputs. Some products may be used as a refinery fuel, evaporation into the atmosphere occurs, light molecules are combined and heavy molecules are cracked which will result in refinery losses or gains. How much a refinery is processing as a percentage of its production capacity is its *run rate*. The *nameplate capacity* of a refinery is the capacity it was originally designed to handle. Refineries can exceed nameplate capacity by de-bottlenecking inefficient processes observed once the refinery is up and running.

Refinery maintenance

Refineries have very high fixed costs. Owners would prefer to operate the plant continuously with no shutdowns. However, because mechanical equipment wears out over time, catalysts have to be regenerated, and new equipment has to be installed, maintenance has to be routinely scheduled.

Refineries tend to plan most maintenance at a similar time each year in the spring and fall known as turnaround season or maintenance season. These are generally the periods in which demand for all petroleum products is least, and are known as the shoulder months, in between winter heating and summer driving season. The main turnaround season is in the spring with a lesser turnaround season in the fall (Fig. 7-4). The dates for scheduled maintenance are often published ahead of time by various reporting agencies and are monitored by traders and factored into market analysts' supply and demand (S&D) models. The turnaround time for maintenance can be from a day to a few weeks.

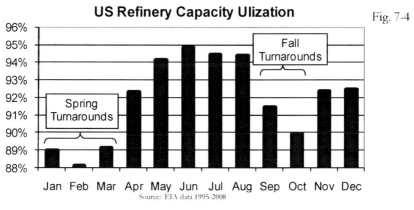

Fig. 7-4

The spring turnaround season starts and peaks at slightly different times in the US, Europe and in the Asia/Pacific region, with each of the three geographies lagging each other by two months (Fig. 7-5). One reason for the regional offset in turnaround seasons is that the independent contractors specializing in carrying out maintenance stagger their crew workload.

Fig. 7-5

Spring Refinery Turnaround Season		
US	Jan-Mar	(peak in February)
Northern Europe	Mar-May	(peak in April)
Southern Europe	Jan-Jun	(peak in April)
Asia/Pacific	Apr-Jun	(peak in May)

Fall Refinery Tunraround Season		
Global	Sep-Oct	(peak in October)

CHAPTER EIGHT
STANDARDS

Without standardization of finished products, it would be difficult to develop devices that run on oil. Standardization of gasoline allows automobile designers in Germany, Japan or the US to independently develop engines which maximize fuel efficiency for a particular grade of gasoline which can be re-fueled at any service station in any part of the world. Standardization also permits travel over wide distances as end users know that a suitable fuel will be available when required. For example, aircraft on long haul flights can re-fuel with the same grade of jet fuel at airports on opposite sides of the planet.

Standardization also facilitates a stable and reliable supply of petroleum products. Shortages in one geographic area can quickly be met with supplies from another area. It increases competition between refineries and distributors, reducing prices for end users. It allows regulators to set common rules and environmental controls for each type of fuel.

Although standards cover a wide range of physical characteristics, there are some which are discussed more in the market in reference to crude oil and various refined products.

Key measures for crude and refined products   Table 8-1

Crude oil: API gravity, sulfur content, TAN, Distillation Temperature Profile (DTP)
Gasoline: octane rating, vapor pressure, oxygenate, sulfur and benzene content, DTP
Jet fuel: flash point, smoke point, freezing point, DTP
Diesel: cetane index, cloud point, sulfur content, DTP
Heating oil: cloud point, sulfur content, DTP
Residual fuel oil: viscosity, sulfur content, flash point, DTP

When setting standards for petroleum products it is important to also define standardized test methods to be applied to determine if a standard is being met.

Enforcing standards
Standards are enforced in most jurisdictions by regulations governing fair representation of the sale of goods. Almost no standards setting agencies test oil before it is sold to consumers and, therefore, consumers are expected to test fuels for themselves or rely on someone to test it for them.

The actual properties of petroleum products found in real life are often much higher than standards require. However, occasionally, due to either an error, oversight or even a deliberate act, an end user may be offered oil which is off-spec, or sub-standard. In order to prevent this from occurring, it is quite common for a sample of fuel to be tested whenever large volumes change hands. Testing can be very important, especially in situations where fuels not meeting specifications, such as specs for jet fuel, can have catastrophic results.

Instead of carrying out a detailed chemical analysis to characterize a petroleum product, it is far easier to define each petroleum product by physical properties which can be quickly tested in the field with relatively basic equipment.

The following are some of the physical properties used for standard setting:

Appearance involves a simple eyeballing of the product.
Visual: a clear and bright visual appearance is desirable with no undissolved water or dirt.
Color: fuels are often colored for tax purposes or safety.
Ash and suspended particulates cause wear to engine parts which can lead to an incomplete burning of fuel and higher exhaust emissions.

Composition ensures that levels of potentially harmful substances such as sulfur, olefins, aromatics, particularly the aromatic benzene, are limited. Minimum and/or maximum oxygenate content is also often specified for gasoline.
Sulfur reduces the life of catalysts in catalytic converters which are used to oxidize harmful emissions. Sulfur also produces sulfur dioxide when burned, which is harmful to animal and plant life and corrodes buildings.
Olefins contribute to exhaust containing both ground level ozone (smog) and butadiene, a known carcinogen.
Aromatics are a good source of octane rating. High levels of aromatics, however, form deposits in the combustion chamber which causes improper combustion and leads to higher exhaust emissions.
Benzene, an aromatic molecule, is a good source of octane rating. Unfortunately, benzene is a carcinogen. Other aromatics, toluene and xylenes can be used as octane enhancing alternatives to benzene.
Oxygen content: oxygenates promote a more thorough burn and less polluting tailpipe emissions.
Mercaptan sulfur, a compound chemically known as thiol, is usually neutralized in a refinery as it creates an unpleasant odor in fuels. Mercaptan is used in trace amounts to create an odor in naturally odorless natural gas for safety reasons.

Volatility refers to how readily a fuel will vaporize (boil into a gas) and the quantity which vaporizes as the temperature is raised. Petroleum fuels contain a range of hydrocarbon molecules which each vaporize at different temperatures.

Distillation temperature profile (DTP): As hydrocarbon molecules evaporate at different temperatures, the distillation temperature profile indicates the range of hydrocarbons a fuel contains.

> Initial boiling point (IBP) temp.
> Fuel Recovered
>   10% vol. temp.
>   50% vol. temp.
>   90% vol. temp.
> End boiling point (EBP) temp.

Distillation temperature profile

Flash point temp is the lowest temperature at which a fuel gives off sufficient vapor to form a flammable mixture with air. While not directly related to engine performance, flash point is an important safety measure for ships and aircraft which cannot be evacuated easily and must store large quantities of fuel. Gasoline has a flash point of around -40°F (-40°C), which is why any small spark at a gas station can be dangerous even when it is extremely cold, whereas Jet-A1 used in jet aircraft has a flash point around 100°F (38°C.) Gasoline and other fuels with flash points below 100°F are referred to as flammable liquids. Kerosene and other liquids with flash points above 100°F are known as combustible liquids.

Density is measured by specific gravity or degrees API. A more dense fuel usually produces more power, but with a higher smoke exhaust. Measurement of density is the first test performed on an unknown fuel. Density can also indicate if a fuel has been contaminated with other fuels.

Vapor lock index: Before fuel pumps in gasoline tanks became common, gasoline was pulled to the engine by a pressure differential between the gas tank (high pressure) and engine (lower pressure.) In hot weather gasoline might vaporize in fuel lines and the engine would be unable to pull the vaporized fuel through the fuel lines. Vapor lock was particularly prevalent at high altitudes. At high altitudes, the pressure differential between the fuel tank and the engine lessens and vapor lock can occur. The vapor lock index specifies how likely a pocket of vaporized fuel, especially gasoline, will obstruct the normal flow of gasoline to an engine. Vapor locks are less common these days as most new cars have fuel pumps located in the gas tank which pushes fuel to the engine therefore making pressure differentials unnecessary.

Reid Vapor Pressure (RVP): Vapor pressure measures how easily a fuel evaporates and is most often measured using an index called RVP. Gasoline burns as a vapor mixed with air, so the more easily the gasoline vaporizes, the better for combustion. RVP has to be high enough to ensure that the gasoline will vaporize when required. However, higher RVP gasoline evaporates more easily during re-fuelling and also from a vehicle's fuel system before it is burned. When the gasoline evaporates, it goes into the atmosphere where it reacts with sunlight to cause smog. Upper (for evaporative emission controls) and lower (for engine performance) limits are therefore placed on vapor pressure.

<u>Low temperature fluidity</u> describes the ability of fuels to flow in a wide range of ambient temperatures. At low temperatures, a pre-heater may have to be used to warm the fuel so it can be pumped prior to combustion.

Freezing point is the temperature at which the fuel solidifies. The freeze point is very important for engines which operate at very low temperatures, such as jet aircraft which fly at high altitudes. The freezing point for Jet-A1 is -47°C.

Cloud point is the temperature at which wax crystals form. Cloud point is important for middle distillates such as diesel which is more susceptible to clouding compared with lighter gasoline.

Pour point is the temperature at which excessive wax crystals form and impede the flow of the fuel. This is important for transportation of oil in pipelines.

Cold Filter Plugging Point (CFPP) is the temperature at which diesel will pass through a fine wire mesh. The CFPP is important as it indicates the temperature at which the diesel will clog engine fuel filters.

Viscosity at low temperature refers to how easily the fuel flows at specified temperatures. This is particularly important for middle distillates (jet fuel, kerosene, diesel, and heating oil) and residual fuel.

<u>Combustion</u> parameters describe the release of energy when a fuel is oxidized.

Specific energy is particularly important for jet fuel, as aircraft have to lift the fuel into the air, it is important that the fuel contain as much energy as possible for its weight and volume.

Smoke point: Smoke, a sign of incomplete combustion, consists of particles of carbon suspended in exhaust. Carbon deposits can damage turbine fans and fuel injectors. By lowering the amount of aromatics, the smoke point can be raised, which is why Jet-A1 kerosene contains 25% vol. max aromatics.

Naphthalenes content: Naphthalene, an aromatic, is particularly responsible for smoke containing unburned carbon.

Cetane index measures how easily diesel ignites on compression. A higher cetane index indicates a shorter ignition lag and smoother running.

Octane rating: octane rating indicates anti-knock performance of gasoline. Knocking, which causes engine damage and improper combustion, is a result of the ignition of fuel before the spark plug heats the mixture. A higher octane rating enables fuel to be compressed more before it is ignited.

<u>Corrosion</u> of storage tanks and piping are caused by elements in certain grades of oil.

Copper strip corrosion: indicates whether the product will corrode copper, brass, or bronze which it may come in contact with.

<u>Stability</u> is important for storage longevity and resistance to high temperatures.

Oxidization stability: prolonged storage of certain fuels such as diesel can result in oxidization of the fuel which forms gums (thick sticky deposits) and sediment.

Thermal stability measures the ability of a fuel to resist decomposing at high temperatures which can cause coke-like deposits to form which may block fuel jets. Thermal stability is especially important for jet fuel as in addition to burning as a fuel, the jet fuel stored on aircraft is used as a heat sink to remove heat from engine oil and hydraulic fluid.

<u>Contaminants</u> are important to limit for efficient running of engines and less polluting exhaust

Existent gum: gum deposits can clog fuel systems.

Water: most fuel contains very small amounts of dissolved water. Water can facilitate corrosion of storage tanks. Free, non-dissolved, water can also enable the growth of organisms in fuel which can clog fuel filters.

Acidity: organic acids can corrode a vehicles fuel system.

<u>Conductivity</u> ensures safety from static while loading and unloading from tankers.

Electrical conductivity: low electrical conductivity is desired to prevent an electrostatic charge building during pumping to and from storage tanks.

<u>Lubricity</u> prevents wear on engine components, fuels, such as diesel, should have a minimum amount of lubricity.

<u>Additives</u>: important limits or requirements are set on the use of additives. For example, gasoline octane rating enhancers may be limited or required such as TEL (Lead); Ethers (MTBE, ETBE, TAME, DIPE) and Alcohols (Ethanol, Methanol, TBA).

Standard setters

There are many organizations which set fuel specifications and test methods.

<u>ASTM International</u>, formerly the American Society for Testing and Materials, sets the most commonly used standards and test methods for petroleum fuels globally.

<u>The International Standards Organization (ISO)</u>, most well known for their marine fuel standards, also sets standards for other oil products.

<u>Individual national standards setting organizations</u> set standards to control polluting emissions resulting from handling and burning fuels.

In the US standards are set by the Environmental Protection Agency (EPA), the California Air Resources Board (CARB) and individual state weights-and-measures agencies.

European standards are set by the Comité Européen de Normalisation (CEN) which issues EN specifications such as EN590 for diesel fuel, EN228 for gasoline, and EN589 for automotive LPG. These specifications are translated into standards by organizations in individual European countries, such as by the British Standards Institution in the UK. For example, some of the countries adopting the CEN gasoline standard (EN228) in Europe are:

Germany:	DIN EN228
U.K.:	BS EN228
France:	NF EN228
Italy:	UNI EN228

In Canada, the Canadian General Standards Board (CGSB) issues CGSB fuel specifications and test methods.

In Japan, the Japan Standards Association (JSA) issues JIS fuel specifications and test methods.

Military organizations, such as the British Ministry of Defense (MoD), NATO, the US Navy, and the US Air Force each set standards for grades of jet fuel, diesel and other fuels used in military operations.

The International Maritime Organization (IMO), which is part of the United Nations, sets standards in its MARPOL, marine pollution, convention affecting fuel burned in international waters and phasing out single hulls in large tankers.

Conseil International des Machines a Combustion (CIMAC), is an organization which focuses on standards for diesel and gas turbine engines.

Individual large petroleum refining and marketing companies usually adhere to ASTM, ISO, and government standards, but some such as BP, Shell and ExxonMobil also set their own standards for certain specialty finished products.

SAE International, formerly the Society of Automotive Engineers, is most well known in the US for its motor lubricant oil standards.

The American Petroleum Institute (API) sets standards for oil industry equipment.

CHAPTER NINE
FINISHED PRODUCTS

Fractions and individual products

The nineteen products of crude oil in approximate order of distillation temperature, from light gases through to the heavy end of the barrel, are:

Table 9-1

Wide fractions/cuts →	Narrow fractions/cuts →	Refined Products
Petroleum gases ▲	Natural gas	1. Methane
	Natural gas liquids (NGLs)…2. Ethane	
		3. Propane
		4. Butanes:
		Normal butane
		Isobutane
Light-ends	Naphthas…………………..5. Light naphtha	
Lighter		6. Heavy naphtha
	Gasolines…………………..7. Motor gasoline	
		8. Aviation gasoline
Middle distillates	Kerosenes…………………9. Jet fuel:	
		Kerosene-type jet fuel
		Naphtha-type jet fuel
		10. Gas turbine fuel
		11. Kerosene
	Light fuel oils/Gas oils…......12. Diesel fuels:	
		Automotive diesel
		Marine diesel
		13. Light fuel oil (burner fuels):
Heavier		Home heating oil
		Industrial light fuel oil
Heavy ends	Heavy fuel oils……..………14. Residual fuel oil:	
		Bunker (Marine) fuel
		Heavy (Industrial) fuel oil
	Specialty products………..15. Base oils & Finished lubricants	
		16. Waxes
		17. Bitumen:
		Asphalt
		Road oil
		Emulsion fuels
		18. Petroleum coke
		19. Carbon black

Individual products are *finished,* if they require no further processing and are ready to be consumed by end users, or *intermediate* if they require further processing. Finished gasoline is the product sold at a retail station. Naphtha is an example of an intermediate product, which has only niche direct end uses, and is most commonly processed further to produce finished products such as gasoline or plastics.

An important consequence of products distilling at similar temperatures in a refinery is that their prices are more highly correlated. This is because products produced from the same part of the barrel have to compete for similar molecules. For example, jet fuel molecules and diesel fuel molecules are both from the middle distillates part of the barrel. Their prices will be more correlated than jet fuel and motor gasoline prices.

1. Methane

Primary uses:	- Main component of natural gas used in home heating, cooking, and utility electricity generation - Petrochemicals, especially fertilizers - Manufacture of MTBE for gasoline outside the US
Synonyms:	CH_4, C1, Major component of: natural gas, natural, nat gas, natty, LNG (liquefied natural gas), CNG (compressed natural gas)
Demand seasonality:	Higher during winter heating and lower in summer

Methane is the simplest and lightest hydrocarbon with one carbon atom in its molecular structure. It is a colorless and odorless gas at room temperature and pressure.

Methane is produced along with other hydrocarbon gases in three ways. Firstly, *non-associated* gas is produced from a gas field and wellhead which does not produce crude oil. In 2008, more than 75% of natural gas produced in the US was non-associated gas. Secondly, *associated gas*, or casinghead gas, is produced from the same wellhead as crude oil, known as live oil if it contains high levels of dissolved gas, and then separated in a gas-oil separation plant (GOSP). Finally, gases are produced in an oil refinery system during the various distillation and cracking processes of crude oil. Refinery produced gas is known as *still gas*, or refinery gas, and is processed at a refinery gas plant.

The gas produced in any of the three ways is called raw gas and may contain:

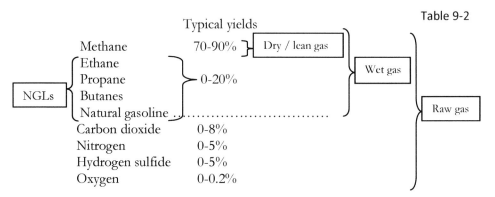

Table 9-2

Sour gas is raw natural gas with high levels of hydrogen sulfide (H_2S), a flammable and poisonous gas, or other sulfur compounds. Gas pipeline operators usually require carbon dioxide, oxygen, nitrogen, and hydrogen sulfide to be removed before transportation. Once undesirable components have been reduced, the remaining gases and natural gas liquids (NGLs) are called wet gas, or greasy gas. NGLs are also known as natural gas plant liquids (NGPL). NGLs are gases at normal room temperature, but because of their higher boiling points are far more easily cooled to liquids than methane. Wet gas is also a term used for raw gases produced with a lot of water.

Methane is separated from the other gases by gas processors in a stripper or fractionator. The resulting methane alone is called dry gas, or lean gas, because of its low energy compared to the other components of wet gas. NGLs can be stripped from wet natural gas with cooling processes (refrigeration and cryogenics), or by passing the wet gas through substances which can selectively remove individual gases via absorption and adsorption.

Methane typically makes up between 75-95% by volume of the consumer product called natural gas. Natural gas is referred to as nat gas, natural, or natty, to distinguish it from motor gasoline. The minor (5-25% by volume) components of natural gas, with much higher energy densities than methane, are the NGLs ethane, propane and butane (Table 9-3). Although the non-methane components can be left in the natural gas stream if it is uneconomical to separate them, there is a maximum amount which can be left in as the equipment designed to burn natural gas may melt if there are too much of the higher energy content NGLs.

Table 9-3

Approximate energy densities and boiling points of hydrocarbon gases

		Btu/cf	Initial boiling point
	Methane	1,011	-161.6°C (-259°F)
	Natural gas (avg. US)	1,030	-161.6°C (-259°F)
Light to "Heavies"	Ethane	1,750	-88.6°C (-128°F)
	Propane } NGLs	2,500	-42.1°C (-44°F)
	Butanes	3,250	-0.5°C (normal) -11.7° C (iso)
			(31°F) (11°F)

US and Canadian natural gas is traded in cubic feet (cf) with a specified energy in British thermal units (Btus). One Btu is the energy required to heat a pound of water by 1°F. Btu measurement always takes place at one standard atmosphere of pressure, but the reference temperature at which measurements are taken varies by region and is commonly 39°F, 59°F, or 60°F.

Table 9-4

Volume, energy and energy density units used for natural gas

Volume:
cf: Cubic foot
Mcf: Thousand cubic feet (used to describe gas production of an individual well)
Bcf: Billion cubic feet (used to describe reserves and large storage quantities)
Tcf: Trillion cubic feet (used to describe reserves and large storage quantities)

Energy:
joule (J)   Metric (SI) measurement of energy
gigajoule (Gj)   1 billion joules (used to describe large consumption)
British thermal unit (Btu) US (English) measurement of energy
therm (th)   100,000 Btu (used for domestic billing)
MMBtu   1 million Btu (used to describe larger consumption)
Dekatherm(Dth)   1 million Btu

Energy density example: MMBtu per cf

One Mcf of nat gas contains approximately one Gj, or approximately one MMBtu, of energy and therefore most natural gas traders use these volume and energy terms interchangeably.

$$1 \text{ Mcf} \approx 1 \text{ Gj} \approx 1 \text{ MMBtu}$$

A price of natural gas commonly mentioned in the general media is traded on the New York Mercantile Exchange and is quoted in US dollars per MMBtu, which in the past few years has been hovering around $8 per MMBtu, and one contract on the exchange contains 10,000 MMBtu.

In financial accounting, a very rough conversion of 6 Mcf per barrel of oil equivalent (bboe or BOE) is used to compare natural gas and crude oil volumes.

Most pipeline operators will specify an acceptable energy density, also known as a Btu factor, for natural gas at a level between 950 and 1,150 Btu per cf. The most common energy density in physical natural gas trading in the US is 1030 Btu per cf, or 1.030 MMBtu per Mcf.

NGLs most often trade at a higher price than natural gas as they can be used for petrochemicals, to make octane rating enhancing gasoline blendstocks, or as refinery fuels. Therefore, there is usually an economic incentive to strip out the NGLs from wet gas. The price spread between each of the NGLs above dry natural gas is referred to as the frac (fractionation) spread. If the frac spread is not wide enough to compensate the gas processors for stripping the NGLs from wet gas, the processor can simply leave the NGLs in the natural gas stream so long as it stays under the maximum Btu factor of a pipeline operator.

For safety, mercaptans are added in very small amounts to the normally odorless natural gas so that leaks can be more easily detected.

When gas is produced at a location where there is no pipeline, it is referred to as stranded gas. Stranded gas sells relatively inexpensively as it is difficult to get to consumers. Stranded gas, if found along with crude oil, is sometimes burned at the site of production, which is called flaring. Although flaring natural gas is wasteful and often illegal, it is much better than simply releasing unburned methane into the atmosphere. Unburned methane is a much more powerful greenhouse gas than carbon dioxide when it comes to trapping the sun's heat.

An alternative to flaring is to reinject associated gas into an oil reservoir to maintain pressure. Reinjected gas can potentially be recovered later. Reinjection involves putting the associated gas through a gas cycling plant where condensates are removed. The remaining gas is then compressed and reinjected back into the reservoir. Stranded associated gas is also used in gas lift pumps to enhance production rates from a crude oil field.

Stranded gas can also be used to fuel lease separators and gas processing plants, where it is known as lease fuel and plant fuel, respectively.

Most natural gas is transported with minimal compression over land via pipeline. Pipelines can handle natural gas under a very small amount of compression but cannot handle natural gas which has been cooled and compressed to liquid form. Due to the difficulty in transporting large quantities of natural gas in anything other than a pipeline, there has been relatively little international trade in gas except between areas with land borders, such as between Canada and the US, or Russia and Western Europe. Natural gas prices in various parts of the world can diverge quite significantly because of this limited physical arbitrage ability. This differs from most other hydrocarbon products, which are more easily physically arbitraged and therefore which have more closely correlated prices around the world.

Winter heating is a very large use of natural gas. Natural gas production is not seasonal, but as demand for it is. Producers fill depleted natural gas reservoirs, underground salt caverns, and aquifers with gas during the summer at locations close to end users, in a process called natural gas stock building. These stocks are then drawn down during cold winter heating months, November through March.

A counter seasonal demand for natural gas is electricity generation, particularly peaker power generation. Peaker power generation is the production of electricity only at periods of very high demand, such as air conditioner use on hot summer days. It is a supplement to baseload power generation, which occurs all of the time.

In the US, oil products account for only 2% of US electrical power production. The oil typically burned is residual fuel oil. Residual fuel oil, which is highly polluting and difficult to pump, is inexpensive and trades at a price significantly under the crude oil from which it was refined. Although residual fuel oil trades at a lower price than crude oil, natural gas typically trades at a price under residual fuel oil. Natural gas is used to produce about 20% of US electricity. If natural gas prices are too high relative to resid, US oil demand increases by around 200,000 barrels per day as a result of utilities burning oil rather than natural gas. The process of utilities moving from one fuel to another due to prices is known as fuel switching. The amount of oil involved in switching is not large in the global or even US oil market, but it does provide a theoretical ceiling for natural gas prices and floor for residual fuel prices.

A spark spread is used to evaluate the relative incentive of a power plant to burn natural gas or oil. The natural gas spark spread is the market price of a megawatt hour (MWh) of electricity minus the cost of natural gas used to generate that electricity (Table 9-5). The oil spark spread is the market price of a megawatt hour (MWh) of electricity minus the cost of residual fuel oil used to generate that electricity (Table 9-6). The spark spread for coal is called the dark spread, but natural gas rarely if ever gets cheap enough to compete on a switching basis with coal.

Table 9-5

Natural gas spark spread calculation example

Natural Gas Price	Heat Rate (in Btu/KWh) …differs for every power plant …the lower the more efficient
$11/MMBtu ÷ 1,000,000 X 7,500 X 1,000 = $82.50/MWh	
To convert gas price into $ per Btu	To convert from KWh to MWh	Subtract this from price of Power in MWh to determine gas spark spread.

Table 9-6

Residual fuel oil spark spread calculation example

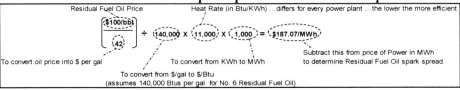

In addition to its primary uses for heating, cooking, and electrical power generation, methane is also a feedstock for the fertilizer industry. The hydrogen in methane is removed in a steam methane reformer (SMR) and combined with nitrogen from air to form ammonia (NH_3), which is the most commonly used petrochemical fertilizer. Further processing of ammonia produces the fertilizers, urea, ammonium nitrate and calcium ammonium nitrate. Demand from the fertilizer manufacturing industry is highest in the summer.

Methane is also used to produce methanol. When combined with isobutylene methanol is used to manufacture gasoline octane raising MTBE.

Finally, methane is used to produce acetylene used for welding, cutting and lighting.

Liquefied natural gas (LNG): When pipeline transportation of natural gas is not possible, the gas can be cooled at atmospheric pressure to a temperature of -259°F (-162°C) at which point the methane condenses into a liquid. This cooling to a liquid achieves an approximate 600-fold reduction in volume. This LNG can then be shipped using a vessel, called an LNG tanker, over long distances (Fig. 9-1).

LNG is traded in metric tonnes, and not the volume measurement of Mcf or the energy measure of MMBtu. One metric tonne of LNG equals 48 Mcf or 48 MMBtu of uncompressed gas.

Shipping LNG is quite a complex and expensive process and requires high natural gas prices in the consuming country to cover costs. The most expensive part of handling LNG is building and running the liquefaction plant, which consists of what are called LNG trains, or liquefaction units. Once cooled, the liquid is transferred to LNG tankers which are easily recognizable due to the large white spherical structures jutting out of their decks. Most LNG tankers are dedicated to particular LNG projects and routes, with very few independent LNG tanker owners. Once LNG has reached its destination port on a tanker, an LNG regasification terminal with storage facilities and connections to consuming region pipelines is required. These LNG regasification terminals are, once again, very expensive to build and operate compared with a crude oil or gasoline terminal.

As a rough guideline, an LNG train adds approximately US$1.10/MMBtu, and the regasification terminal adds a further US$0.30/MMBtu to the cost of natural gas. One must then add shipping costs. LNG shipping costs to the US increase with the distance involved and can range from US$0.35/MMBtu from Venezuela and Trinidad and Tobago to US$0.60/MMBtu from Algeria and Norway, US$0.80/MMBtu from Nigeria, US$1.40/MMBtu from Qatar in the Middle East, and US$1.75/MMBtu from Australia.

LNG shipments are relatively common in Asia where production of natural gas in the Middle East, South East Asia and Australia, is quite a distance from the large consuming regions of Japan, South Korea and China. In the US, LNG imports arrive through only four ports: Cove Point, Maryland; Elba Island, Georgia; Everett, Massachusetts; and Lake Charles, Louisiana. Imports come from many countries, primarily Algeria, and Trinidad and Tobago. A large number of new LNG importing terminals are under construction in the US, and will be in place over the coming years to supplement domestic natural gas production.

LNG tanker   Fig. 9-1

Source: EIA

<u>Compressed natural gas (CNG)</u>: Natural gas is sometimes compressed into a storage tank for use as a transportation fuel. The gas is not compressed into a liquid which would require extremely high pressures or low temperatures, but instead is merely compressed to a small degree (under 3,000-3,600 psi) above atmospheric pressure so a higher volume of gas can be stored on the vehicle.

Most gasoline-powered vehicles can be modified to run on natural gas and many urban bus and taxi fleets use CNG. CNG has never really taken off with general consumers due the higher initial vehicle purchase cost (currently around $2,000-$3000 per vehicle). Additionally, the gas cylinder takes up a larger amount of space than a gasoline tank, vehicle range being is less than if the vehicle had an equivalently sized gasoline tank, as well as the lack of re-fueling stations. There are currently 1,600 retail stations which sell CNG in the US, compared with 190,000 stations selling gasoline.

A significant benefit of CNG is environmental. Natural gas, because it contains simple hydrocarbon molecules, burns thoroughly in air. Water vapor and carbon dioxide are almost the only emissions. CNG powered vehicles also have acceleration comparable with gasoline or diesel powered vehicles.

CNG is mainly used in fleet vehicles, such as city buses, which are required to emit a low level of pollutants. In fact, one in every five new transit buses in the US is powered by natural gas.

As CNG is purchased as a gas and not a liquid, a gasoline gallon equivalent (GGE) is used in the US at the pump to enable consumers equate the energy content of the CNG purchased to gasoline.

CNG should not be confused with Autogas, or LPG, which is a propane/butane fuel mix also used in cars.

Unconventional sources of methane
Along with the non-associated, associated and refinery produced methane, there are three sources of unconventional gas: tight gas sands, coalbed methane, and shale gas. Unconventional gas requires reservoir stimulation or other high cost recovery techniques in order to remove the gas. Until the late 1990s, unconventional gas accounted for a small amount of production in the US. With the higher recovery costs of unconventional production being offset by the rise in natural gas prices since 2000, unconventional production has quickly risen to almost 50% of US dry gas production. Unconventional gas production is still quite rare outside of the US.

Methane hydrates are a fourth source of unconventional gas currently untapped, but with potential in the future.

Tight gas is methane locked in impermeable sandstone rock which prevents the gas flowing to a well. The reservoir has to be fractured with explosives, acid, polymers and water to enhance permeability of the rock. Tight gas extraction is becoming more common with higher natural gas prices. Tight gas dominates unconventional production. Currently the San Juan Basin in New Mexico/Colorado is the location of most tight sands production.

Coalbed methane is methane found along with coal. Coalbed methane has traditionally been vented into the atmosphere as coal is removed from a mine. It is only recently that this methane has begun to be extracted commercially.

Shale gas is methane gas found sandwiched between fine-grained shale deposits with low permeability which makes the gas difficult to remove. As with tight gas, artificial fracturing of the reservoir is required in order to remove the gas. Most shale gas in the US is found in the Barnett Shale of the Fort Worth Basin, Texas in addition to Appalachian Basin Devonian shale.

Coalbed methane and shale gas deposits are unusual in that a single rock is the source, reservoir and trap for hydrocarbon deposits.

<u>Methane hydrates</u> are molecules of methane surrounded by, but not bonded to, molecules of water. Methane hydrates are crystalline in appearance and have traditionally been perceived as a nuisance as small amounts occasionally clog natural gas pipelines. The estimated naturally occurring quantities of methane available from methane hydrates are enormous, more than all coal, oil and non-hydrate natural gas combined. Naturally occurring methane hydrates are most often found under ocean floors.

Producing large quantities of methane from methane hydrates in a commercially viable manner is not currently technically possible. One danger with with methane hydrates is that an accidental release of large quantities of methane into the atmosphere could rapidly exacerbate global warming. Methane is up to ten times more effective at trapping the suns heat than carbon dioxide.

Methanol
Methanol (CH_3OH) is a volatile, colorless, flammable, and poisonous alcohol used in the production of formaldehyde. Formaldehyde is used in wide variety of plastics. Methanol is also used to produce MTBE, anti-freeze, and is used by itself as a niche automotive fuel. Methanol is also used in the manufacture of biodiesel.

The most frequently used method of producing methanol is called steam-methane reforming (SMR.) The first stage in methanol production via SMR involves producing synthesis gas by reacting methane with steam at very high temperatures and moderate pressure in the presence of a nickel catalyst. Synthesis gas is a mixture of carbon monoxide and hydrogen. Methanol is subsequently created when synthesis gas is then heated at higher pressure in the presence of a catalyst comprising of copper, zinc oxide, and alumina.

2. Ethane

Primary uses: - Petrochemicals, particularly ethylene production
     - Minor use as a refinery fuel and refrigerant gas
Synonym: C2, C_2H_6
Demand seasonality: Demand not seasonal

Ethane is a colorless and odorless gas at room temperature and pressure. It is often referred to by traders as C2 because it has two carbon atoms in its molecular structure, C_2H_6.

Ethane is the lightest of the four natural gas liquids, which are, from lightest to heaviest: ethane, propane, butane and natural gasoline. Although the NGLs are gases at room temperature and pressure, they are far more easily compressed or chilled into liquids than methane.

Ethane's primary use is as a feedstock for the petrochemicals industry where it is converted to ethylene, C_2H_4, by steam cracking in a petrochemical plant. Ethylene can be subsequently polymerized, which involves linking a series of the molecules, to produce polyethylene. Polyethylene, also called polyethene, polythene, or PE, is the most commonly used plastic. Variations of PE are, in reducing order of density: HDPE (high density PE), used in plastic soda bottles; LDPE (low density PE), used in plastic bags; and LLDPE (linear low density PE), used in flexible tubing. Ethylene can also be combined with benzene to produce ethylbenzene, which is used to manufacture polystyrene for hot food containers. In addition, ethylene can be chlorinated to produce 1,2-dichloroethane, which is used to produce the plastic polyvinyl chloride (PVC). Ethylene can also be used to hasten the ripening of fruits such as apples and oranges.

Ethane is flammable and is sometimes used to fuel the gas processing plant or refinery it is produced in. Ethane can also be used as a refrigerant gas. A refrigerant gas is simply a gas which is easily compressed using an electrical, or other powered, motor. Gases generate heat when compressed, which is why the back of a refrigerator is always warm, and then absorb heat when decompressed, which is what happens in the inside walls of a refrigerator.

Ethane is produced either through either gas processing of raw gas feedstock or from crude oil at a refinery.

Gas processing takes place if the prices of individual NGLs are high enough to cover costs involved, or if the BTU of natural gas is too high. The NGLs are first stripped from natural gas (leaving a higher percentage of methane) and then the NGLs, having been removed from natural gas, are subsequently fractionated to their four components. The price spreads between each of the individual NGLs and methane is referred to as the frac spread or stripper spread, and is actively traded.

Refinery production of ethane results from various oil refining processes beginning with basic distillation in the ADU. NGLs are stripped from refinery produced natural gas and then individually fractionated at a refinery's gas plant.

If the prices of the individual NGLs are not high enough relative to methane, a refinery may simply use the ethane as a cheap fuel to power the refinery or leave it in the natural gas stream and sell it onto natural gas consumers, so long as it doesn't raise the BTU energy content of the natural gas above tolerable specs. Also, if it is uneconomical to separate ethane and propane, a mixture of the two hydrocarbons called EP mix, is separated from NGLs as a single stream. EP mix is then used as a feedstock for petrochemicals.

3. Propane

Primary uses:	- Petrochemicals
	- Primary component of liquefied petroleum gas
Synonym:	C3, C₃H₈
Demand seasonality:	High during winter and low during summer

Propane has three carbon atoms in its molecular structure, C_3H_8. Propane is a colorless and odorless gas at room temperature. As with natural gas, a foul-smelling material called mercaptan sulfur is added to propane when sold to retail consumers so that leaks can be more easily detected. One of the primary advantage of propane is that it remains a liquid, and thus in a dense form, until it boils into a gas much closer to room temperature and pressure than methane. The boiling point of propane is approximately -42°C (at 1.013 bar pressure), enabling it to be easily compressed into a liquid for transportation and storage, whereas methane's boiling point is approximately -164°C.

Propane has four major uses: petrochemical production, heating and cooking fuel, farming use and transportation.

Almost 47% of propane in the US is used for petrochemical products such as plastics. Similar to ethane, propane is cracked in a steam cracker to produce propylene and ethylene, which are used to make plastics. On a price basis, propane has to compete with ethane, normal-butane, natural gasoline and naphtha, all of which can be used to varying extents as a substitute for propane in petrochemical production.

A further 39% of propane in the US is used for home and commercial heating and cooking, where it is marketed as liquefied petroleum gas (LPG). LPG typically contains 90% or more propane and 10% or less normal-butane.

Farm use, crop drying in particular, accounts for 8% of US propane demand. Finally, transportation accounts for a very small amount of propane demand, where it is often marketed as a fuel called Autogas.

Propane prices are closely tied to both natural gas prices and oil prices. In the US, natural gas processing is a source of approximately 60% of propane produced and oil refining is the source for the other roughly 40%. The ratio between gas processing and refining produced propane varies around the world. As the US has a large petrochemicals industry, the country currently imports about 10% of its propane needs annually.

Since residential and commercial heating is seasonal, propane demand in winter is almost double that of summer. Propane is stored in underground salt domes during the summer low demand period and then de-stocked during winter heating demand periods.

4. Butanes

Primary uses:	- Gasoline blendstock
	- Minor component of liquefied petroleum gas (LPG)
	- Refrigerant gas
Synonyms:	nC4 (normal butane), iC4 (isobutane), C_4H_{10}
Demand seasonality:	Higher in summer driving season and low in winter

Butane is a colorless and odorless gas at room temperature and pressure. Butane can exist in the form of one of two structural isomers: normal butane (n-butane) and isobutane (i-butane) - hence the plural butanes. Normal butane and isobutane are often referred to as nC4 and iC4, respectively. Isomers of hydrocarbon molecules begin to exist once a molecule contains four or more carbon atoms, as is the case with butanes. Isomers are molecules with the same chemical formula and numbers of atoms but with the atoms arranged in a different structure and thus have differing chemical and physical properties. For example, at standard atmospheric pressure (1.013 bar), normal butane boils at approximately -0.6°C, whereas isobutane boils at the lower -11.7°C. Even more than propane, due to its relatively high boiling temperature, butane can be easily stored and transported in liquid form in low pressure containers.

Most consumers are familiar with normal butane which is a minor component of the household fuel LPG. Normal butane is also used as a cigarette lighter fuel, refrigerant gas, as an aerosol propellant, and in the manufacture of synthetic rubber (as opposed to natural rubber which comes from the sap of the rubber tree). Normal butane also can comprise up to 50% of the automotive fuel Autogas, with the remainder being propane. Finally, normal butane has a high octane rating and thus can be used directly as a gasoline blendstock. The high RVP of normal butane usually means that butane is used as a gasoline blendstock during the winter, but not in the summer.

Although normal butane can be cracked to form a feedstock for petrochemicals, the high profit margin of gasoline means that refineries try to turn most normal butane into isobutane in a refinery process known as isomerization. Isobutane is then subsequently converted to isobutylene for use in the manufacture of octane rating enhancing MTBE, or three alternative octane rating enhancing gasoline blendstocks: alkylate, polymerate, and isooctane. Blending MTBE into gasoline raises the octane rating without having the high RVP issues which limit the use of normal butane.

Butane demand in the US has been hit in recent years as MTBE bans spread across the US. MTBE readily dissolves in groundwater and imparts an unpleasant taste, forcing gasoline blenders to use ethanol as an octane enhancer.

MTBE is still the octane enhancer gasoline blendstock of choice in most areas outside of the US. Legislative limits on octane raising blendstock benzene, in European gasoline, has driven increased demand for MTBE in that region, which has offset somewhat the reduced demand in the US for MTBE.

Up until now, in describing methane, ethane, propane, normal butane and isobutane, we have been dealing with single hydrocarbon molecules. The heavier products of crude oil outlined next, such as naphtha, gasoline, and so on, are composed of ranges of hydrocarbon molecules. For example, gasoline may contain a small amount of methane, some normal butane, heptanes, octanes, and so on, blended together,

Naphtha overview:

Naphtha is a generic term applied to petroleum molecules with an approximate boiling range between 30°C (the boiling point of pentanes, C_5) and 200°C. Naphtha range molecules are used to improve the octane of gasoline, in petrochemicals such as plastics, and in the production of fertilizers.

When naphtha is initially produced from a refinery tower it is called whole cut naphtha. Whole cut naphtha is divided into two categories: light naphtha and heavy naphtha (Fig. 9-2).

Light naphtha (boiling range of C_5 through 80-90°C) is also known as low aromatic naphtha (LAN), or naphtha petrochemical. Light naphtha contains relatively low levels of aromatic molecules, and high levels of paraffins, which are used in petrochemical industries. In particular, paraffins (which tend to have relatively low octane ratings) are cracked to produce olefins, especially ethylene used to make plastics.

Light naphtha is considered to be paraffinic naphtha when it contains 65% or more paraffins. Three grades of paraffinic naphtha commonly traded are: light grade naphtha (approx. 92% paraffins); full range naphtha (approx. 85% paraffins); and open spec naphtha, OSN (min. 65% paraffinic content).

Heavy naphtha (boiling range 80-90°C through 150-200°C) is also known as reforming naphtha, naphthenic naphtha, N+A (naphthenes and aromatics) naphtha, aromatic naphtha, or high aromatic naphtha (HAN). Heavy naphtha is used as feedstock for a refinery reformer which produces octane enhancing gasoline blendstock, in an aromatics plant to make benzene, and in fertilizer manufacturing industries. US and Asian companies tend to prefer heavy naphtha to have an N+A content of at least 40% for use as a reformer feed. European companies often consider an N+A content of 30% acceptable.

Naphthas summary

Fig. 9-2

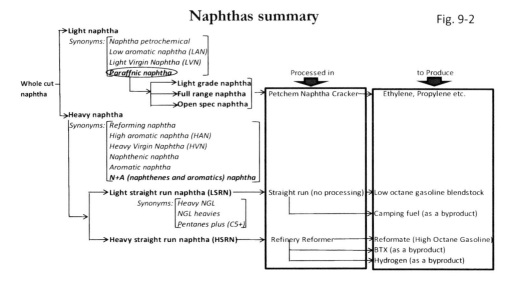

5. Light naphtha

Primary uses: Petrochemical feedstock

Synonyms: Naphtha petrochemical; low aromatic naphtha (LAN); paraffinic naphtha; light grade naphtha; full range naphtha; open spec naphtha.

Demand seasonality: Demand is not seasonal

Light naphtha competes with NGLs for use in petrochemicals. Just as with cracking NGLs, steam cracking light naphtha in a petchem plant known as a naphtha cracker produces ethylene and propylene, which are the most common petrochemical starting points.

Light naphtha is the feedstock of choice in the Asian petrochemicals industry. Due to the ease of transporting liquids compared with gases, and the fact that Asian petchem plants tend to be located thousands of miles from gas fields, it is less expensive in that region than NGLs.

NGLs are used as a petchem feedstock in the US, rather than light naphtha, as US gulf coast petrochemical plants are located close to gas production fields in the Gulf of Mexico and transportation of NGLs via pipeline to the petrochemical plants is relatively inexpensive.

High US natural gas prices in the early 2000s caused US Gulf Coast petchem manufacturers to lose dominance in global petrochemicals markets. High nat gas prices were the result of the construction of many new natural gas powered electrical power generation plants in the US. Ethane and propane prices soared along with natural gas and their values as petrochemical feedstocks became overwhelmed by their value as BTU enhancers for natural gas and as alternative refinery fuels. Petchem producers using NGLs rapidly became uncompetitive.

Competitive light naphtha prices and the growth of China as the world's plastic goods manufacturing base is such that today, Asian petchem and plastic markets have become the benchmark against which global petchem prices are set instead, rather than the US Gulf Coast markets of the past.

6. Heavy naphtha

Primary uses:	- Gasoline blendstock
	- Fertilizers
	- Solvents
	- Camping fuel and lantern fuel
Synonyms:	Reforming naphtha; naphthenic naphtha; N+A (naphthenes and aromatics) naphtha; aromatic naphtha; high aromatic naphtha (HAN); Light straight run naphtha (LSRN); Natural gasoline; Straight-run gasoline; Heavy straight run naphtha (HSRN)
Demand seasonality:	Higher in summer driving season and lower in winter

There are generally two types of heavy naphtha: light straight run naphtha (LSRN) and heavy straight run naphtha (HSRN).

LSRN, is also called natural gasoline, or straight run gasoline. As LSRN is often recovered along with NGLs, it is sometimes called the heavy NGL, or NGL heavies. LSRN is also occasionally referred to as pentanes plus, or C5+, as it contains mostly pentanes and heavier molecules.

Despite its name, natural gasoline does not have a high enough octane rating to be used directly in most modern automobile gasoline engines. However, it is used as a low octane blending component in gasoline without requiring any treatment.

HSRN containing molecules which can be modified in a reformer to produce high octane gasoline blendstock called reformate, as well as the aromatic molecules benzene, toluene and mixed xylenes, known together as BTX, which are used in petrochemicals.

Hydrogen is a major byproduct of reforming HSRN. Hydrogen can be combined with nitrogen from air to produce ammonia and other fertilizers. Hydrogen can also be used in hydroprocessing, which is a refinery process to remove sulfur from products. Hydrogen is additionally used in hydrocracking, which is a method of increasing the yield of gasoline and middle distillates.

Heavy naphtha, in addition to being used to produce gasoline, is used to produce solvents, and also as a camping stove and lantern fuel.

Naphtha solvents are made out of the aromatics benzene, toluene and mixed xylenes. Solvents are materials capable of dissolving another substance, and are used in many applications from fabric cleaners to paint thinners. There are several specifications for naphtha solvents and often the three primary aromatic hydrocarbon molecules, benzene, toluene, and xylenes, are used separately as solvents instead of as a group.

Naphtha camping and lantern fuel in the US is sometimes referred to as white gas or water white gasoline. White gas is produced from LSRN, has a low octane rating, and does not contain any of the additives which motor gasoline usually has.

Another liquid fuel sold in the US referred to as camper stove fuel, lantern fuel, or Coleman fuel™, is a blend of naphtha solvent, aliphatic hydrocarbons (non-aromatic hydrocarbon molecules), in addition to a small amount of toluene and xylenes.

7. Motor gasoline

Primary uses: Motor car spark-plug ignition internal combustion piston engine fuel
Synonyms: Mogas (motor gasoline); Gas (US and Canada); Petrol (Ireland, Australia, New Zealand, and the UK); Benzol; Benzine; Motor spirit; Automotive gasoline
Demand seasonality: Higher in summer driving season and lower in winter
Volume energy content: US Conventional: 114,500Btu/gal (summer blend);
                       US Conventional: 112,500Btu/gal (winter blend)
                       Ethanol (E100):   76,100 Btu/gal

Motor gasoline is usually the most profitable petroleum product, so refineries will crack, combine or modify the molecules of most other petroleum products that come off the atmospheric distillation unit to try to maximize gasoline production at the expense of almost everything else.

Finished gasoline is a mix of primarily naphtha-range (C_5 to C_{12}) hydrocarbon molecules blended together to meet a tight range of physical specifications. Key specifications for gasoline are octane rating, volatility (how easily the fuel evaporates as measured by RVP), distillation temperature profile, and content of sulfur, aromatics (particularly benzene), and olefins. There are also often restrictions on the content of specific octane rating enhancers, such as lead and MTBE.

Since January 2006 all gasoline sold in the US must be under 30ppm (parts per million) sulfur, so called ultra low sulfur gasoline, which is one tenth the level of sulfur previously allowed. The European Union moved to 10ppm sulfur gasoline, which is called sulfur free gasoline, in January 2008. Japan, Australia, and many other nations, have also been moving toward ultra low sulfur and sulfur free gasoline. The most lax environmental regulations worldwide generally allow for gasoline with 1000ppm (0.1% by weight).

Aromatics such as benzene, toluene, and xylenes are particularly good at increasing octane rating; however, their content in gasoline is usually limited by legislation as they are highly reactive chemicals and can cause health issues. The aromatic benzene by itself is known to be a carcinogen and is often required to be limited to under 1%-2% of gasoline by weight.

Olefinic molecules are also prized for their high octane rating value. However, olefins can leave deposits on engine surfaces which result in increased emissions as the deposits prevent an engine from burning fuel efficiently. Many parts of the world, including the US, limit olefin content in gasoline for this reason.

Octane rating is the most important and generally the most expensive characteristic for a refinery to produce. Gasoline engines generate power by compressing vaporized gasoline with air before igniting the mixture by means of a high temperature electrical spark. As gases heat up when compressed, the process within a gasoline engine has to be carefully controlled such that ignition should take place only as a result of the spark plug as a piston reaches the top of the engine cylinder, and not before due to compression. If ignition occurs out of sequence, because of the heat of compression and not a timed spark, the effect is known as autoignition. Autoignition in a gasoline engine is known as knocking, because of the pinging sound which it often makes. Knocking is also called pinking, pre-ignition, or detonation. As knocking is an uncontrolled burning of gasoline in the piston, the engine can be damaged, energy wasted, and more polluting exhaust emitted from an incomplete burning of the fuel. Severe knocking can quickly rip an engine apart. The measure of how well gasoline will resist autoignition from the heat of compression is called octane rating.

Octane ratings are not correlated with energy content. In fact, some high octane rated fuels, such as pure ethanol, have a much lower gallon for gallon energy content compared with regular octane gasoline. High octane rated fuels allow more fuel, more molecules, to be compressed into the cylinder without autoignition occurring. The extra fuel which can be placed in the cylinder can be used to offset any lower energy content of the fuel. High octane gasoline is therefore often less efficient (less miles per gallon) but can be more powerful on each engine stroke, which is why race cars use high octane gasoline.

Individual gasoline engines are tuned to compress a specific quantity of fuel before a spark plug ignites the mixture. The tuning is set for a certain octane rated gasoline outlined in a car owner's manual. Modern gasoline engines will usually continue to run fairly smoothly by using a computer to change the quantity of fuel and spark timing when given a slightly lower or higher octane fuel than recommended. However, in general, a lower octane than the engine is tuned for may risk damage from knocking, and a higher octane than the engine is tuned for may merely be a waste of money as higher octane fuel often contains less energy. As the engine is not configured to compress additional fuel before igniting it, there would not be any power gain.

The measure of resistance to ignition from compression is called an octane rating because the anti-knock property of gasoline is measured relative to a mixture of isooctane (C_8H_{18}), a hydrocarbon molecule with excellent anti-knock properties, and normal heptane (C_7H_{16}), a hydrocarbon with a very low anti-knock property.

A 100 octane rated gasoline doesn't contain 100% isooctane, instead it means that it has anti-knock properties in a test engine of a blend of 100% isooctane and 0% n-heptane fuel. If gasoline has an octane rating of 90, this means the gasoline has the anti-knock properties of a mixture of 90% isooctane and 10% n-heptane in the same test engine. Gasoline can also have octane ratings over 100 which means it has a greater resistance to knocking than 100% isooctane.

There are three commonly referred to octane ratings: research octane number (RON), motor octane number (MON) and road octane number (RdON.) RON and MON are both calculated in a laboratory using a single cylinder, variable compression ratio test engine. RON uses less severe test conditions and is generally around 10 points higher than MON in finished gasoline. RdON is a measure of actual anti-knock as experienced in real-world, on road, conditions. RdON is mostly used for racing fuels and measurement takes place with a dynamometer attached to a multi-cylinder test engine.

The number printed on the pump at a retail station, the pump octane number (PON), is different depending on which part of the world the station is located.

In the US, the PON is also called the US federal octane rating, the anti-knock index (AKI), or simply (R+M)/2. The US PON is a simple average of RON and MON. Canada uses the same octane rating system as the US for display on retail pumps.

Under the US PON rating system of (R+M)/2, regular gasoline is between 85 and 88 octane, midgrade gasoline is between 88 and 90, and premium gasoline has a greater than 90 octane rating. In 2008, regular gasoline accounts for around 80% of gasoline sold in the US, with midgrade and premium accounting for approximately 10% each, down due to higher prices from around 15% each in the 1990s.

Refineries often only produce regular gasoline and premium gasoline. The midgrade gasoline that is found at many gas stations is blended at a terminal or at the gas station itself from the regular and premium gasoline storage tanks.

Octane rating is altitude sensitive such that in mountainous areas, octane rating can be as much as two points lower and still possess the same anti-knocking properties of a higher octane rated gasoline.

In Europe, Australia, Singapore and many other places outside of the US and Canada, the octane rating listed on a pump is usually RON. As a rough approximation, one could subtract 5 points from RON to get the equivalent US federal octane rating. Thus, for example, a retail pump which indicates 92 octane rated fuel in Europe is approximately equivalent to 87 octane rated gasoline on a retail pump in the US.

Most cars built after 1971 use 87 and higher US federal octane gasoline. Cars built before 1971 and cars with high compression engines tend to use 93 US federal octane and higher. Cars built since 1984 typically have knock sensors and computers in the engine which can adjust engine performance for temperature and pressure to reduce knocking.

Straight-run gasolines have a federal octane rating in the 50-70 range. In order to meet the high octane requirements of automobiles, refineries must add high-octane rated blendstocks to straight-run gasoline.

High octane blendstocks are expensive and often trade separately in the wholesale global oil market between refineries, blenders and others. Choosing which high octane blendstocks to add to gasoline is based on availability and price, also ensuring that the finished gasoline will meet end user and environmental specifications.

A lot of research into anti-knocking blendstocks has been carried out. During the 1920s it was discovered that adding small quantities of inexpensive tetra ethyl lead (TEL) to gasoline dramatically reduced knocking and resulted in a much smoother running engine. The use of TEL in gasoline became widespread, until leaded gasoline accounted for over 90% of the global gasoline market by the 1970s. Unfortunately, lead ingestion from the environment, such as from automobile emissions, can result in serious brain damage, particularly in children. Surprisingly, this is not the reason unleaded gasoline is now the required standard in the US, Europe and many other parts of the world. It so happens that because of smog there was a major push in the US for lower emissions in the mid 1970s. New US clean air standards required the use of catalytic converters on exhaust tailpipes. Catalytic converters use the precious metals platinum, palladium and rhodium as catalysts to oxidize carbon monoxide into carbon dioxide, nitrogen oxides (NOx) into nitrogen, and unburned hydrocarbons into carbon dioxide and water. Catalytic converters cannot operate with leaded gasoline as lead poisons the catalysts. So, gradually over the next 20 years, unleaded gasoline became standard. Leaded motor gasoline is used in very few countries today and the additive TEL is only manufactured at a single plant in the UK.

When lead was phased out in the 1970s and 1980s, refineries had to come up with new inexpensive ways to generate octane rating enhancement. One solution was to add more aromatic and olefinic molecules. However, because aromatics and olefins create a disproportionate amount of pollutants compared with other types of hydrocarbon molecules such as paraffins and naphthenes, regulators have placed limits on their use.

One alternative to increasing octane rating without aromatics and olefins is to use oxygenates. Oxygenates are molecules containing oxygen atoms in addition to hydrogen and carbon atoms. Along with raising the octane rating of gasoline, the oxygen allows for a more thorough burn (oxidization) of the gasoline and thus fewer unburned hydrocarbon pollutants in exhaust emissions (although oxygenates do create other pollutants). The two groups of oxygenates used to increase the octane rating of gasoline are: ethers (MTBE, ETBE, TAME, DIPE); and alcohols (ethanol, methanol, TBA). The ether, MTBE, followed by the alcohol, ethanol, are the two most popularly used oxygenates globally today. The other ethers and alcohols have environmental or cost issues which prevent them from being used widely.

Ethers
MTBE (methyl tertiary butyl ether) is manufactured from methanol, which is made from methane, and isobutylene, which is produced from butane. The first MTBE production plant in the US was built in 1979.

Despite its benefits in raising the octane of gasoline and reducing exhaust pollutants, traces of MTBE have on very rare occasion been found in groundwater. The full health effects of consuming water with MTBE are not fully known at this time; however MTBE does cause water to have a foul taste and odor, often to an extent that it is undrinkable. For this reason, a majority of states in the US, including the large gasoline consuming states of California and New York, have, within the past few years, banned the use of MTBE as an oxygenate. A US nationwide ban on MTBE use is required by January 1, 2015 as a result of the Energy Policy Act of 2005 (EPACT). The oil industry sought protection from MTBE contamination lawsuits to be included in the EPACT. Immunity was not granted and so, in order to protect themselves from lawsuits, the US oil industry voluntarily ceased using MTBE nationwide such that by 2007 MTBE was effectively no longer used in US motor gasoline. MTBE continues to be used in most areas outside the US, such as in Europe and Asia.

Other oxygenate ethers: ethyl tertiary butyl ether (ETBE), tertiary amyl methyl ether (TAME), and disopropyl ether (DIPE) all face similar groundwater pollutant issues as MTBE and cost either the same or more to produce, so there is little economic motivation for widespread use of these alternatives.

Alcohols
Ethanol, also known as ethyl alcohol (EtOH), is an alcohol produced by fermenting starch-based agricultural products, principally corn in the US, wheat in Europe, and sugarcane in Brazil.

Ethanol is the same alcohol found in beverages such as beer and wine. Ethanol is relatively easy to produce, using a similar yeast fermentation process to manufacture alcohol for human consumption. The major physical byproduct from ethanol production from corn is distillers dried grains (DDGs), which is sold as a feed for farm animals.

Bacteria break down ethanol quickly in groundwater, which means it can be used to raise the octane of gasoline without having the same pollution issue as MTBE.

When ethanol is used in gasoline, the blend is sometimes known as gasohol. There are three commonly sold gasohol blends in the US which can be used in gasoline engines: E10, E85 and E100.

E10 is 10% by volume ethanol and 90% gasoline. E10 is the most common blend of ethanol within gasoline. Ethanol can usually be blended with gasoline as high as 15% by volume without requiring adjustments to most conventional gasoline engines or fuel handling systems.

E85 is 85% by volume ethanol and 15% gasoline. Blends this high require modifications to an automobile as ethanol can damage over time rubber components of the fuel handling system. In addition, because the octane rating of ethanol is so high, above 110 (R+M)/2, adjustments may need to be made in an engines ignition spark timing and compression ratio. Finally, as ethanol does not contain as much energy as gasoline (E85 has only 73% of the energy content of conventional gasoline), larger fuel jets may be required. In the US, vehicles capable of handling high levels of ethanol in addition to conventional gasoline are called flexible fuel vehicles (FFVs). The 15% gasoline in E85 allows the fuel to operate in cold climates, where 100% ethanol may make it difficult to start an engine.

E100 is 100% pure ethanol. Ethanol, because of its high octane rating, is used neat in extremely high compression race car engines where immediate power is more important than fuel efficiency. As E100 contains only 67% of the energy content of gasoline, consumers of such a fuel have to refuel more often to drive the same distance.

Other alcohols, such as methanol, also known as methyl alcohol, and TBA (Tertiary-Butyl Alcohol) can also be used to raise octane rating of gasoline. However, due to energy efficiency and environmental reason, neither is commonly used. Methanol contains only 50% of the energy of an equivalent volume of conventional gasoline. In addition, methanol is poisonous to humans in relatively small volumes. Ethanol is also toxic, but at much higher quantities.

Methylated spirits, sometimes referred to as meths, is over 90% ethanol with an additive, up to 10% by volume of the highly toxic alcohol methanol, to render the product undrinkable. The methanol, added to render the ethanol undrinkable, can kill or create blindness when consumed. Methylated spirit is used as a camping stove fuel or as a solvent used to remove stains on fabrics.

The process of rendering ethanol undrinkable is necessary so as to avoid taxes levied on alcohols consumed as beverages. This is called denatured ethanol. Adding small amounts of gasoline or methanol to ethanol produces denatured ethanol.

The ethanol debate
Oxygenates have several benefits. They add to the supply of liquid oil at a time conventional petroleum supply is struggling. They raise the octane rating of gasoline which allows for smoother running engines. They reduce tailpipe exhaust pollutants by displacing environmentally reactive olefins and aromatics in gasoline. As oxygenates contain oxygen atoms, a more thorough burn results in less volatile organic compounds (VOCs) in tailpipe exhaust.

In spite of their advantages, oxygenates such as ethanol face some issues that makes their replacement of gasoline a challenge.

Ethanol production is difficult to scale up to become a major gasoline substitute. Ethanol production requires a large amount of corn to produce relatively little gasoline. Ethanol production requires 2.8 bushels of corn per gallon of ethanol[4]. At present, approximately 25% of US grown corn produces 7 billion gallons, or 3.3% of the energy content of gasoline (and only 1.5% of all US finished oil products) consumed in the US. If the entire corn crop was used to produce ethanol, it would only offset the energy in 6% of total US oil consumption.

To address the scalability limit of corn-based ethanol, there is ongoing research into producing ethanol using non-food cellulosic portions of crops such as corn stalks, switch grass and wood chips in addition to using genetically modified yeast.

It is often asked why other countries don't follow Brazil's example in growing sugarcane rather than corn to produce ethanol. The reason is Brazil has unique rainforest climate and soil suitable for growing large volumes of low cost sugarcane which is difficult to replicate elsewhere.

Lower efficiency: Oxygenates are comprised of molecules which contain oxygen in addition to hydrogen and carbon. As oxygen does not contain any energy by itself (it is the reaction of oxygen and hydrocarbon that releases energy), a higher volume of oxygenated gasoline has to be burned to generate the same power than if the gasoline did not contain such oxygenate.

When blended with conventional gasoline (CG) in small amounts, the improved burn of the CG molecules due to additional oxygen compensates somewhat for the lower energy content. For example, oxygenated gasoline in the US is often a blend of 2% by weight ethanol or 2% by weight MTBE, which is 5.71% and 11% by volume. This addition only reduces energy content by approximately 2%:

Table 9-7

2% oxygenate by weight blended into CG

5.71% vol. ethanol + 94.29% vol. CG.........111,836 Btu/gal (98.1% of CG)
11% vol.   MTBE + 89% vol. CG...........111,745 Btu/gal (98% of CG)

---

[4] The corn crush is the price spread in $/gallon between the price of ethanol and the price of corn. The calculation is | (ethanol price in $/gal) – (price of corn in $/bushel) / 2.8 |.

However, as one raises the oxygenate content in conventional gasoline, the energy content of the gasoline relative to CG will fall toward 67% and 82% for neat ethanol and MTBE, respectively (Table 9-8).

Table 9-8

Conventional Gasoline (CG).......114,000Btu/gal
100% MTBE..........................93,500Btu/gal (82% of CG)
100% Ethanol.........................76,100Btu/gal (67% of CG)

Ethanol use increases certain pollutants. Adding ethanol to gasoline increases evaporative volatile organic compounds (VOCs) emissions from gasoline. Ethanol, with an RVP of approximately 19psi, is much more volatile than conventional gasoline, which in the US has an RVP typically under 11.5psi. In addition, nitrogen oxides (NOx) emissions also increase linearly on combustion of increasing amounts of ethanol blended into conventional gasoline. VOCs and NOx are two primary causes of smog.

The role of ethanol in smog formation is the main reason the state of California, which is aggressive in tackling automobile pollution problems, initially objected to a federal mandate requiring ethanol use before being required to consume it.

Ethanol storage and transportation is challenging. Ethanol is more corrosive than conventional gasoline and special storage and handling facilities have to be built to withstand it. In addition, ethanol cannot be transported easily across long distances as it absorbs water during pipeline transit and must therefore be transported at high cost via railcar, truck tanker, or barge. The difficulty in transporting and storing ethanol had been one of the reasons it never became popular outside of the US Midwest, the corn belt of the country, until bans on MTBE use came into effect. Ethanol is blended with gasoline close to the market where it is sold. The blending of ethanol at a tanker truck loading dock just before the truck leaves for the retail gas station limits the opportunity it has to pull water into the gasoline.

US gasoline grades

Motor gasoline around much of the world is fairly similar, typically meeting the ASTM's D 439 or D 4814 standards. Limits on sulfur content and the choice of either MTBE or ethanol as an octane replacement for lead being are the most common variation within the standard in addition to octane rating. MTBE is the most common octane enhancing oxygenate apart from in the US and Brazil, where ethanol is used.

Although almost all US grades of gasoline will meet ASTM D 439 or D 4814, these standards allow for a large subset of so called boutique gasolines created by the EPA and individual states for pollution control.

The boutique nature of the gasoline market in the US is a contributor to higher prices as it results in small, almost monopolistic, markets within the nation instead of a single large competitive one. The EPACT tried to address the boutique nature of the US gasoline market by setting a limit on the number of variations of gasoline sold in the US, and preventing any new blends being sold unless the new blend fully replaces an existing blend.

Finished gasoline is a fuel which is ready for sale to an end user in a retail station. In the US there are four categories of finished gasoline: federal fuels; California fuels; state alternative fuels; and, seasonal pollution control fuels (winter oxygenated fuels and summer low RVP fuels.) There is also a wholesale market in unfinished pre-oxygenated gasoline in the US.

Federal fuels: There are two types of federal gasoline sold across the US: the default gasoline is called conventional gasoline (CG), and an alternative gasoline mandated in certain areas called reformulated gasoline (RFG). Both CG and RFG are defined by the US Environmental Protection Agency (EPA) with a goal to reduce vehicle emissions of toxic and ozone (smog)-forming compounds. RFG has to meet more strict anti-pollution standards than CG. Today, RFG accounts for 30% of gasoline consumed in the US.

As a result of the 1990 Clean Air Act, RFG is required to be sold year-round in areas of the US with the worst pollution as defined by the US EPA, called nonattainment areas, in addition to some areas which voluntarily opted into the RFG program. RFG had, until May 2006, to contain a minimum of 2 percent oxygenate (MTBE or ethanol usually) by weight. As a result of the EPACT, the minimum oxygenate requirement for RFG was phased out in May 2006. Despite the removal of a minimum oxygenate requirement, when burned, RFG must still meet a strict battery of vehicle emission performance standards.

Among the most significant pollutants which RFG is used to control are volatile organic compounds (VOCs), which are a precursor to ozone (smog). The continental US is divided by the EPA into Southern (VOC control region I) and Northern (VOC control region II) areas. During the summer so called VOC control season (June 1 – Sep 15 in retail outlets and May 1 – Sep 15 upstream of retail) RFG must enable vehicles to meet VOCs emission standards in each of these two regions.

California fuels: The state of California led the nation in controlling the type of gasoline sold because of severe problems with automobile pollution in that state beginning in the 1960s and 70s. Although individual states can require a unique fuel, those fuels have to meet federal emission standards, whereas California, in addition to creating its own fuel standards, is the only state permitted to set its own tighter than federal emission standards. California is a large consumer of gasoline. Approximately 12% of US motor gasoline is consumed in the state. The California Area Resource Board (CARB) issues its own requirements for gasoline sold in California which are more restrictive than those issued by the federal EPA. CARB clean burning gasoline (CBG) contains 2% oxygenate by weight (5.71% ethanol by volume).

State alternative fuels: In addition to California, individual states are permitted by the EPA to require gasoline which is more restrictive than federally required standards. State alternative fuel standards commonly require a minimum amount of ethanol to be added to gasoline. For example, Arizona Clean Burning Gasoline contains 10% ethanol by volume during winter months, November through March. The state of Arizona follows California gasoline standards in part because a major pipeline supplying the state comes from California. Another example is Minnesota Oxygen Minimum Gasoline is require to contain 2.7% oxygenate by weight. Minnesota has an oxygenate requirement because it is an agriculture-dependent state and ethanol usage provides employment for local corn-growing farmers and agribusinesses.

Winter oxygenated fuels: Winter oxygenated gasoline contains 2.7% minimum oxygenate by weight (7.8% ethanol by volume). It is required by the US EPA, in what is sometimes referred to as the winter oxyfuels program, in areas around the nation having particularly bad carbon monoxide (CO) levels during winter, because cold vehicle engines are not capable of producing a thorough fuel burn. Additionally, stagnant winter weather conditions, particularly temperature inversions, trap CO close to the ground. The CO nonattainment areas subject to the winter oxyfuels program use federal conventional or reformulated gasoline (whichever is required by the EPA) during the remainder of the year. Winter is usually defined legally as at least four winter months of the year, typically November 1 through February 28, although the definition of winter used in the oxyfuels program varies in each area.

Summer low RVP fuels: Gasoline evaporates more easily during the heat of summer than during the rest of the year. Evaporated gasoline molecules, when they react with sunlight, are a primary cause of ground level ozone, commonly referred to as smog. Evaporative controls placed on fuel pumps in addition to thicker fuel lines and evaporative recovery devices on an automobile's fuel system help to capture some of the evaporating gasoline. In many states, you may have noticed circular flanges, called a vapor recovery boot, surrounding the gasoline dispenser nozzle. The vapor recovery boot is connected to vacuum tubes within the gas station fuel line, or perhaps as a second line attached to the fuel pump line, to suck vapor back into an underground tank. This vapor is too volatile to be blended back into gasoline in storage at the retail station and is instead collected by the refueling tank truck and returned to a refinery for re-processing.

In the US, limits are placed on the propensity of gasoline to evaporate, as measured by Reid Vapor Pressure (RVP) in pounds per square inch (psi.) The limits are mandatory nationwide on all gasolines:

Table 9-10

Low RVP gasoline.......(low RVP limits on gasoline during summer months only):

Nationwide:	May 1 – Sep 15	9.0psi RVP max.
Low RVP areas:	June 1 – Sep 15	7.8psi RVP max.*
		(* some areas require a 7.0 or 7.2 RVP max)

Winter grade gasoline.................(RVP limit for remainder of year):

Nationwide:	11.5psi RVP max.

Summer blend gasoline is more expensive to refine than winter blend because relatively inexpensive octane rating enhancers such as butane cannot be added as it has too high an RVP. The transition from winter into summer blends each year often coincides with price volatility as brief shortages can occur while refineries, pipeline, and storage operators manage the switch. Ethanol has a very high RVP and the EPA permits the RVP limit of gasoline to be exceeded by 1.0psi if the gasoline contains 10% ethanol, in what is called the one pound waiver.

Instead of having a summer RVP limit and a winter RVP limits, some parts of the world, such as Australia, have different volatility limits for gasoline for each month of the year.

Pre-oxygenated gasoline in the US: Oxygenate is added at the very last stage in delivery due to the difficulties involved in transporting oxygenated gasoline, particularly if the oxygenate is ethanol. Ethanol is usually blended into the gasoline at the truck rack where is it loaded onto tanker trucks just prior to going to retail stations.

There are two methods of blending ethanol into gasoline: splash-blending and match-blending. Splash blending involves adding ethanol to conventional gasoline. The pre-splash-blended conventional gasoline by itself meets end user finished gasoline specifications. Splash blending gives away octane value, as ethanol raises the octane rating of the conventional gasoline, and although it is possible to charge a little extra for midgrade and premium octane fuels, the market for high octane fuels is limited. Also, as conventional gasoline has to meet strict volatility controls and ethanol increases RVP, splash-blending can only carried out on a limited basis.

Match-blending of ethanol into gasoline is more common than splash-blending. Pre-oxygenate-blended gasolines suitable for match-blending are not the same as conventional gasoline. Pre-oxygenated gasolines are blended by a refinery/blender such that they will meet retail specs for octane rating, RVP and so on, only after ethanol or another oxygenate has been added. This process results in wholesale trade in several pre-oxygenate-blended gasolines:

CBOB: Pronounced "C Bob", *Conventional gasoline Blendstock for Oxygenate Blending* is unfinished gasoline which is oxygenated to bring octane rating up to within deliverable specs in areas which use conventional gasoline.

RBOB: Pronounced "R Bob", *Reformulated gasoline Blendstock for Oxygenate Blending* is pre-oxygenated RFG.

CARBOB: Pronounced "Car Bob", *California Reformulated Gasoline Blendstock for Oxygenate Blending* is pre-oxygenated California Clean Burning Gasoline. Finished gasoline, after the oxygenate ethanol has been added, is referred to as CARB gasoline.

OPRG: *Oxygenated fuels Program Reformulated Gasoline* is pre-oxygenated gasoline which is oxygenated during the winter oxygenated gasoline program in areas which use RFG during the remainder of the year.

8. Aviation gasoline

Primary uses: Aviation gasoline for use in sparkplug-ignition piston engines of light general aviation (GA) aircraft
Synonyms: Avgas; 100LL
Demand seasonality: Higher in summer and lower in winter

Prior to the development of the jet engine in the late 1930s, almost all aircraft were powered by high-compression sparkplug-ignition engines similar to those used in motor cars.

Today aviation gasoline, or avgas as it is commonly referred, is a minor fuel used only in high compression sparkplug-ignition piston engines of small private 4-seater propeller airplanes and helicopters in lightweight aviation situations.

High compression aviation engines typically require a higher octane rated fuel compared with motor gasoline. In order to inexpensively obtain the required high octane rating, tetra ethyl lead (TEL) is added to avgas in the US. Avgas is produced using similar blendstocks to motor gasoline, apart from the use of lead, which motor gasoline no longer contains.

The US Federal Aviation Administration (FAA) does not permit any alcohol, such as ethanol or methanol, to be used in avgas. Alcohols raise the volatility of gasoline which may cause vapor lock. Also, aircraft often have rubber fuel handling components which alcohols will quickly corrode. In addition, phase separation of gasoline and alcohol occurs a low temperature, such as when an aircraft climbs to higher altitude, and the separated alcohol may absorb any water previously held in solution which then cannot be handled by the engine sediment bowl. Finally, and most importantly, alcohols have much less energy release potential than avgas and will significantly reduce aircraft range.

Avgas is produced on a very small scale, with 19,000 barrels per day produced in the US; compared with production of 8.7 million barrels per day of motor gasoline and 1.6 million barrels per day of jet fuel in 2008.

By far the most popular avgas blend in the US is Avgas 100 octane LL. LL indicates low lead with 2 grams of lead per gallon. Avgas 100LL is dyed blue in the US so that private pilots can ensure they are filling with the correct fuel, and especially so they don't accidentally fill up at an airport with jet fuel, which is usually not dyed, and will destroy a spark-plug ignition piston engine.

The LL reference in Avgas 100LL is used to distinguish the fuel from a less commonly used Avgas 100 which contains 4 grams of lead per gallon.

Avgas 100/130 Octane, which is dyed green in the US, and Avgas 80/87 Octane, dyed red in the US, are also produced, but in much smaller quantities than the popular Avgas 100LL. An older form of avgas, called Avgas 115/145 Octane, dyed purple in the US, was used during WWII and the Korean War in high-performance military piston aircraft and is only produced in tiny quantities today.

Avgas is typically US$1-$2 per gallon more expensive than motor gasoline for several reasons. It is produced by a very small number of refineries (such as ConocoPhillips in the US) in batch runs and then stored for distribution throughout the year - this storage cost adds to the price. Furthermore, TEL the lead octane-rating enhancer is only produced by a single manufacturer and, as lead is a banned additive in motor gasoline, avgas cannot be moved in the same inexpensive pipelines that transport motor gasoline.

Middle distillate products overview:

Before we begin to discuss the next product, jet fuel, which is a middle distillate, it is useful to explain why middle distillate products are grouped together.

Middle distillates boil at similar temperatures as they contain many of the same hydrocarbon molecules. In refinery terms, the tail ends of their cuts cross significantly.

As middle distillate products contain many of the same molecules, they compete with each other, in refinery processes, to use those molecules in finished products. This results in middle distillates prices being very highly correlated which can result in unusual price situations.

For example, Jet A-1, the most commonly used form of fuel for jet aircraft engines, is usually a lot more expensive during the northern-hemisphere winter compared with the summer, despite the fact that jet fuel demand is lower during the winter. The reason for this is that during winter, a portion of the part of the barrel which normally would have gone to make jet fuel is steered toward making home heating oil and winterized automotive diesel. The supply of hydrocarbon molecules for jet fuel is squeezed in winter by these other middle distillates.

Another point to note in relation to the similarity of the middle distillates is that some finished distillate molecules can be reassigned to different end uses so long as they meet the minimum standards required by standard setting organizations and governments.

On-road diesel (diesel sold at a retail station), for example, which has a low sulfur content in the US, could be substituted for home heating oil, which currently allows much more sulfur. Because of its lower sulfur, on-road diesel is usually more expensive than home heating oil so it is unlikely that anyone would swap one for the other. However, if there was a severe and prolonged cold spell during winter, home heating oil should not rally in price above on-road diesel for very long. If it did, storage operators and refineries would divert low sulfur on-road diesel molecules into the heating oil pool.

Finished home heating oil doesn't have low enough sulfur or necessarily meet other specifications such as a minimum cetane index, to be used as in automobile diesel engines.

Middle distillates are burned in three devices: *Turbine engines* are used for aircraft jet engines, some heavy military transportation equipment, such as US M1 Abrams tanks, and utility electrical power generation with gas turbines; *Diesel (compression-ignition) engines* use the heat of compression and do not have spark plugs; and, *fuel oil burners,* which are relatively simple devices involving burning the fuel in a controlled manner with a small open flame without compression or high temperature spark plugs. Fuel oil burners are used for residential heating, with home heating oil, lighting with kerosene lamps and cooking with kerosene stoves.

Following are the various types of middle distillate:

Middle distillates

Table 9-11

Finished product		Method of energy release	Standard grades
Kerosenes	Jet fuel	Turbine engine	*Civilian:* Jet A, Jet A-1, Jet B *Military:* JP 8, JP 5, JP 7
	Gas turbine fuel	Turbine engine	No.1 GT, No.2 GT
	Kerosene for heaters and lamps	Open flame burner	1-K, 2-K
Light fuel oils/ Gasoils	Diesel (land-based engines)	Diesel compression engine	1-D, 2-D, 4-D
	Diesel (marine-based engines)	Diesel compression engine	DMX, DMA, DMB, DMC
	Light fuel oil (home heating oil)	Open flame burner	No.1 Fuel Oil, No.2 Fuel Oil
	Light fuel oil (industrial fuel)	Open flame burner	No.4 Fuel Oil

9. Jet fuel

Primary uses: Jet propulsion aircraft fuel
Synonyms: Jet A, Jet A-1, JP8, JP5, Jet B, JP4, Avtur (Aviation Turbine Fuel), ATF (Aviation Turbine Fuel), Kero, Kerosene
Demand seasonality: Higher in summer vacation season and lower in winter

Jet fuel is a clear, or very slightly yellow, low-sulfur middle distillate used to power aircraft jet engines. Jet fuel is relatively easy to refine, especially compared with the process associated with gasoline. Often straight-run jet fuel from the ADU can be sold after minor treatment to bring it within acceptable end user specs.

There are two types of jet fuel: kerosene-type, also called narrow cut, jet fuel; and naphtha-type, also called wide cut, jet fuel.

Kerosene-type jet fuel

Kerosene-type jet fuels are used for commercial and military turbojet and turboprop aircraft engines. The most commonly used kerosene-type grade in the US is referred to as Jet A. The slightly different Jet A-1 is more common outside the US for commercial use. JP-8 is the kerosene-type grade used in US military aircraft (Table 9-12). JP, in the military spec, stands for Jet Propulsion. Jet A-1 has a slightly lower freezing temperature of -47°C versus -40°C for Jet A and is therefore more suitable for flights over polar routes during the winter.

US domestic commercial airlines use Jet A, and not Jet A-1, as refineries can produce a couple of percentage points more volume of Jet A than Jet A-1, and it is therefore marginally more profitable for refineries to produce. Apart from the use of additives, military and commercial jet fuels only differ slightly.

Table 9-12

Jet A:	US domestic commercial
Jet A-1:	International commercial
JP-8:	US Air force
JP-5:	US Navy

Flash point is an important safety characteristic of jet fuel. The flash point of a fuel is the lowest temperature at which it will form sufficient vapor to be ignited. The flash point of Jet A and JP-8 is 38°C.

The flash point is especially important on aircraft carriers which store large quantities of fuel onboard and have restricted evacuation options. For the US Navy and other marine uses, jet fuel must have a minimum flash point of 60°C. In other words, US Navy jet fuel should not burn unless it is first heated to above 60°C. Since the flash point of Jet A and JP-8 is 38°C a separate blend of jet fuel is produced specifically for use on Navy aircraft carriers. The blend of high (60°C) flash point jet fuel used in the US navy is called JP-5.

Smoke point is another important property of jet fuel. Smoke in engine exhaust is typically a sign of unburned hydrocarbons. Soot from smoke can coat engine surfaces and reduce efficiency. As one has to lift the fuel in addition to the aircraft, passengers and cargo, engine efficiency is extremely important in aviation and smoke point is a key characteristic in determining efficiency. In general it is desirable for the smoke point of a fuel to be as high as possible. In other words, one only wants soot (unburned hydrocarbons) to form once very high temperatures have first had a decent chance to burn all of the hydrocarbon molecules. Low smoke points are correlated to high aromatic content in a fuel.

Smoke point is measured by burning a sample of fuel in a special kerosene lamp. The flame is allowed to get higher until visible smoke appears. The number of millimeters which the flame reached is known as the fuel's smoke point. The smoke point minimum for Jet A-1 is 25mm.

Naphtha-type jet fuel
Naphtha-type jet fuels are used for commercial and military turbojet and turboprop aircraft engines where a lower freezing point than kerosene-type jet fuels is required for high altitude operations and for use in cold climates. The grade is referred to as Jet B for commercial use and JP-4 for military use (Table 9-13). Jet B is only used by commercial airlines in some parts of Canada and Alaska, but is otherwise not commonly used by commercial airlines. Since the 1970s, the US military has also moved away from JP-4. Since 1996, it is primarily using JP-8 (kerosene-type jet fuel) as it is more widely available and thus less expensive.

Table 9-13

Jet B:	Commercial
JP-4:	Military

Naphtha-type jet fuel is known as wide-cut jet fuel. In addition to kerosene-range molecules, it is comprised of hydrocarbon molecules spanning the more volatile gasoline and naphtha boiling ranges. Naphtha-type jet fuels were originally the most commonly used specification for the military. Because of the wide cut, it was easy to produce in large quantities. However, as gasoline is usually more profitable and naphtha-type jet fuel wide-cuts eat into the gasoline pool, kerosene-type jet fuels have steadily become more popular. Another reason naphtha-type jet fuel is less popular is that naphtha-type jet fuels are more volatile and have a lower flash point than kerosene-type. This makes it more dangerous to handle on the ground and it makes crashes less survivable.

Jet fuel price volatility
Jet fuel prices are usually the most volatile of all oil product prices for several reasons. Firstly, jet is the most strategic of fuels essential for most military aircraft. On the announcement of large scale military actions, such as the Gulf War, jet fuel prices rally hard. Secondly, absolute production and storage of jet fuel is not large relative to other products such as gasoline and diesel therefore it doesn't take much incremental demand or supply to affect prices. Thirdly, jet fuel consumer demand is more elastic. In other words, consumers can do without it by foregoing vacations and business trips compared with other products such as gasoline. Terrorism events, such as September 11 or biological events, such as the spread of SARS in Asia during the second quarter of 2003 had a depressing effect on jet demand and prices collapsed.

10. Gas turbine fuel

Primary uses: Power (electricity) generation
Synonyms: Kerosene, 0-GT, 1-GT, 2-GT, 3-GT, 4-GT
Demand seasonality: Higher in summer

Gas turbine fuel is very similar to jet fuel and kerosene.

A gas turbine operates on the same principle as a jet engine. Air is compressed and then burned by a flame with fuel in a confined area. The resulting hot gas expands rapidly through a turbine which turns in a generator to produce electricity (or produces thrust in a aircraft) and pulls more air into the front of the engine to be compressed. As the hot exhaust gases exit, its pressure will quickly return to close to atmospheric pressure although the temperature of the exhaust may still be over 500°C. The gas compression, combustion and expansion process is called the gas cycle.

The hot exhaust gas leaving the gas-turbine cycle has a significant amount of energy remaining as heat. This heat is sometimes used to boil water and produce steam which drives a steam–turbine connected to a second electrical generator. This second stage of electricity production is called the steam cycle.

The combination of the gas and steam cycles in a single facility is referred to as combined cycle gas turbine (CCGT) plant.

In the US there have been many combined cycle gas turbine plants constructed in recent years due to their relatively low construction cost, their improved efficiency in generating power, their flexibility to be switched on and off rapidly, and relatively low pollution compared with coal. Micro-turbines, which are small turbines used to generate electricity for factories or even large retail stores, have also become popular recently.

Most gas turbine plants burn natural gas. Oil based gas turbine fuels are rarely used for electrical power generation and usually only in peak power demand periods. In the US oil accounts for a tiny amount of electrical power generation and residual fuel is more commonly used than gas turbine fuel. The use of gas turbine fuel is limited because of its high cost relative to other power generation fuels, such as hydro-gravity, coal, nuclear and natural gas.

11. Kerosene

Primary uses: Heating, lighting and cooking
Synonyms: Kerosine, Kero, No.1-K, No.2-K, Furnace oil No.1, K1, Diesel fuel No. 1, Range oil, Stove oil, Lamp oil, Paraffin oil (UK), Liquid paraffin, Dual Purpose Kerosene (DPK)
Demand seasonality: Higher during winter

Kerosene, originally the most commonly produced petroleum product, is a light petroleum distillate burned in portable space heaters, stoves, water heaters and as a light source when burned in wick-fed lamps.

Kerosene is very similar to the material used for jet fuel and gas turbine power generation. In fact, a grade of kerosene called dual purpose kerosene (DPK) can be used as jet fuel or illuminating kerosene (IK) as it meets specifications for both uses.

No. 1-K kerosene is water-clear low sulfur kerosene often recommended for indoor use in the US. There is also a grade named No.2-K which is recommended only for applications where a flue, or chimney-type exhaust, is directly connected to the burning device as 2-K contains more sulfur.

Due to the fact that kerosene and most other hydrocarbon fuels generate carbon monoxide when burned, ventilation to outside air is essential regardless of the grade burned. To avoid a dangerous buildup of volatile fumes, a kerosene heater or lamp should only be re-filled with fuel outdoors when the unit is cold. Gasoline, diesel, heating oil or any other petroleum product should never be used as a substitute for, or mixed with, kerosene for domestic use.

12. Diesel fuels

Primary uses: Automobile and marine diesel compression-ignition engines
Synonyms: Diesel; 1-D; 2-D; 4-D; Light diesel oil (LDO); Number 1 oil; Number 2 oil; Number 4 oil; Marine diesel; Distillate fuel oil; Automotive gasoil (AGO); Automotive diesel oil (ADO); Diesel engine road vehicle (DERV); Industrial diesel oil (IDO)
Demand seasonality: Higher in the summer driving season and lower during the winter

Diesel fuel is the term used for any fuel suitable for use in a diesel engine. Diesel engines, also known as compression engines, were developed in 1892 by Rudolph Diesel. A diesel engine has no spark plugs. Instead, diesel engines rely on the piston in the engine cylinder to compress and heat air. As with any gas, air heats when it is compressed because molecules increasingly bounce off each other in a more confined space. When diesel fuel is finally injected into the cylinder containing the super-heated compressed air it burns immediately. This differs from gasoline engines where the air and fuel are compressed together at lower pressure levels in the combustion chamber before a spark ignites the mixture.

Depending on the grade of diesel fuel and how the engine is configured, diesel fuel and engines can offer 20% to 40% more mileage per gallon compared with gasoline fuel and engines for two reasons: energy density and high compression.

Diesel fuel contains heavier hydrocarbon molecules than gasoline. This is why diesel doesn't evaporate as easily as gasoline and is noticeably oilier or waxy to the touch. The oilier touch of diesel is the reason it sometimes referred to as diesel oil, distillate fuel oil, or gasoil. Heavier hydrocarbon molecules have greater energy densities as there are more hydrogen and carbon atoms available to react with oxygen.

Because air is first compressed without fuel, much higher compression rates, typically double, can be achieved relative to gasoline engines. Because of the higher compression, when the diesel fuel is finally added, there is much more oxygen in the chamber to burn the diesel. Temperatures in the chamber are higher which results in a more complete and thus efficient burning of the fuel compared with gasoline engines. As the burn is more efficient, a smaller amount of diesel fuel has to be added to the combustion chamber to achieve a desired power output compared with gasoline engines.

Because of a higher compression ratio, diesel engines offer much more power at lower engine speeds (lower RPMs) than gasoline engines. As diesel engines compress more air, diesel pistons generally cannot move as fast as gasoline engine pistons, which have less compression to deal with. Because diesel compresses more air, the power of each piston stroke is much higher, even though it is slower than a gasoline piston. This feature of a diesel engine is useful for commercial applications which do not require high engine speeds, but instead require the ability to pull heavy loads.

Diesel is easier and less expensive for a refinery to produce than gasoline as most crude oil naturally contains more diesel range molecules than gasoline molecules. Refineries do not have to employ expensive processes to the same degree which crack, combine or modify molecules to generate diesel compared with gasoline.

In addition to being cheaper to produce, governments have tended to charge less tax on diesel than gasoline in order to support industry and farming.

Because of the lack of an electrical ignition system, diesel engines tend to require less maintenance. In addition, diesel engine blocks are built to withstand much higher combustion pressures than gasoline engines and tend to have a longer lifespan. The downside of being able to withstand higher stress is that diesel engines are heavier than gasoline engines. The weighty engine block of a diesel engine is one of the reasons few aircraft have them. However, continual advances in fuel efficiency and performance of diesel engines is starting to offset this load disadvantage and a very small number of light general aviation aircraft have started to use diesel rather than expensive avgas.

Diesel engines have a reputation for not starting well in cold weather. Gasoline engines use a high temperature spark for ignition, which is relatively unaffected by ambient temperatures. On the other hand, because diesel is ignited from compression-heated air alone, it sometimes needs a helping hand to start on very cold days. In order to assist a diesel engine when starting on cold days, diesel engines have glow-plugs which pre-heat the cylinders, or a heater which warms the engine block.

Diesel engines also have a reputation for not having the acceleration capabilities of gasoline engines, as well as being more expensive and noisier. The higher expense of purchasing a diesel engine is offset by the efficiency of the engine.

The development of turbocharged direct injection (TDI) diesel engines has reduced, and in many cases eliminated, a lot of the performance differentials between gasoline and diesel engines. TDI engines have turbochargers, intercoolers and direct fuel injection. Turbochargers compress air before it enters the combustion chamber and results in more power being generated as there is more oxygen available to burn the fuel. As cold air is denser than hot air, cooling the air after it has been compressed, in what is known as a turbo intercooler allows even more air to be delivered into the combustion chamber. Direct injection involves a fuel injector nozzle spraying diesel directly into the combustion chamber, instead of a pre-chamber which had been used in older diesels. Direct injection reduces heat loss, which improves efficiency, and facilitates easier cold engine starts. A Mercedes variation of direct injection called common-rail direct injection (CDI) which uses electronic controllers to equalize the pressure in each of the engines injector valves, further reduces diesel engine noise and provides a more efficient burn.

Diesel engines, because of the more efficient burn, produce 20-30% less carbon dioxide and much less carbon monoxide than gasoline. However, diesel engines also have a reputation for being more polluting than gasoline engines, especially producing more soot and other particulate matter, as well as more nitrogen oxides (NOx).

Government standards requiring the introduction of ultra lower sulfur diesel fuels have permitted the use of oxidizing catalytic converters which bring NOx emissions from diesel exhaust down to similar levels as gasoline engines. In addition, Catalytic Diesel Particulate Filters (CDPFs), which also require ultra low sulfur diesel fuel, dramatically reduce the problem of soot and other particulate matter emissions from exhaust. There are currently four categories of diesel depending on sulfur content (Table 9-14):

Table 9-14

High sulfur diesel (HSD)	> 500ppm	(> 0.05% by weight)
Low sulfur diesel (LSD)	<=500ppm	(<= 0.05% by weight)
Ultra low sulfur diesel (ULSD)	<=15ppm	(<=0.0015% by weight)
Sulfur free diesel	<=10ppm	(<=0.0010% by weight)

In recent years, the US, the EU and many other regions have introduced legislation to reduce the sulfur content of diesel. Sulfur is removed from diesel in a refinery hydrotreater unit.

US diesel sulfur limit reduction schedule
Table 9-15

	On-road diesel	Off-road diesel (non-road, locomotive, and marine diesel)
Originally	500ppm (0.05% by weight)	5,000ppm (0.5% by weight)
Sep 2006	15ppm (0.0015% by weight)	-
June 2007	-	500ppm (0.05% by weight)
June 2010	-	15ppm (0.0015% by weight) - non-road diesel
June 2012	-	15ppm (0.0015% by weight) - locomotive and marine

The California ARB had a slightly more aggressive sulfur reduction schedule than the federal EPA.

CARB diesel sulfur limit reduction schedule
Table 9-16

	On-road diesel	Off-road diesel
Originally	500ppm (0.05% by weight)	5,000ppm (0.5% by weight)
June 2006	15ppm (0.0015% by weight)	15ppm (0.0015% by weight)

In addition to sulfur, the State of California limits aromatics content of diesel to 10% by weight, or an equivalently low emission fuel. This is known as CARB diesel. Diesel fuel meeting the sulfur content restriction but not the aromatics limits is known as EPA diesel.

European diesel sulfur limit reduction schedule
Table 9-17

	On-road diesel	Off-road diesel
Current	10ppm (0.001% by weight)	50ppm (0.005% by weight)
Jan 2009	-	10ppm (0.001% by weight)

Diesel has become especially popular in Europe and Asia over the past ten years. Diesels often account for over 25% of new European and Asia car sales compared with less than 1% of all new non-commercial passenger vehicle sales in the US. The rapid so called dieselification of the European automobile fleet has been one of the primary reasons for the lack of any growth in European per capita oil consumption over the past few years. In Europe, diesel engines now power over 40% of all road vehicles compared with around 4% in the US.

In the US, fuel taxes are low on both gasoline and diesel, and the efficiency of diesel has not convinced consumers to switch from the historically quicker acceleration of gasoline-powered vehicles. Only 42% of filling stations in the US in 2006 offered diesel. Although diesel vehicles account for a small number of US vehicles, diesel consumption is approximately 16% of total oil demand in the US as those vehicles, such as commercial trucks and trains, burn large amounts of fuel and run quite frequently.

Cetane number is a very important characteristic for diesel fuel performance in an engine as it measures how easily the fuel ignites when it is added to compression-heated air. A higher number is desirable as the higher the cetane number the more easily the diesel will ignite. A low cetane number fuel will have a longer ignition delay period.

In a refinery, high cetane number fractions tend to go diesel, and low cetane fractions go to gasoil/heating oil, which has no minimum cetane requirement.

Diesel does not actually contain much cetane, which is a paraffinic hydrocarbon molecule. The hydrocarbon molecule cetane is used as a reference as it ignites relatively easily under pressure. Pure cetane is assigned a cetane number of 100 in a test engine and diesel fuels are rated as to how easily they ignite relative to this in the same engine.

Maintaining and operating a cetane number engine is expensive and difficult. In order to make the process of determining a diesel fuel's cetane number easier, a method of estimating the cetane number is often used. The estimate of the cetane number is called the cetane index, and uses the measured density of the fuel as well as its distillation temperatures at 10% vol., 50% vol. and 90% vol. recovery to generate the cetane index estimate.

Octane rating is not used for diesel fuels. Cetane numbers, for diesel fuels, and octane rating, for gasoline fuels, both measure the propensity of a fuel to autoignite from the heat of compression (Table 9-18). Autoignition can result in knocking, which is not desirable in gasoline engines and the higher the octane number the more resistant a gasoline is to spontaneous combustion. In diesel engines, the opposite is true as it is desirable for the fuel to readily combust.

Table 9-18
Difference between cetane number and octane rating

Diesel cetane number:
- o   <u>Propensity</u> to autoignite on compression.
- o   Higher number is better.
- o   Applies only to diesel compression engine fuels.

Gasoline octane rating:
- o   <u>Resistance</u> to autoignition on compression.
- o   Higher number is better.
- o   Applies only to gasoline spark-plug engine fuels.

The lower the cetane number, the more difficult the diesel will be to use in cold conditions. The engine will run rough due to a delay in the ignition of the fuel. Rough running reduces engine performance and may damage an engine.

There are three types of diesel engine: high-speed (>1000 RPM), used for mobile transportation, such as trucks, buses and cars; medium-speed (450-1000 RPM), used for electrical power generators, diesel locomotives, and auxiliary ship engines; and low-speed (100-450 RPM), used for electrical power generation and primary ship engines. Low speed diesel engines use residual fuel oil, which is discussed later in this chapter. The diesel fuels we are discussing here are burned in high and medium-speed diesel engines. High speed engines tend to use high speed diesel (HSD) and medium speed engines use light diesel oil (LDO).

Usually the cetane number of diesel is between 40 and 55 for medium to high speed engines. For automotive fuel, the cetane number of diesel in Europe is required to be at least 51 whereas in the US most diesel has a cetane number in the low 40s. The lower cetane number of diesel in the US results in slower engine starts, higher emissions, but contains more energy when oxidized.

<u>High speed diesel</u> is the most commonly used type of diesel. There are two grades: No.1 diesel fuel (ASTM 1-D), which is used in high-speed engines where there is a wide variation in speed and load; and No. 2 diesel fuel (ASTM 2-D) for use in high-speed engines where there is a less variation in speed and load. 2-D is the standard grade for US automotive diesel engines and is much more commonly used than 1-D.

The energy densities of high speed diesels compared with gasoline are approximately:

Table 9-19

114,000Btu/gal	Typical US conventional gasoline (CG)	
130,000Btu/gal	No. 2-D diesel fuel	(114% of CG)
124,000Btu/gal	No. 1-D diesel fuel	(109% of CG)

Diesel grade 2-D, also called number 2 oil, typically has a cetane number of between 40 and 43, and is often labeled as regular diesel in North America. 1-D usually has a cetane number of 44 or above, and is sometimes called premium diesel in the US. Although it is easier to use in cold weather, it contains less energy than 2-D. The diesel fuel named 3-D is no longer produced as a separate product. The standard diesel grade in Europe, EN590, has a minimum cetane index of 51.

In the US and other parts of the world where there are large differences between winter and summer temperatures there are generally two types of diesel, a standard version and a winterized version. Winterized diesel is regular 2-D with some 1-D blended in along with some other low-temperature fluidity improver additives.

Winterizing diesel grade No.2 reduces engine performance slightly because No.1 diesel fuel is less dense and has approximately 95% of energy density of No. 2 diesel fuel. Winterized diesel is more resistant to clouding or gelling and ensures the diesel engine will start and run smoothly in cold weather. The degree of diesel winterization varies from state to state depending on the expected range of temperatures in that region. Non-winterized No.2 diesel fuel will generally be fine to use if temperatures stay above -15°C. Adding 20% diesel grade 1-D to diesel grade 2-D will reduce the cloud point of the diesel by 3°C. In Europe, winterized diesel is called arctic diesel.

Light diesel oil is a blend of distillate fuel with a small amount of residual fuel oil. LDO, such as No. 4 diesel fuel (ASTM 4-D), has a higher viscosity compared with HSD. LDO is used in low and medium speed engines, and is used in diesel engines which operate in a sustained manner with no major changes in engine speed or load, particularly in marine diesel engines.

Marine diesel
Marine fuels are often collectively referred to as bunker fuels. Refuelling a ship is called bunkering, which is a throwback to the early 20th century when steamships had coal bunkers.

When discussing bunker fuels, we are talking about oceangoing vessels, not cruisers or small non-commercial fishing boats which burn motor gasoline or automotive diesel and carry just one type of fuel. Bunker fuels are used in oceangoing vessels such as large fishing vessels, bulk carriers, oil tankers, and container ships, which have two engines: an auxiliary and a primary, and burn two types of bunker fuel: marine diesel, and residual fuel oil

The auxiliary engine burns marine diesel. It is used when a vessel is close to shore, as it produces relatively few pollutants and can be used to maneuver the ship into port. Tight maneuvering requires a fuel which can be used to quickly add and reduce power. The auxiliary engine also generates electricity for the ship while in port.

The primary engine burns a much less expensive residual fuel oil on the long steady journey across oceans which will be discussed later of this chapter.

A key criterion for all marine fuels is the flash point of the fuel. Because of the confined area onboard, the fuel obviously has to be stored relatively close to the crew and engine room. The flash point, as mentioned earlier, is the lowest temperature at which sufficient vapor will form to sustain a flame. While this can be as low as 38-55°C for some grades of automotive diesel, on a ship the minimum flash point specified is usually 60°C. This is to prevent accidental ignition of the fuel.

There are four types of marine diesel:

<u>Synonyms</u>   Table 9-20

Diesel marine A (DMA): [Marine gas oil (MGO), DA, M1]
Diesel marine B (DMB): [Marine diesel oil (MDO), Distillate Marine Diesel, DB, M2]
Diesel marine C (DMC): [Blended marine diesel oil (BMDO), DC, M2]
Diesel marine X (DMX): [DX]

DMA is 100% diesel and is commonly used fuel for small fishing boats, ferry boats, tugboats, and other marine vessels that operate close to the shore or spend a relatively large part of their time at port. It is very similar to diesel fuel No.2 and fuel oil No.2 (heating oil) in specifications. In fact, diesel fuel No.2 or fuel oil No.2 are sometimes used as an alternative to DMA when DMA supplies are short.

DMB is 99% diesel (DMA) and 1% residual, or heavy, fuel oil. DMB is not actually produced as a separate refinery product. It is usually DMA which has been transported in a pipeline or barge which has previously carried DMC or a heavy residual fuel oil. Because it is really a contaminated version of DMA, not all ports regularly supply DMB and shippers do not request it often. DMB is usually slightly less expensive than DMA.

DMC can be up to 20% residual fuel oil and 80% diesel (DMA). DMC is similar to diesel fuel No. 4 and fuel oil No. 4. DMC is used in primary engines as a cleaner alternative to residual fuel and is usually the least expensive marine diesel.

DMX is a form of low operating temperature marine diesel which is 100% diesel. It differs from DMA in that it can be used when ambient temperatures are as low as -15°C, the cloud point of DMX. Because DMX has a lower flash point (43°C) than that normally required for safety in maritime use (60°C), it is often only used in lifeboat motors and emergency generators, when all-temperature performance is key and storage safety is less of a concern.

Biodiesel

Fats and oils are some of nature's mechanisms for storing surplus energy for future use. Biodiesel, also referred to as green diesel, or diester, is a fuel derived from the oils of agricultural vegetation, such as soybean, rapeseed, sunflower, palm and coconut oils, or animal byproducts, such as beef tallow and other animal fats. Biodiesel typically burns in unmodified diesel engines.

Before it can be used as a fuel, the animal or plant oil is first mixed with an alcohol such as methanol. After a period of time, which can be quickened by the addition of caustic soda (anhydrous sodium hydroxide), the mixture separates into a small amount of glycerin, which sinks to the bottom of the production tank, and biodiesel, which forms above the glycerin. The glycerin and methanol are drained away, leaving biodiesel.

The process of producing biodiesel from vegetable oil and animal fat using alcohol is called transesterification. It is so simple that it is performed in garages by enthusiasts, some of whom collect and process used vegetable oil from restaurants.

Sulfur from diesel fuel reacting with nickel from engine block steel has traditionally been an important lubricant in diesel engines. Biodiesel adds lubricity without the need for sulfur. As biodiesel contains virtually no sulfur, it can meet new ULSD (Ultra Low sulfur diesel) restrictions on diesel fuel. Biodiesel also contains fewer aromatics than petroleum diesel.

The primary environmental downside to biodiesel is that it produces more NOx (nitrogen oxides) than petroleum diesel when burned.

Biodiesel has a cetane number of between 46 and 57, with vegetable oils producing a lower cetane number than animal fats.

The two grades of biodiesel available in the US are B100 (100% non-petroleum) and B20   (80% petroleum diesel).

Biodiesel manufacture is currently a cottage industry with just under 1 billion gallons produced in the US during the year, which is approximately 65,000 barrels per day. It takes about 7.3 pounds of soybean oil to produce one gallon of B100 biodiesel. Soybean oil is the most commonly produced oil in the US with approximately 17 billion pounds of soybean oil currently produced each year. Bean oil accounts for nearly 80% of edible oil consumption in the US.

Bean oil is also a primary feedstock for biodiesel production. Distillates (jet fuel, kerosene, diesel and heating oil) consumption in the US is approximately 4 million barrels per day. If the entire US soybean oil production was diverted to replace middle distillate then it could replace only 13 days of current annual US distillate demand. One has to be realistic, therefore, about the current capacity of soybean oil production and accept that biodiesel is not a realistic alternative for mass scale implementation.

Biodiesel is sometimes confused with ethanol, which is discussed in connection with gasoline earlier in the chapter. Ethanol is a form of alcohol, the same alcohol contained in wine, beer and spirits, whereas biodiesel is derived from animal and plant fats. The use of pure ethanol requires modifications to a gasoline engine, whereas pure biodiesel can be used in a diesel engine without any modifications. Biodiesel is a very good solvent and may dissolve petroleum deposits in the fuel system which may clog filters when it is first used.

Biodiesel is different from unprocessed vegoil, or straight vegetable oil (SVO), discussed next.

Straight vegetable oil (SVO)
Unprocessed vegetable oil, often referred to as vegoil or SVO, is fat which has not been converted into an ester-based fuel. Whereas biodiesel can be used in unmodified diesel engines, vehicle modification is required to burn SVO. Modification kits range from US$ 250 to $2000, depending on the diesel engine manufacturer. Among other changes, the modification kits is required to heat the vegetable oil which has a much higher viscosity than petroleum diesel, before it reaches the combustion chamber.

In 2005, during the run up of oil prices, several thousand car drivers with modified diesel engines in France began buying human consumption vegetable oil from the supermarket and blending it with petroleum diesel, thus avoiding the large tax on diesel fuel.

Diesel emulsion fuel
Diesel emulsion fuel is a mix of diesel fuel and 10% to 20%, by mass, of water. The water is suspended in the diesel as tiny droplets by the use of a surfactant. The mixture can be burned in unmodified diesel engines. Emulsion fuels are less efficient in a gallon per gallon comparison with regular diesel as the water produces no energy. The primary benefits of diesel emulsion fuel are the 25% nitrogen oxides (NOx) and 60% particulate matter (PM) reductions in exhaust. For this reason, diesel emulsion fuels are currently used in some mass transit applications in Europe.

Water vapor increases fuel dispersion in the combustion chamber by enabling the production of smaller droplets of fuel. Smaller droplets of fuel mean a larger surface contact between the fuel and oxygen, which results in more efficient combustion and fewer pollutants.

Water is usually undesirable in petroleum fuels as it can promote the growth of bacteria and other organisms which metabolize the fuel, potentially knocking it off spec, and creating biological debris which can clog filters and pipelines.

13. Light fuel oil (burner fuel)

Primary uses:	- Domestic/industrial heating with open flame burner
	- Electrical power generation
Synonyms:	Home heating oil; Gasoil; Fuel oil No.2; No.2 fuel oil; Furnace oil
Demand seasonality:	Higher due to winter heating and lower in summer

The most common use for light fuel oil is home heating oil, which is referred to as gasoil in parts of Europe and Asia. Large heating oil consuming regions are Northern Europe, the Northeastern US, Japan and South Korea.

Home heating oils are used primarily as space heaters. The fuel is burned in a boiler room to heat water which is then pumped around a building to radiators. In newer buildings, air is heated in the boiler room and pumped around a building in air conditioning ducts, a process termed forced air heating.

There are two types of heating oil burner in common use: atomizing-type burners, and vaporizing pot-type burners. Atomizing-type burners are the most commonly used burner type. The fuel is vaporized by passing it through a high pressure nozzle, which mixes the fuel with air. The fuel-air mix is then burned. Vaporizing pot-type burners are the simplest of burner types and are only used in small-scale heating applications. Heat is applied to a small pool of oil, and the vapors are burned once they have mixed with the necessary amount of air.

Home heating oil, also known as fuel oil No. 2, contains very similar hydrocarbon molecules to diesel fuel 2-D, and therefore their prices are highly correlated. The fact that there are two uses, heating and transportation, of the same hydrocarbon fraction in winter months, means that the price for both diesel and heating oil tends to be higher in winter and lower during summer.

Light fuels oils in the US, such as home heating oil, tend to have more than 500ppm (0.05% by weight) sulfur. Individual US states regulate the sulfur content of heating oil such that it may, depending on the state, contain up to 5,000ppm (0.5%) compared to the 15ppm (0.0015%) limit which applies to US automotive diesel. Most European countries limit heating oil sulfur to 1,000ppm (0.1%), with Germany permitting only 50ppm.

As home heating oils are burned by igniting fuel with an open flame, cetane numbers are not important for light fuel oil as they are for diesel.

Light fuel oils are categorized in a similar manner to diesel with three grades (Table 9-21). The grades are numbered in order of increasing viscosity and density – fuel oil No.1 being the lightest and fuel oil No.4 the heaviest. Fuel oils No.1 and 2 contain 100% distillate. Fuel oil No.1 contains some of the higher volatility kerosene fraction. Fuel oil No.2 is the most commonly used grade. Fuel oil No.4 can contain up to 20% residual fuel, with 80% distillate. As with diesel, fuel oil No. 3 is no longer produced as a separate product.

Table 9-21

Fuel oil No.1........used in vaporizing pot-type burners
Fuel oil No. 2 ....standard grade for home heating with atomizing-type burners
Fuel oil No. 4.....lower cost and used primarily for industrial burners

Power utility consumption
Utilities sometimes burn light fuel oil to generate steam which in turn powers a turbine to produce electricity. As light fuel oil products are relatively expensive compared with hydro-gravity, coal, nuclear and natural gas, utility consumption of light fuel oil tends to be restricted to peaker power generation during short periods of high power demand.

14. Residual fuel oil

Primary uses:	- Primary marine engines of oceangoing vessels for operations well away from the coast as this oil is very high in polluting sulfur - Electrical power generation
Synonyms:	Resid; Fuel oil; High-sulfur fuel oil (HSFO); Low sulfur fuel oil (LSFO); Heavy fuel oil (HFO); Fuel oil No.6; Bunker fuel; Bunker C; Bunker A; Navy special; Low sulfur waxy residual (LSWR); Low sulfur heavy stock (LSHS)
Demand seasonality:	Slight seasonal peaks in both winter and summer for electrical power

Residual fuel oil, also called resid, or fuel oil, is frequently seen as the ugly duckling of liquid fuels because it is an undesirable fuel. Resid is from the bottom of the barrel remaining after the production of more profitable gasoline and middle distillates. Fuel oil is a sludgy, highly viscous fuel which must often be heated in order to pump through fuel systems. Residual fuel also contains large amounts of sulfur.

Straight-run residual fuel oil is a product of just the atmospheric distillation unit, which is the first and simplest separation facility in a refinery. In a complex refinery this straight-run resid can be used as a feedstock for a vacuum distillation unit (VDU), a visbreaking unit, or a coker to distill and crack the fuel into higher value petroleum products such as gasoline and distillates. If a refinery does not have a VDU, visbreaker, or coker, they may sell the straight run to another refinery which does. Straight run resid, because high value products can be squeezed out, trades at what is called the straight run premium to cracked resid.

Cracked residual fuel oil is what remains once all economic possibilities of processing the fuel into higher value products have been exhausted. Cracked residual fuel oil is sold to end users as a shipping bunker fuel or for electrical power generation.

Refineries sell resid for less than the crude oil feedstock that was used to produce it. In other words, the refinery margin for resid, the crack spread, is almost always negative. The refinery takes a loss on residual fuel production which is subsidized by the much larger profits they are making on other petroleum products such as gasoline and middle distillates. Some refineries take advantage of the cheap cost of resid to burn it as a refinery fuel.

Heavy crude oil feedstock with a low API and high sulfur content will produce relatively large amounts of residual fuel oil compared with light sweet crude oils.

Crude oil and residual fuel, both referred to as black oils, can be much more damaging when involved in oil spills than lighter petroleum products such as gasoline and diesel, which are referred to as white oils. This is because crude oil and resid are not as volatile and therefore do not evaporate as easily as gasoline, diesel and similar light products which simply float on the water surface and evaporate in the sunlight to disperse over a couple of days. Resid is often more dense than water and may sink to the ocean floor or form slicks on shorelines where it can slowly cause damage for many years.

In recent years, the refinery margin loss for producing residual fuels, the fuel oil crack, has been becoming ever larger primarily due to several factors. Firstly, much of the global spare refinery capacity since the 1970s, particularly in Asia, has been in basic topping refineries which do not have the facilities, lacking cokers in particular, to crack residual fuel into lighter products. This results in more resid being produced. Secondly, Saudi Arabia increased production to full capacity between 2003 and 2008 because of high crude oil prices. The incremental Saudi oil was heavy and sour crude. Heavier crude has higher resid yields. The quality of crude oil being produced by non-OPEC producers is also becoming heavier and sourer. Additionally, demand for residual fuel oil is not growing as much as products such as gasoline or distillates because of environmental regulations and restrictions on burning high sulfur fuel oil.

The two most important characteristics of fuel oil are viscosity and sulfur content.

Viscosity
Centistokes (cSt) is the unit used to define how viscous fuel oil is. The higher the number of centistokes the higher the viscosity, the more slowly the material flows. Viscosity is temperature dependent. If one heats a material it will flow more easily.

A key specification of fuel oil dependent on viscosity is pour point. Pour point is the lowest temperature which the fuel still behaves as a liquid and can be poured. This is important for fuel handling purposes in cold climates.

If residual fuel is too highly viscous then more expensive distillates such as light fuel oil or kerosene are blended in so that the resid flows more easily. Light fuel; oil and kerosene are referred to as cutter stock when used for this purpose. The price spread between different viscosities of resid, such as 180cSt and 380cSt, is called the viscosity spread, or visco diff, and is primarily due to the additional cost of this cutter stock.

In order to reduce the amount of expensive cutter stock required to meet viscosity standards, a process called visbreaking, also called viscracking, is used. The visbreaking process is a milder form of thermal cracking than coking.

Sulfur
Because of its high sulfur content, demand for residual fuel is limited to oceangoing shipping. The fuel can be burned in international waters where pollution laws are lax. It can also be used in electrical power generating utilities which may install sulfur scrubbers on exhaust flues.

Residual fuels with greater than 1% sulfur by weight are referred to as high sulfur fuel oils, and those with sulfur contents of 1% or less are referred to as low sulfur fuel oils. The word low in this usage is relative, considering that even 1% sulfur by weight is a huge amount of sulfur compared to the new US EPA ultra low sulfur diesel regulations that specify the sulfur content of diesel be under 0.0015% by weight.

Marine use residual fuel
Residual fuel oil, when used as a marine fuel is called bunker fuel and the process of refueling is called bunkering. Bunker fuel is burned in massive two-stroke low speed (<450rpm) compression ignition engines. The engines rely on compression-heated air, rather than a high temperature spark plug to ignite the fuel. This is similar in concept to on-road automobile diesel engines. Because resid contains quite a large percentage of sulfur, most commonly from 1% to 4.5% by weight, these fuels are typically only used in oceangoing ship's primary, or main, engines once in international waters.

ISO standards
The International Standards Organization (ISO) marine fuels standards (ISO 8217), which have also been codified in a similar manner by the ASTM (ASTM D 2069), attempted to bring uniformity to international marine fuel markets. The format used by the ISO for distillate marine oils (DMA, DMB, DMC and DMX) where the D indicates distillate and the M indicates marine has been discussed earlier. For residual marine fuels the ISO uses a similar format with 15 RM specifications: RMA, RMB, RMC....through RML used for residual marine fuels. The R indicates residual and the M indicates marine.

In addition to the three letter designation, such as RMA, for each residual fuel specification, there is a viscosity limit. For example, RME-25 is residual marine fuel E with a maximum viscosity, at 100°C, of 25 centistokes.

Although most other parts of the oil market will use ASTM or ISO standards, the marine fuel market also continues to uses somewhat broad categories called Intermediate Fuel Oils, or IFO.

IFO grades
The most commonly used marine residual fuels are colloquially referred to as Intermediate Fuel Oils, or IFO. They are intermediate in that they may contain up to 6-7% of middle distillates, used as cutter stock to lower the viscosity of the otherwise heavy residual fuel. IFOs are named after their viscosity at the handling temperature for marine-use residual fuel oil, which is commonly 50°C. The 50°C handling temperature is above normal atmospheric temperatures because residual fuel is so viscous it has to be pre-heated in order to pump it into fuel tanks and to the engine room. The IFO grades refer to viscosity measured at 50°C, which differs from ISO standards which measure viscosity in centistokes at 100°C.

The three most commonly used intermediate fuel oil grades are called IFO-180, IFO-380 and IFO-460. Oil traders simply refer to the grades as 180, 380 and 460. Of the four IFO grades, by far the most common residual fuel used for bunker is 380, sometimes referred to as Bunker C, and followed in popularity by 180. The shipping industry prefers to use higher viscosity residual fuels, as the higher the viscosity, the less expensive the fuel is.

Light Marine Fuel Oils (LMFO):

> IFO-180 can be either of the ISO specifications RME-25 or RMF-25. The differences between the two ISO specs are primarily carbon, ash and vanadium content.

> IFO-380 can be either of the ISO specifications RMG-35 or RMH-35. Once again the differences between the two ISO specs are primarily carbon, ash and vanadium content. 380cSt, as it is more viscous, is less expensive than 180cSt.

Motorship Marine Fuel Oils (MMFO):

> IFO-460 can be either of the ISO specifications RMH-45 or RMK-45. The difference between the two ISO specs is primarily density. IFO-460 is usually the highest viscosity residual fuel grade available at ports. 460cSt is less expensive than 180cSt and 380cSt and not widely used.

Comparing IFO grades to ISO standards   Table 9-22

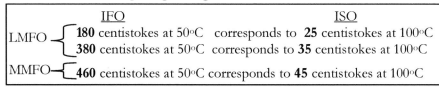

	IFO	ISO
LMFO	**180** centistokes at 50°C	corresponds to **25** centistokes at 100°C
	380 centistokes at 50°C	corresponds to **35** centistokes at 100°C
MMFO	**460** centistokes at 50°C	corresponds to **45** centistokes at 100°C

180cSt marine residual fuel oil is usually more expensive than 380cSt marine residual fuel oil as it will have more middle distillate cutter stock content and thus will require less heating to pump. 180cSt contains approximately 6-7% middle distillates, whereas 380cSt will have only about 3%. 180cSt (RME-25 and RMF-25) can contain up to 5% sulfur by weight. 380cSt (RMG-35 and RMH-35) may contain up to 4.3% sulfur by weight. In practice, almost all 180cSt and 380cSt purchased by shipping companies contains less than 4% sulfur.

As with marine diesel, for safety purposes, all marine-use residual fuels must have a flash point minimum off at least 60°C.

The compression engines in ships are not choosy and can tolerate some pretty heavy fuels. It doesn't happen frequently, but occasionally in recent years, unscrupulous residual fuel oil blenders have used residual fuel as a dumping ground for unwanted pollutants and low value products from other parts of the oil industry, including the petrochemical polypropylene, which because it is comprised of microscopic long strands, can block fuel filters. Ship owners frequently sample fuel being pumped into their tanks and send samples to fuel testing centers which ensure that the fuels meet ISO standards.

Marine fuel sulfur restrictions
In international waters, where domestic laws do not apply, the UN sets minimum pollution standards.

The International Maritime Organization (IMO), an agency of the United Nations, has issued emission limits on marine engines as part of MARPOL (Marine Pollution) regulation, Annex VI. Annex VI specifies emission limits on NOx (Nitrogen Oxides), SOx (Sulfur Oxides), and VOCs (Volatile Organic Compounds) among others.

In order to meet emission limits, Annex VI sets limits on sulfur content in marine fuels. The MARPOL regulation sets a maximum of 4.5% by weight sulfur on bunker fuel globally moving to 3.5% by 2012 and 0.5% by 2015. Most residual fuel used in bunkers contains 2% to 3.5% sulfur and so the initial 4.5% limit did not make much difference in practice. However, in what are known as Emission Control Areas (ECA), sulfur content in residual fuel is limited to 1.5%, moving to 1% by March 2010 and 0.1% by 2015. The first ECA was the Baltic Sea area, as defined in MARPOL, from May 2006, with the North Sea and an area west of the UK becoming the second ECAs in 2007. Vessels entering an ECA will either have to constantly use fuels with less than the ECA sulfur limit or will have to blend the sulfur levels of bunker fuel down to under the ECA sulfur limit before entering the area.

Utility power-generation residual fuel use
Power generation utilities generally have a portfolio of power sources, some of which, such as coal and nuclear plants, are used constantly, night and day, and some which are used just when power demand surges.

The demand ranges which are met all of the time are termed base load, and the high demand periods are termed peak demand. Utilities separate their fuels into what is called a stack. Generators which are low cost, such as hydro-gravity, and are difficult to shut down and re-start quickly, such as coal and nuclear, form the bottom of the stack. Fuels at the bottom of the stack are used to continuously produce electricity and meet base load power demand. Fuels which are relatively more expensive and which can be burned in generators which can be started up and shut down quickly, such as natural gas, residual fuel oil and even home heating oil and kerosene/jet fuel, form part of the top of the stack. Fuels at the top of the stack are often only burned at what are called peaker stations during the relatively short periods of high power demand. Technology and efficiency improvements in recent years have moved natural gas from a peaker fuel and into the base load part of the stack in many areas.

In the US, petroleum products account for less than 5% of power generation. The percentage varies by region in the US, but petroleum is generally a minor component in most utility stacks. The primary reason for the dwindling use of petroleum for power generation in the US has been the availability of vast quantities of inexpensive and domestically available coal, nuclear, natural gas and hydro-gravity power.

Outside the US, in Japan for example, petroleum accounts for a much higher percentage of power generation because of the lack of inexpensive alternatives.

When burning oil, utilities will use the following grades, in order from most expensive No.1, to least expensive, No. 6:   Table 9-23

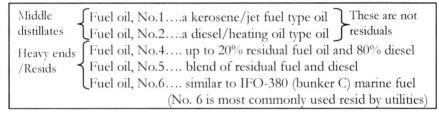

By far, the most common fuel burned by utilities for power generation is the cheap No.6 fuel oil. Fuel oil numbers 4, 5 and 6 often need to be pre-heated before they can be handled.

Due to local government environmental regulations, utilities often specify much lower sulfur content for residual fuels than permitted in ASTM and ISO standards. In many countries where there are lax environmental regulations, residual fuel oil may contain up to 5% sulfur by weight. In the US, the highest sulfur content is 3% by weight for fuel oil No.6, however many states and regions within the US, especially in large metropolitan areas, specify sulfur maxima of 2.2%, 1%, 0.7% or 0.3%. Lowering sulfur usually involves blending cutter stock.

ASTM International, and not the ISO, is responsible for designating fuel oils with the number 1 through 6 descriptors. Hence the terms fuel oil numbers 1 through 6 are rarely used outside the US. The term bunker C is frequently used in the US to describe fuel oil No.6 when it is used for marine fuel. Bunker A and bunker B, are very rarely used terms to describe fuel oils number 2 and 5, respectively. Fuel oil No.5 is also referred to as Navy Special Fuel Oil (NSFO), or simply Navy Special, and is the preferred residual fuel of the US navy.

A grade of residual fuel oil frequently exported from Russia is called Mazut-100, or M100. M100 is typically purchased by international trade houses to be blended with cat slurry oil, which is a heavy byproduct produced in a refinery FCC unit, and this blend is then used as a high sulfur marine fuel or fuel oil to be used by utilities.

In Asia there is a grade of residual fuel called Low Sulfur Waxy Residual (LSWR). LSWR is a low sulfur, approximately 0.2% by weight, residual fuel oil produced in Indonesia primarily for use in the Japanese power generation market. Occasionally LSWR will, because of its low sulfur level, be blended with a high sulfur crude oil and used as a refinery feedstock. LSWR, when used as a refinery feedstock is referred to as Low Sulfur Heavy Stock (LSHS.)

Some Japanese utilities burn unrefined crude oil rather than residual fuel oil to produce electricity. These so called direct-burning crudes or ready-to-burn crudes tend to be primarily from Indonesia, Vietnam and Sudan.

15. Base oils and Finished lubricant formulations

Primary uses:	- Prevents friction and thus reduces heat and wear in mechanical devices
	- Personal care products
	- Candles
Synonyms:	Base stock, Lube oils, Motor oils, Greases, White oil, Paraffin wax, Mineral oil, Paraffinic base oils
Demand seasonality:	not seasonal

Base oils, also known as base stock, are used mostly to manufacture finished lubricants and greases, which prevent friction between moving parts in machinery.

It is important for lubricating oils not to thicken at low temperatures, while starting an engine on a cold day, or thin at high temperatures, such as running an engine at high loads on a hot day. Because of this, a key factor with base oils is its Viscosity Index (VI) which is a measure of thermal stability. The higher the VI the more stable viscosity will be over a wide range of temperatures. Unstable aromatic molecules are particularly undesirable in base oil stock as they have a very low VI. Paraffinic hydrocarbon molecules tend to have the highest viscosity index compared with other hydrocarbon molecules.

The reason base oil can sustain a relatively constant viscosity over a range of temperatures is because they contain long chain paraffin molecules which coil up tightly at low temperatures so they don't impede flow. These molecules unwind into long chains at high temperatures, ensuring the oil doesn't thin too much and can form a film even at high temperatures.

In addition to VI, another quality used to categorize base oils is the absolute viscosity of the base oil. Viscosity of most base oils in the US is measured at 100°F in Saybolt Universal Seconds (SUS). Viscosities of some base oils, bright stock for example, as well as those defined by the ISO, are measured in centistokes (cSt) at 100°C.

Base oil production
Vacuum gasoil (VGO) produced from straight-run resid in a refinery's Vacuum Distillation Unit (VDU), is sent to the refinery's lube plant where aromatic molecules are removed to produce raffinate (Table 9-24). Raffinate can be used as a low octane gasoline blendstock. Alternatively, wax is removed from this raffinate to produce base oils.

Refinery lube plant processes   Table 9-24

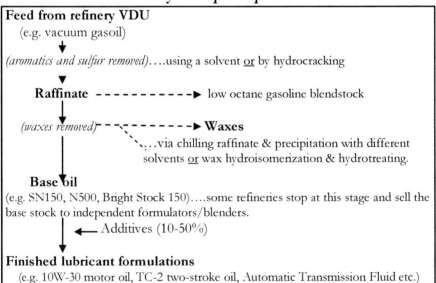

Processing the feedstock from the VDU with hydrogen produces higher quality base oils than the older method of using solvents.

Base oil categories

Base oils are classified as either solvent neutral, neutral, or bright stock.

Solvent neutral (SN) is base oil produced using solvents. The neutral designation is a throwback to earlier days when acid was used to produce base oils. The acid had to be neutralized before the base oil could be used. Examples of Solvent Neutral base oil names are: SN60, SN150, SN500, SN600, and SN700. The numbers refer to the SUS viscosity. The higher the SUS, the more viscous the oil is.

Neutral (N) is base oil produced using hydrogen instead of solvents. Examples of hydroprocessed base oils are N60, N150, and N500. The numbers again refer to the SUS viscosity.

Bright stock (BS) is the heaviest base oil and is defined by its viscosity at 100°C (212°F), which is a higher temperature than that used for the neutral base oils which are measured at 100°F (38°C.) The most common grade of bright stock is BS150, where 150 is the SUS viscosity at 100°C. At the lower temperature of 38°C, BS would have an SUS viscosity of 2,500, which shows how heavy it is relative to the neutral base oils. Bright stock has a golden color, whereas most neutral base stocks have a water white appearance. Bright stock is almost always produced using solvents, not hydroprocessing, and is therefore sometimes referred to as Solvent Bright Stock (SBS.)

API categories of base oils

In addition to the Solvent Neutral, Neutral and Bright Stock classifications, the American Petroleum Institute (API) also has another categorization method for base oils, in order of lower to higher VI performance: Group I, Group II, Group III, Group IV and Group V (Table 9-25).

Base Oil Categories

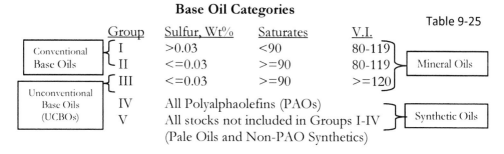

Table 9-25

	Group	Sulfur, Wt%	Saturates	V.I.	
Conventional Base Oils	I	>0.03	<90	80-119	Mineral Oils
	II	<=0.03	>=90	80-119	
	III	<=0.03	>=90	>=120	
Unconventional Base Oils (UCBOs)	IV	All Polyalphaolefins (PAOs)			Synthetic Oils
	V	All stocks not included in Groups I-IV (Pale Oils and Non-PAO Synthetics)			

Group I base oils are typically produced by solvent-refining and de-waxing, and contain higher levels of undesirable aromatics and sulfur than Group II or III. Group II and III are usually produced by hydroprocessing and typically offer higher performance characteristics, such as a higher VI and lower levels of impurities than the solvent-refined Group I base oils.

There are two other base oil groups defined by the API: Group IV which are Polyalphaolefins (PAOs) and Group V, which is a catch all for any other type of base oils, such as those manufactured using esters, polyisobutylene, polyalkylene glycol, phosphate esters, and poly-internal olefins. Group IV and V tend to have the highest VI of base oils. PAOs, for example, have a VI of around 150.

Two sub-groups referred to in industry are:

	V.I.	
Group II+:	110-119	Table 9-26
Group III+	135+	

Synthetic base oils

One of the drawbacks of base oils manufactured from petroleum is that they are composed of a wide spectrum of hydrocarbon molecules which have differing viscosity properties at various temperatures. This range of hydrocarbon molecules tends to reduce thermal stability. Lubricant manufacturers have tried to get around this drawback by using base oils which are comprised of a single molecule type which can withstand a wide range of temperatures. Such base oils are called synthetic base oils and offer higher performance than hydrocarbon-based base oils, though sometimes at an increased cost.

Most synthetic oils are Group IV PAOs. The most well known brand name of motor oil which is made with synthetic base oil is Mobil 1™.

The definition of synthetic originally simply referred to just PAO base oils. However, as a result of a decision by the US National Advertising Department of the US Better Business Bureau, Group III oils which meet most performance features of early PAOs are now also permitted to be marketed as synthetic.

Castor oil

Castor oil is a motor oil which is not derived from crude oil at all, but is produced by crushing castor beans and distilling the resulting oil. Castor oil is an excellent lubricant. Unfortunately, castor oil degrades quickly and is only used in extreme high performance race cars that have their engine pulled apart and new oil added for every race. Castor oil results in a very sweet engine exhaust odor. Castor oil is not recommended for use in most modern engines. A British company, Castrol, now owned by BP, originally made motor oil from castor oil, but now concentrates on producing petroleum based lubricants.

Finished lubricant formulations

Although refineries do produce finished lube oils for the end user market, many refineries sell their base stock oil to lubricant blenders. These small manufactures formulate lubricants for specialized uses. Lube base oils (LBOs) are used in the manufacture of finished formulations for several uses: automotive engine and hydraulic oil, industrial machine oils such as axle lubes, gear oils, greases, turbine oils, electrical oils, drilling fluids; and food grade and medicinal oils.

Finished lube oil formulations are base oil plus an additive package. The additives package is usually minor, but can be up to 50% of the volume of the finished formulation. The nature of the lube oil market, where large refineries sell to thousands of small independent finished oil formulators, results in very aggressive marketing. Small blenders try to differentiate themselves with colorful advertising, resulting in lube oil formulations having high levels of brand loyalty compared with other parts of the oil industry.

Finished lubricant formulation: motor oil

Motor oil specifications are set by the Society of Automotive Engineers (SAE) in its standard SAE-J300. The SAE motor oil specifications have names such as: 0W, 5W, 10W, 15W, 20W, 25W, 20, 30, 40, 50, 60, through 120. The W in some specifications indicates that the motor oil passes low temperature test that measure cold winter performance.

A higher specification number indicates a higher viscosity. A high viscosity means thicker lubricant, which means less wear and tear on moving parts, but less fuel efficiency as the engine has to overcome the resistance of heavier oil.

Originally the SAE measured viscosity only at 100°C, a so called mono-grade rating, and the motor oil scale was approximately the number of seconds the oil took to pass through a viscometer.

The scale used today, although still linked to viscosity, no longer corresponds to the exact number of seconds the oil takes to pass through a viscometer. The numbers used today such as SAE5W and SAE30 are simply specification names, in other words the numbers 5 and 30, in these examples, do not refer to 5 or 30 viscosity seconds, as they originally did. For example, SAE30 indicates that an oil has a kinematic viscosity of greater than 9.3cSt and less than 12.5cSt at 100°C. W specifications outline acceptable viscosities of oil at various points between -40°C and -10°C in addition to viscosity at 100°C.

Most motor oils today have what are called multi-viscosity ratings, in that the oil meets both a low temperature W specification, and a high temperature, non-W, specification. Multi viscosity ratings indicate motor oils viscosity index (VI).

There are four commonly formulated multi-viscosity rating grades of motor oil used in automobile engines:

Table 9-27

| Oil meets this low temp. spec. | Oil also meets this high temp. spec. |

SAE 5W - 30: viscosity with good cold weather performance.
SAE 10W - 30: viscosity for all-weather performance most popular in US.
SAE 5W - 40: preferred viscosity in Europe provides broad protection range.
SAE 15W - 50: viscosity for use in climates which are warm all year around.

In addition to the SAE specifications, in order to distinguish gasoline and diesel engine motor oils, an API designation of S is used as a prefix for gasoline motor oils and C for diesel engine motor oils.

Finished lubricant formulation: mineral oil
Mineral oil, also known as white mineral oil, white oil, or paraffin oil, is a colorless, odorless, tasteless base oil used in the manufacture of industrial polymers, personal care lubricants such as baby oil, sun tan lotion, laxatives, food applications such as bakery pan oils, in addition to being used as a grain dust inhibitor. The feedstock for the manufacture of white oil is simply high quality hydroprocessed (not solvent processed) base oil. Mineral oil consists of chemically stable saturated paraffinic and naphthenic hydrocarbon molecules. Aromatic molecules, sulfur, nitrogen and oxygen will have almost all been removed from the lube base oil feedstock.

In the US, the Food and Drug Administration (FDA) sets standards for food grade oils.

In the UK and South Africa, the term paraffin oil is sometimes used to refer to kerosene even though kerosene and mineral oil are two very different products.

16. Waxes

Primary uses:	- Candles, waterproofing packaging, fruit and cheese coating, portable fuel, chewing gum, floor polish, crayons
Synonyms:	Petroleum wax, Mineral wax
Demand seasonality:	not seasonal

Waxes are combustible hydrocarbon solids at room temperature, which can be heated and reformed many times. In essence they are thermoplastic. Petroleum waxes compete with non-petroleum waxes, such as beeswax. The largest applications of petroleum waxes are in the manufacture of candles and producing waxed paper. The US FDA sets standards for food grade waxes.

Important characteristics of waxes are melting point and a needle penetration test, which measures the hardness of the wax, with higher values indicating a softer wax.

The petroleum waxes discussed here are referred to as natural waxes. Synthetic waxes are produced from petrochemicals, which are complex chemicals made from petroleum, such as polyethylene (PE), often used in applications such as in the manufacture of paint.

There are four categories of wax depending on how much room temperature liquid oil has been removed from the wax (Table 9-28). Lower liquid oil content means less smoke if the wax is burned. Oil is removed from wax in a process known as de-oiling. The non-wax oil liquids removed during de-oiling are called foot wax, or sweat oil.

Table 9-28

Wax type	Liquid oil content
Slack wax/Crude wax	3-50%
Scale wax	1- 3%
Semi-refined waxes	0.5-1%
Fully refined paraffin wax	< 0.5%

In addition to labeling waxes according to how much liquid oil they contain, there is another categorization method for petroleum waxes used in industry: paraffin wax, microcrystalline wax, and petrolatum wax.

Paraffin wax is non-toxic, water resistant, colorless and clean burning. The feedstock for the manufacture of paraffin wax is neutral lube base oil. Paraffin wax is primarily used in the manufacture of candles, waxed paper, and as a clean-burning solid fuel such as camping stove fuel.

Microcrystalline wax is also known as micro wax. Micro wax is produced from heavy hydrocarbon molecules such as bright stock. Micro wax is darker than paraffin wax with a higher viscosity and melting point. Micro wax is often used as a coating for cheese wheels and as a base for chewing gum.

Petrolatum waxes, often referred to as mineral jelly, or petroleum jelly, is a mix of paraffin wax, white mineral oil and micro wax. Petrolatum is semi-solid at room temperature. Petroleum jelly ranges in color from white to yellow and is used as a formulation base in a range of personal care products such as ointments, lotions, and creams. Vaseline®, a trademarked product since 1870, is one of the most well known petroleum jelly products.

17. Bitumen

Primary uses:	- Asphalt and road oil for road surfacing and roofing
	- Emulsion fuels for power generation
Synonyms:	Incorrectly: tar/pitch
Demand seasonality:	Higher demand during summer road construction season

Bitumen is one of the heaviest, most viscous liquid refinery products of petroleum. Bitumen, which must be heated above 150°C to maintain a liquid state, is primarily used in the manufacture of asphalt and emulsion fuels. Bitumen used in the production of road paving asphalt is seasonal and accounts for over 85% of bitumen consumption.

Bitumen consists of heavy asphaltic petroleum molecules and hence it is sometimes referred to as asphalt, but as you will see below, bitumen is a small part of the product known as asphalt.

Tar, also referred to as pitch, is commonly and incorrectly confused with bitumen and its product, asphalt. Tar is a product of coal, not petroleum, and is no longer used in the manufacture of roads.

Asphalt
Asphalt is a mixture of approximately 5 percent bitumen, referred to as a binder in this use, and 95 percent aggregates, which are crushed stone, gravel, and sand. The aggregates are typically sourced locally, close to the site of final use, and mixed with bitumen to form asphalt at an asphalt plant.

Asphalt can be hot mix or cold mix. The vast majority of asphalt used in North America is hot mix asphalt (HMA) and requires the bitumen binder to be mixed with aggregate at 250-325°F (121-163°C). Then, the resulting asphalt is quickly rushed to the paving site before the mixture cools and sets. Road manufacture in the US is seasonal, with most construction during the summer in order to take advantage of additional cooling time. Cold mix asphalt (CMA), on the other hand, remains workable for up to 2 weeks in ambient temperatures as it contains a small amount of more volatile material such as diesel or kerosene, called cutback when used in this manner. Cold mix is often used in pothole repair or other such emergency situations throughout the year.

Bitumen is typically defined by a penetration (PEN) test whereby a sample is heated to 25°C and a standard needle with a 100g weight is placed on the sample for 5 seconds. The number of tenths of millimeters which the needle penetrates gives a good indication of the viscosity and hardness of the bitumen. Often the PEN measurement of a grade of bitumen is given as a range such as 80/100PEN. Some common grades are:

Table 9-29

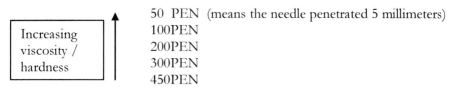

50 PEN (means the needle penetrated 5 millimeters)
100PEN
200PEN
300PEN
450PEN

Increasing
viscosity /
hardness

The harder the bitumen, the more difficult it is to work with, but the more durable and longer lasting the road surface will be. Usually, road crews will use differing specifications of bitumen for high and low traffic areas.

Cutback bitumen

Cutback bitumen, also referred to as road oil, or deferred set bitumen, is bitumen mixed with a small amount of cutback material such as diesel or kerosene. Cutback bitumen is defined by the number of seconds it takes for 50cc of the material to run through a standard hole in a viscometer at 40°C. The larger the number the more hard the bitumen is – the opposite of the PEN scale.

Table 9-30

200 SECS (means it took 200 seconds)
100 SECS
50 SECS

Increasing
viscosity

Surface dressing

In addition to its use in making cold mix asphalt, cutback bitumen is used for surface dressing of a road. Surface dressing is designed to prevent water from entering the road, which results in quick disintegration of the asphalt integrity. Surface dressing also provides a rough surface texture which reduces skidding. The surface dressing process involves spraying the cutback bitumen from the back of a tanker and then applying surface chips.

Roofing oil

Bitumen is used to make roofing oil which is typically spread on a roof surface from a bucket with mops.

Emulsion fuels

Emulsion fuels are a combination of bitumen, water (around 30% by weight) and a surfactant to allow the water and oil to mix. Emulsions are used as alternatives to coal by electrical power generation utilities. One of the most well known emulsion fuels was produced from bitumen in Venezuela using the trademarked name Orimulsion.

18. Petroleum coke

Primary uses:	-Fuel in steel smelting, cement manufacture and power generation
	- Anodes in electric furnaces to manufacture aluminum and steel
	- Electrodes in many applications
	- Furnace lining
Synonyms:	Sponge coke, needle coke, shot coke, fuel grade coke, specialty coke
Demand seasonality:	not seasonal

Coke is mostly carbon, with a minor amount of hydrogen, nitrogen, sulfur, and oxygen. As volatile components (light hydrocarbon molecules) have mostly been removed, coke burns at a very high temperature with very little smoke. The lack of volatile components makes coke especially useful for smelting steel, where the coke heats the iron ore. At the same time, the coke impregnates the iron with carbon atoms to produce carbon steel, which is harder than iron by itself. Coke, if it is pure enough, can be also used as anodes in the production of aluminum. Anodes, which are one of two electrodes (the other being the cathode) used in making aluminum and steel, are eventually oxidized and must be replaced regularly.

Traditionally, there have been three sources of coke: wood, coal and petroleum.

Wood-based coke is called charcoal, and is produced by heating wood in the absence of oxygen. The wood is heated to remove volatile components, and the lack of oxygen prevents the carbon from burning. Wood-based charcoal was originally used to smelt steel, but is used today primarily as a barbeque fuel.

Coal-based coke can be produced from bituminous coal by heating the coal at a very high temperature in the absence of oxygen until all components that easily vaporize have been removed.

Petroleum-based coke is mainly produced in a refinery coker unit. The feed to a coker is most often straight-run residual fuel produced from atmospheric and/or vacuum distillation. Coke is a byproduct of attempting to maximize the production of higher value products such as gasoline and diesel.

There are two categories of petroleum coke depending on the level of carbon purity: green coke and calcined coke.

Green coke is a solid industrial fuel used in the manufacture of cement and also by power generation utilities as an alternative to coal. Green coke may contain up to 15% hydrocarbon molecules contained in pores within the coke, the remainder being predominantly carbon. Sulfur content of green coke is typically between 4% and 7%. Green coke can also contain up to 10% water from the quenching operation in the coker (which rapidly stops coke formation at a desired point.) Green coke is used mostly to make fuel-grade coke and calcined coke

Calcined coke is green coke which has been further heated to above 1,000°C which removes almost all hydrocarbon molecules to under 0.25% by weight. It is used to make what are called specialty cokes: anode-grade coke and graphite needle grade coke, which is a more pure carbon than anode-grade.

In addition to being classified by its level of carbon purity, coke is also classified by its physical appearance as sponge coke, needle coke, and shot coke.

Sponge coke is primarily used as a fuel and is in the form of visibly porous, irregular shaped lumps. The vast majority of sponge coke is used as a fuel in power plants, cement kilns and steel manufacturing plants, although some sponge coke can be used to make anodes used in the production of aluminum and steel.

Fuel grade coke must be ground into fine particles to ensure a more thorough and rapid burn. The Hargrove Grindability Index (HGI) measures the coke resistance to grinding. A HGI below 35 is considered very hard and will require more intense grinding. A HGI above 65 is considered soft.

Because of its end use, fuel grade coke prices tend to be more correlated to coal prices than to oil prices. Despite the fact that it has a higher energy content, coke typically is less expensive than coal because it has higher levels of sulfur and metals, which are pollutants, and lower levels of volatile hydrocarbon molecules, which makes ignition a challenge.

Needle coke has a parallel ribbon, or needle shape, structure. Because of its high level of carbon purity, needle coke is used to make electrodes for the manufacture of aluminum and steel. Needle coke is usually the most expensive form of coke and usually results from aromatic crude.

Shot coke takes the form of small spherical structures and is a byproduct of the manufacture of either sponge or needle coke. Shot coke is blended with sponge coke to be used as a fuel. Shot coke is very hard and has a low HGI.

19. Carbon black

Primary uses:	- Reinforcing agent in rubber tires
	- Manufacture of ink
Synonyms:	Soot; lampblack
Demand seasonality:	Not seasonal

Carbon black, commonly known as soot, is a black powder produced by subjecting residual fuel oil to extremely high temperatures in a carefully controlled combustion process. Carbon black is essentially finely dispersed carbon. Carbon black's primary uses are a reinforcing agent in rubber tires and in the manufacture of ink.

Lampblack has similar properties to carbon black but it more oily and bluish. Lampblack is produced by subjecting kerosene to extremely high temperatures and is used to produce contact brushes for electrical applications.

CHAPTER TEN
PETROCHEMICALS

Petrochemicals are chemicals derived from hydrocarbon molecules and account for approximately 6% of total global crude oil use.

Unlike most finished petroleum products of crude oil, petchems are not used as fuels directly. They are instead used to enhance certain properties of fuels, or as a component in the manufacture of plastics, synthetic fibers, agricultural fertilizers and hundreds of other niche industrial and consumer products.

Petrochemicals can theoretically be manufactured from any part of the oil barrel or natural gas stream. However, simpler hydrocarbon molecules with less carbon atoms are more commonly used as feedstocks as their chemical reactions can be more easily controlled. The end results are similar regardless of the feedstock used, although the ratio of outputs differs.

In the US and Middle East, the primary feedstocks used to manufacture petrochemicals are the NGLs ethane and propane. Heavy naphtha and light fuel oils/gasoils are more commonly used in Asia. Petrochemical feedstocks can come from either an oil refinery or a gas separation plant (Table 10-1).

Petrochemical feedstocks   Table 10-1

Oil refinery <u>or</u> Gas separation plant sourced feedstocks:
 	
- NGLs { - Ethane
   - Propane
   - Butanes
   - Light naphtha
 	
Oil refinery sourced feedstocks:
   - Heavy naphtha
   - Light fuel oils/Gasoils (diesel and heating oil)

Steam cracking is the initial process used to produce petrochemicals. Steam cracking involves using very high temperature steam to break apart the molecules of ethane, propane, light naphtha, heavy naphtha and light fuel oil/gasoil. Hydrogen (H) in the steam (H_2O) bonds with the cracked molecules to form new hydrocarbon molecules.

The main petrochemicals produced in steam cracking are ethylene and its co-product, propylene. Byproducts include butanes, pygas (pyrolysis gasoline - used to make benzene and as a gasoline blendstock), light fuel oil, hydrogen and methane. Ethylene, propylene, and benzene are the principal petrochemicals used in products such as consumer plastics.

Steam cracker outputs can be divided into two molecular structure groups: olefins and aromatics (Table 10-2). Olefins and aromatics have double bonds between some or all of the carbon atoms in their molecular structure. Double bonds in a molecule are unstable and it is this instability which makes olefins and aromatics useful as petrochemicals as they are relatively easy to break and rejoin in new ways.

Steam cracker outputs are also called monomers. The most commonly produced monomers are ethylene and propylene. The term monomer is used as monomers can be combined in a repeating fashion, usually in the presence of a catalyst, to form longer-chained derivatives called polymers, the most commonly produced of which are polyethylene (PE) and polypropylene (PP). The vast majority of consumer plastics and other petchems are made from these two polymers.

A homopoly, or homopolymer, is a polymer formed from only one type of monomer. Polyvinyl chloride (PVC), produced from the monomer ethylene, is an example of a homopoly. A copoly, or copolymer, is a polymer manufactured from two or more different monomers. Acrylonitrile butadiene styrene (ABS), a light rigid plastic, is an example of a copolymer made from the monomers butadiene, benzene, propylene as well as ammonia.

Table 10-2

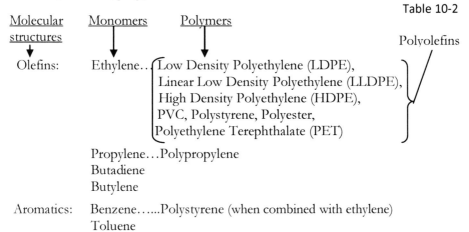

As was mentioned briefly in the chapter on the chemistry of oil, in an attempt to make nomenclature of chemicals simpler and standardized the International Union of Pure and Applied Chemistry (IUPAC) uses a naming convention which has removed the "yl" from many petrochemical names (Table 10-3). The oil industry continues to use the older chemical names, with the "yl".

Old name	IUPAC name	Table 10-3
Ethylene……………….Ethene		
Propylene……………...Propene		
Butylene……………….Butene		
Polyethylene…………...Polythene		
and so on…		

Olefins

Ethylene

Ethylene is the most produced petrochemical and is used to manufacture a variety of plastics and other chemicals.

Steam cracking is the principal method of olefin production. In steam cracking, the gas or liquid hydrocarbon feed is mixed with steam and then heated to a very high temperature for a brief time before being cooled rapidly to quickly stop any further chemical reactions. The products of steam cracking depend on the feedstock used, the feedstock to steam ratio, the cracking temperature and the residence time in the furnace.

Olefin feedstocks, in the descending order of the amount of ethylene they produce when cracked are: ethane, propane, light naphtha, heavy naphtha and light fuel oil/gasoil. Cracking each of these feedstocks, in addition to producing ethylene, also produces the co-product propylene and byproducts butanes, pyrolysis gasoline (pygas), light fuel oil, hydrogen and methane. Pygas is used to produce aromatics. Pyrolysis is any reaction which requires heat alone to occur.

The average size of ethylene plants has grown over the past ten years such that most new plants have a capacity of over 1 million mt per year. Annual global demand is just over 100 million mt per year.

Ethylene is a volatile and extremely flammable gas. Because it cannot easily be transported, it is usually used in a petrochemical manufacturing process in the immediate vicinity of where it is produced.

Propylene

Following ethylene, propylene is the second most commonly produced petrochemical.

Propylene is mainly produced as a co-product of steam cracking in the manufacture of ethylene. There are three additional sources of propylene: as a byproduct of oil refinery fluid catalytic cracking (FCC) of gasoil; propane dehydrogenation; and ethylene/butylene metathesis.

Propane dehydrogenation and ethylene/butylene metathesis are the two on-demand methods available for producing propylene – in other words, where propylene is the primary desired product, and not merely a byproduct.

In addition to its use in making petrochemical plastics, propylene can be sent through a refinery's alkylation process to produce a high octane, low sulfur gasoline blendstock. Propylene prices are therefore often highly correlated to gasoline price. The blendstock value provides a floor for propylene prices.

Butadiene/butylene

Butadiene and butylenes are produced as byproducts of steam cracking in the manufacture of ethylene.

Butadiene is also produced by solvent extraction of a butane feedstock. Butadiene is used to manufacture (acrylonitrile butadiene styrene) ABS resin and synthetic rubber.

Butylene, in addition to being a byproduct of steam cracking, can be produced by either solvent extraction of a butane feedstock or as a byproduct of an oil refinery's fluid catalytic cracking (FCC) unit. Butylene is used to produce MTBE, and can also be sent through a refinery's alkylation unit to produce high octane alkylate for gasoline blending.

Aromatics

The aromatics benzene, toluene and mixed xylenes, are referred to together as BTX. As with many petrochemicals, refineries and petrochemical manufacturers never really set out to make aromatics. Aromatics occur as byproducts of other processes.

Almost all aromatics are obtained from pygas and reformate, which are highly aromatic byproducts in the production of ethylene and the reforming of naphtha to produce gasoline blendstock, respectively. A third source of aromatics is as a byproduct of producing raffinate, another gasoline blendstock, from vacuum gasoil.

Aromatic molecules are removed from pygas, reformate and raffinate via either solvent extraction or a hydrocracking process. The removed aromatics are sent to a fractional distillation unit, where benzene, toluene and mixed xylenes are separated. The non-aromatics remaining are typically blended as a low octane component of gasoline.

Although most aromatics production is from byproduct sources, it is possible to produce additional benzene by turning toluene into benzene and xylenes in a process called hydro dealkylation (HDA). Such production is referred to as on demand benzene manufacture and is used to balance the market in periods of benzene shortage.

Both toluene, which is a primary component of fuel in Formula 1 cars, and mixed xylenes have high octanes and low RVP, which make them ideal gasoline blending components. Such gasoline blending provides a floor for toluene and xylenes prices. Petrochemical feedstock prices should be above this floor in order to pull the aromatics into those products.

Petrochemical applications   Table 10-4

Used to manufacture

Ethylene……...Polyethylene (HD, LD, LLD)
               ⌠ HDPE (High Density Polyethylene)
            →⎨ LDPE (Low Density Polyethylene)
              ⌞ LLDPE (Linear Low Density Polyethylene)
              Ethylene oxide
                 ↳(Ethylene Glycol for antifreeze, coolant and fibers)
              Ethylene dichloride (PVC - Polyvinyl Chloride)
              Ethyl benzene (for styrene)

Propylene…….Polypropylene (PP)
              Acrylonitrile
              Propylene oxide
              Acrylic acid
              Gasoline blendstock (alkylate)

Butadiene…….Styrene-Butadiene (SB) rubber
              Polybutadiene
              Styrene-Butadiene (SB) latex
              ABS resin

Butylene………Gasoline blendstock (alkylate)
              MTBE (gasoline octane enhancer/oxygenate)

Benzene……...Styrene (Polystyrene, ABS)
              Cumene/Phenol (Polycarbonate)
              Cyclohexane (NYLON)

Toluene………Benzene
              Xylenes
              Solvents and Urethane
              Gasoline octane enhancer

Xylenes………Paraxylene used for Polyester and PET
              Orthoxylenes (Plasticizer)
              Solvents
              Gasoline octane enhancer

Methanol……...Formaldehyde
              MTBE (gasoline oxygenate)

Plastics terminology

The term plastic is a generic term for any substance capable of being molded, extruded, or cast into various shapes such as films or fibers. Petrochemical-based plastics are often delivered from petrochemical producers in resin format. Plastic resins are most often in either solid bead or semi-solid powder-like form. Plastic processors referred to as converters buy the resin and form it into finished products such as soda bottles.

The main methods converters use to shape plastic polymers are: *injection molding*, used to make garbage cans and garden furniture; *die extrusion molding*, used for pipes, threads of fabric and insulation for electrical cables; *blow molding*, used for soda bottles; *calendaring*, used for making large sheets of plastic; and *compression molding*, to shape plastics in a heated mold cavity.

Plastics recycling

Plastics can be either thermoplastic or thermoset. Thermoplastics can be re-cycled relatively easily because they can be heated and formed, then heated again and reformed, which is called curing, without any change to their chemistry. Polyethylene, polypropylene, PVC and polystyrene are the most common thermoplastics.

Thermoset materials cannot be easily re-cycled as they undergo a chemical change when they are heated and cannot be re-heated or re-formed the same way. An example of a thermoset plastic is polyester.

In order to facilitate recycling of plastics, the Society of the Plastics Industry (SPI) created a system of recycling codes in 1988 (Fig. 10-1). The number corresponding to the type of plastic is contained within three arrows, often with the acronym for the plastic at the bottom. Categories one and two are the ones most commonly recycled.

Plastic resin identification coding system

Fig. 10-1

The seven plastics re-cycling categories defined by the SPI are:

1. PET or PETE (polyethylene terephthalate): for clear containers such as soda and water bottles.
2. HDPE (high-density polyethylene): stronger than PET, used in detergent bottles, milk cartons, fuel tanks, and toys.
3. V or PVC (Polyvinyl Chloride): high strength material used for shrink wrap, cold food wrapping, and blister packages.
4. LDPE (low-density polyethylene): strong, flexible and transparent, used for plastic grocery bags, cling film, squeezable bottles and dry cleaning wrapping.
5. PP (polypropylene): high melting point and waxy to the touch, resistant to grease and oil, used in straws, bottle tops, refrigerator containers, microwaveable containers, carpets, and garden furniture.
6. PS (Polystyrene): used in a foam as lightweight disposable cups and plates, meat packing, and shipping packing; or rigid as cutlery, plates and cups.
7. Other: combinations of 1 through 6 plastics or plastics not included in other categories.

CHAPTER ELEVEN
TRANSPORTING OIL

Oil can be transported by one of five methods: tanker ship, pipeline, tank truck, railcar tanker and aircraft tanker.

Tanker ship

There are two categories of commercial load-carrying vessels: dry cargo vessels and wet cargo vessels.

Most dry cargo vessels are container ships, that can carry standard 20 and 40 foot long containers, or dry bulk carriers which carry loose iron ore, grains, and coal. A bulk carrier, or bulker, is simply a vessel in which the material being carried is in one or more large holds and not in containers.

Wet cargo vessels, also known as wet bulk carriers, carry oil and fluid chemicals. There are two types of wet bulk carriers, clean product tankers and dirty tankers.

Clean product tankers, sometimes called white oil tankers, carry intermediate or finished products such as naphtha, gasoline, and middle distillates. Product tankers are typically smaller, have shallower water drafts, and shorter ranges than those which carry crude oil, known as dirty tankers. Water draft is the depth which the hull sits below the waterline. Clean tankers, unlike dirty tankers, usually have zinc or epoxy based paint coating the inside of the cargo tanks to prevent corrosion of the steel tank walls. Clean tanker owners try to carry the same clean oil all of the time, however, clean tankers can switch between various clean products if they have to. Clean tankers cannot easily switch between clean oils and dirty oils.

Dirty tankers carry crude oil, residual fuel oil, or bitumen, and are often are referred to as black oil tankers. Just over half of global daily crude production is transported by dirty tankers at various stages on its way from oil fields to refineries. Some black oil tankers which carry very heavy crude oil or bitumen will have mechanisms for heating the oil via steam passed through coils at the bottom of cargo tanks. The oil can then be pumped freely when it needs to be discharged.

Petroleum tankers have three types of tanks: cargo, ballast and slops, all of which are contained within a hull (Fig. 11-1). The bottom of the hull is usually flat.

A tanker will typically have several individual cargo tanks (Fig. 11-2 and 11-3). The individual tanks are structurally isolated such that a tanker can carry several different grades of cargo at the same time. A tanker can have up to three rows of cargo tanks running the length of the vessel. Large supertankers are generally the only vessels with a three across configuration.

Any space in a cargo tank not occupied by oil can be filled with a blanket of inert gas, such as nitrogen or carbon dioxide, to reduce oxygen content in the empty space and prevent a buildup of combustible vapors. Often, the inert gas is taken from the exhaust gas, called flue gas, of the ship's engine. The depth of empty space in a tanker not occupied by oil is referred to as outage, or ullage. Conversely, the depth of liquid in a tanker is known as innage.

Once cargo has been offloaded at a destination, the tanker cannot travel onwards completely empty. Ballast, usually sea water, lowers the vessel further into the water which stabilizes it and keeps it on an even keel in rough waters. Sea water ballast is not required if a much sought after, and difficult to find, return cargo is available - referred to as a backhaul cargo.

After discharging, many tankers have a system of high pressure water jets within the tanks which can be activated to remove any oil clinging to the internal tank walls. In crude oil tankers, these jets are called crude oil washers (COWs.) Slop tanks are used to collects the mixture of oil and water which results from washing cargo tanks. It is illegal to dump slops at sea. The slop tanks are periodically emptied to companies specializing in the safe disposal of this waste.

A tanker will normally be loaded with cargo by shore pumping equipment, but will use her own pumps to unload the cargo at discharge ports.

Petroleum tanker layout

Fig. 11-1

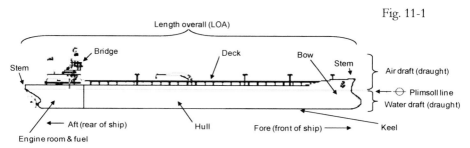

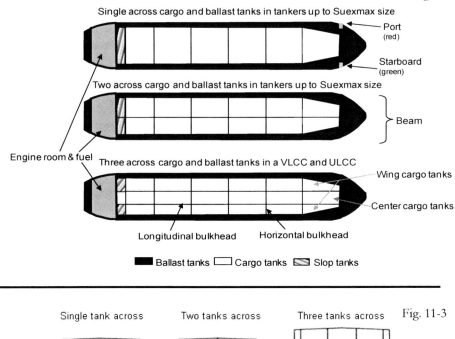

Fig. 11-2

Single across cargo and ballast tanks in tankers up to Suexmax size

Port (red)

Starboard (green)

Two across cargo and ballast tanks in tankers up to Suexmax size

Beam

Engine room & fuel

Three across cargo and ballast tanks in a VLCC and ULCC

Wing cargo tanks

Center cargo tanks

Longitudinal bulkhead   Horizontal bulkhead

■ Ballast tanks ☐ Cargo tanks ▨ Slop tanks

Single tank across   Two tanks across   Three tanks across   Fig. 11-3

Tanker vessel size

Ship size is described in terms of its displacement, volume, and dimensions.

Displacement is the weight of a vessel and any cargo, as indicated by the amount of water the vessel displaces. Displacement is measured in either long tons (2,240 pounds) or metric tonnes (1,000 kilograms or approximately 2205 pounds). Lightship displacement is the weight of the ship itself without any cargo, fuel, ballast, crew, passengers, water, or provisions. Deadweight tonnage (DWT) is the displacement of a fully loaded ship, minus the lightship displacement. Subtracting, fuel, water, ballast, crew and provisions from DWT gives a vessels deadweight cargo carrying capacity (DWCC). DWCC averages around 95% of DWT for an oil tanker. SDWT is DWT measured at summer conditions. DWT measurements are also made at winter and tropical conditions. A fully loaded vessel will sit in the water up to its Plimsoll mark (load line), which is painted on the side of the vessel and indicates the legal safe load limit.

There are several common sizes of ocean-going tanker defined by DWT:

Table 11-1

Coastal .... 3-10,000 DWT........................

Small (SR=short range) .... 10-20,000 DWT

Handy size (general purpose) tanker ..... 20-55,000 DWT.................

    Handymax .... 45,000 DWT

    Super Handymax .... 50,000 DWT

    Ultra Handymax .... 55,000 DWT

    (**MR** is a specific type of medium range Handy size with tanks coated to carry clean products)

Panamax.........................60-80,000 DWT (beam <32.3meters)

    Largest tanker with a full cargo capable of passing through the Panama Canal. **LR1** is a specific type of long range Panamax with tanks coated to carry clean products

Product (clean) tankers

Aframax.................................80-120,000 DWT

    AFRA (Average Freight Rate Assessment) is a price assessment published by the London Tanker Brokers Association since 1954. Although Aframax is now a general size term, Aframax tankers used to be the largest tankers included in the assessment. **LR2** is a specific type of long range Aframax with tanks coated to carry clean products.

Suezmax.................................120-180,000 DWT

    Largest tanker with a full cargo which can pass through the Suez Canal. The Suez Canal has no locks and so beam is not as important as in the Panama Canal which has locks.

Crude (dirty) tankers

Supertankers:

    VLCC (Very large crude carrier) tanker   200-320,000 DWT

    ULCC (Ultra large crude carrier) tanker   320-560,000 DWT

Notes on naming conventions for vessels: The approximate number of barrels can be calculated using a rough conversion of 7.5 barrels per DWT. In practice, different oils have different conversions depending on the density of the oil. In the oil trading world, the shorthand term bargeload is used to refer to 50,000 barrels, and a boatload refers to 500,000 barrels of oil. Capesize is a term used for vessels too large for either the Suez or Panama canals. Supertankers are capesize.

Traders sometimes refer to a VLCC as a V and clean long-range 1 or long-range 2 tankers as LRs. Vs and LRs are most often used to move oil on the high seas east of Suez. Vs and MRs are more commonly used west of Suez.

Volumetric measurements used in shipping refer to the term tonnage, although, a ton in this usage is somewhat confusingly a volume and not a weight reference. Volumetric tonnage is defined by formulae in the United Nations (UN) International Maritime Organization (IMO) International Convention on Tonnage Measurement of Ships, 1969 (London-Rules). Gross tonnage (GT) is the volume available in all enclosed space in vessel. Net tonnage (NT) is the space within a vessel available for cargo. Suez Canal Net Tonnage (SCNT) and Panama Canal/Universal Measurement System Net Ton (PC/UMS NT) are modified net tonnage volume measure for each canal respectively. A PC/UMS ton is equivalent to 100 cubic feet.

Although volume measurements will give some indication of vessel capacity, most oil traders refer to the displacement measurement of DWT rather than volume tonnage, when discussing vessel size.

The outline dimensions of a vessel are also important, particularly when a vessel needs to pass through a canal, or dock at a shallow port. Vessel length, as measured by length overall (LOA), width, called the beam, and height, referred to as the draft, are usually noted on a tanker description.

VLCCs and ULCCs supertankers are usually too large to pass through either the Suez or Panama canals when carrying a full load and must travel around the Cape of Good Hope in South Africa or Cape Horn in South America. On the return voyage, many VLCCs and ULCCs can pass through the Suez Canal (not the Panama Canal) as they are not carrying any oil and are sitting high in the water.

The largest supertankers, because of their massive size and deep draft, often cannot approach port at many delivery locations. Offshore discharging terminals must be used. In the US, for example, VLCCs and ULCCs can only discharge crude oil at the 1 million barrel per day Louisiana Offshore Oil Port (LOOP) near Clovelly, Louisiana. The LOOP facility, which consists of transportation pipelines and storage caverns, opened in 1981. Large vessels dock and discharge at buoys 18 miles off the coast of Louisiana. VLCCs can discharge at Long Beach in California, but that port cannot handle ULCCs.

Non-ocean-going tanker vessels
Non-ocean going vessels are small, broad, flat-bottomed tankers used to transport oil between ocean-going vessels and the shore. These vessels are known as lighter vessels or barges – both are usually physically similar, the distinction being in their use.

Lightering vessels offload oil from large vessels and then deliver the oil to locations onshore. Lightering is quite common as many large tankers are not able to dock at a port because of their size, and also the oil they carry may have several delivery locations onshore. Lightering vessels may also be involved in transshipment, which is the discharging of oil from one tanker for delivery to and loading onto another tanker. Lightering occurs when both vessels are parallel and underway at up to 6 knots. The vessels pull alongside one another with large inflated tire covered lightering fenders placed between the vessels. Oil is pumped from the larger vessel to the lightering vessel via rubber hoses.

Barges are used to move oil between onshore storage terminals. Barges may also be used to offload oil from large vessels. Oil industry journals, such as Platts and Petroleum Argus, list both barges and cargoes prices for the same grade and location of oil. The barges price is always a little more expensive. It includes the cost of lightering the oil from a large vessel, say a 500,000 barrel tanker, onto a smaller vessel such as a 50,000 barrel barge.

Bunkering

Re-fuelling a ship is called bunkering. Small re-fuelling barges pull up alongside a tanker to load fuel onto the vessel. Although almost all ports offer bunkering facilities, there are some benchmark refueling ports, such as Rotterdam, Los Angeles, and Singapore which are commonly referred to in pricing contracts for bunkering fuels.

Tanker charterers will try to manage their bunkering in order to only load at a port with less expensive bunker fuel. This fundamental bunkering arbitrage tends to tighten somewhat the spread between global bunker fuel prices.

The majority of ocean-going vessels burn two types of fue. Clean marine diesel fuel is burned close to shore in auxiliary engines. More polluting residual fuel oil is burned in primary engines out in international waters during ocean crossings. Both fuels are burned in diesel-type compression ignition engines. The most commonly used grade of marine diesel is called Diesel Marine C (DMC) or Blended Marine Diesel Oil (BMDO.) The most commonly used grade of residual fuel oil is 380cSt bunker fuel.

Chartering, or fixing, a vessel

While many large companies involved in the oil industry own their own vessels, of the global tanker fleet a large percentage are owned by individuals and individual corporations with no physical crude oil production or oil refinery system. Greek and Italian nationals in particular are very prominent as vessel owners in the international tanker industry. In addition to having a rich maritime tradition, they purchased a large amount of surplus US merchant vessels as a result of the US Merchant Ship Sales Act of 1946, which decommissioned the Allies World War II sealift capacity.

Individual owners usually use brokers to find cargoes to move in their tankers. The process of hiring a vessel is called fixing a charter.

Within trade houses, refineries and other oil trading organizations, the individuals responsible for ensuring that all freight is managed and that the process of moving oil goes smoothly are called operators. A group of operators are called the traffic department.

Firstly, the potential charterer, such as a refiner or oil trader, ensures that there are product or crude avails, in other words that there is material available for lifting at a port or terminal. If there is little material available at that location then it is said that there are tight avails, the opposite being plenty of avails.

When securing avails, there is usually an operational tolerance (op tol), sometimes referred to as volume variance. This is permitted in the trading contract because the volume occupied by oil varies with ambient temperature. This also takes into account the fact that volatile oil hydrocarbons may evaporate during transportation. The contract will state that the volume or weight is to be within 1% to 5% of the contracted amount.

The minimum loading volume which terminal operators require per vessel is known as a stem size. A buyer may have formed a stem by cobbling together several partial contracts.

If there are sufficient avails to be purchased as a stem, the potential charterer contacts a broker or a ship owner directly and asks if there is any tonnage available. A tanker waiting to be chartered is referred to as being laid-up, or in lay-up. The potential charterer will have in mind the product or crude to be carried, the size of vessel required, the route, and possibly the actual ports of loading and discharging at either end of the route.

A port is a general area where a tanker waits to discharge, and a berth/anchorage where a tanker actually discharges. The berth can be an offshore sea berth, or at a specially designed terminal/jetty.

Tankers don't necessarily have to traverse a route when hired. Often when freight rates are low, and freight price slumps can last for many years, tankers can be pressed into service as stationary floating storage tanks. Using a tanker as floating storage is a relatively expensive undertaking and therefore is usually only used when storage costs onshore are unusually high or onshore tanks are full, referred to as storage being at tank tops.

The charterer will have an idea of the time period, called the lifting window, in which they will need to have the vessel at a port ready for loading cargo. In addition the charterer will know the time period in which the vessel should be discharging its cargo at the destination port. The lifting window and the discharge window are together known as laydays. Laycan (laydays commencement and canceling) is a clause in a charter agreement which defines when laydays begin and end. If a vessel arrives after the end of the laydays, the charterer often has the option to cancel the contract.

Once a tanker reaches a port during the assigned laydays, the crew will alert the port authority with a notice of readiness (NOR.) The actual lifting and discharge of oil will take place in a defined number of running hours known as laytime within the laydays. If the laytime is exceeded then there are additional costs termed demurrage.

Finally, the charterer and owner will agree a price for the transaction. The vessel owner will make a firm offer, allowing the potential charterer a short period of time to consider it. The charterer and owner may then agree to place a vessel on subjects, which is a provisional booking of the vessel, subject to minor details being agreed, with no firm commitment by either party. Depending on the ports which the vessel is expected to load and discharge at, the charterer will have to ensure that the vessel is approved, such as having double hulls or being under a certain age, for calling on those ports. Requesting approval from a port authority is known as nominating a vessel, and placing a vessel on subjects while awaiting approval is known as subject receivers.

Once the charter, or fixture, is firmed up and agreed formally by both parties the contract which describes the agreed terms of the charter, or fixing, is called a charter party.

The charterer and owner may agree on one of three types of charter: a spot charter/voyage charter; a contract of affreightment (COA) charter; or a period charter (Table 11-2).

Spot charters, also known as single voyage charters, are for a specific vessel in the immediate future to move cargo between a specified loading and discharge port. A spot charter is for a one off journey. Once the journey has been completed, each party can go their own way. The agreed spot charter rate covers all expenses in the voyage such as bunkering, port charges, crew costs, repairs, insurance, and canal tolls.

The charterer must load and discharge within the defined laytime and has to pay the costs of loading and discharging the cargo.

Spot charter prices are agreed as a lump sum in US$ per metric tonne with reference to annually revised Worldscale nominal prices.

Contract of Affreightment (COA) charters are agreements to transport a specified quantity of product or crude at a specified rate per metric tonne between specified loading and discharging ports. The primary difference between a COA and a spot charter is that no particular vessel is specified under a COA. Because of this the price of a COA is usually slightly less than either spot or period charters.

Period charters: Unlike spot charters and COAs, which refer to worldscale for pricing, period charters are charged out in US$ per day, called a dayrate, for a vessel. Sometimes spot charters are compared to period charters as Time Charter Equivalents (TCE) which converts the spot charter lump sum cost into a dayrate. There are three types of period charters:

1. Consecutive voyage charter: A consecutive voyage charter is two or three consecutive spot charters with the same vessel.
2. Time charter: this is a charter of a specific vessel for up to ten years with the vessel owner providing and paying for the crew, provisions, insurance and repairs. The charterer is responsible for bunker fuel, canal tolls, port charges and crew overtime. Time charter fees are paid in advance to the owner at regular pre-agreed intervals. If the vessel breaks down then the charterer does not have to pay for down time.
3. Bareboat (Demise) charter: A bareboat charter is basically a time charter except the charterer assumes the full costs of a captain, crew and maintenance and effectively assumes full responsibility as owner for a period usually more than ten years. The vessel remains the property of the owner. Fees are paid at regular intervals to the vessel owner.

Charter pricing summary

Table 11-2

1. Spot charter 2. COA	3. Period charter
↓	↓
One-off lump sum fee negotiated with reference to Worldscale with some minor adjustments.	Fee agreed as a dayrate for a certain number of days. Fees paid at regular intervals.

Worldscale

During World War II, the US and UK governments requisitioned large portions of their private merchant marines for military transport. The allied governments would occasionally temporarily lend vessels back to private industry. However, assessing a free market price for each hire was quite difficult because the governments, as relatively new owners of the vessels, didn't have any sense of what was cheap or expensive. The governments came up with a table of standardized rates which took into account the typical costs to move a metric tonne of cargo over each route. The costs in US$ per metric tonne (mt) of cargo were indexed to 100 to be referenced in the following calendar year. The governments could then use this table as a guideline in negotiating spot vessel charters.

After the War, the shipping industry recognized the value of negotiating freight deals with reference to such standardized cost-based rates and several private organizations prepared tables which were used. By 1969, Worldscale became the predominant industry standard.

The not-for-profit Worldscale Association (London) Limited and Worldscale Association (NYC) Inc. jointly publish a book available on a subscription basis for a fee at the end of each year, listing nominal voyage costs for routes in US$/mt of cargo. The annually published Worldscale prices are used in the following calendar year, such that 2012 Worldscale prices are published at the end of 2011. The Worldscale book is revised annually and the nominal, or standardized (WS100), prices are reset to take account of changes in bunkering prices, canal tolls and port charges since the previous year.

These published US$/mt flat rates are the benchmarks against which the spot market charter parties negotiate. Worldscale flat is also referred to as WS100.

The worldscale nominal cost assessment (WS100) is based on how much it would cost a hypothetical vessel with 75,000 DWT cargo capacity, performing a round-trip voyage, loading and discharging cargo, back to loading port via the shortest route at 14.5 knots on 55 tonnes of bunker fuel oil per day during the voyage and 5 tonnes of fuel per day at port, including canal fees and allowing for four days at port and a standardized return on capital.

In addition to the Worldscale Annual containing rates for voyages between individual ports, the book also contains rates for voyages between regions. This allows parties negotiating a charter party to agree to a range of ports at either end of a route if the tanker charterer wants the option to load from and deliver to a port where he/she can obtain a higher profit. The regions included in the Worldscale book include the US Gulf Coast (USGC), US Atlantic Coast

(USAC), Caribbean, Baltic, West Africa (WAF), UK/Continent (UK/Cont), Mediterranean, Black Sea, Arab Gulf (AG), South East Asia, and Korea/Japan. Each region will have a port defined in the Worldscale book as a so called connecting point.

When a charterer finally decides on which port to use in each region, the worldscale rate (in $/mt) between the chosen port and the connecting point port is added to the inter-regional worldscale rate (in $/mt). The total Worldscale rate (in $/mt) is then multiplied by the Worldscale points which the parties agreed to in the charter party.

Assuming WS100 for hypothetical route Z from the Worldscale Annual is US$10/mt and the charter parties agreed to charter a vessel on that route at WS150, 150 Worldscale points, then the price of the charter is US$10 x 1.50 = US$15/mt. This $15/mt is multiplied by the vessel DWT to determine the total charter cost.

If the charter parties agreed on WS80 then the price for the vessel would be US$10 x 0.80 = US$8/mt. If the charter parties agreed to WS100 then they would have agreed to US$10/mt x 1.00 = US$10/mt.

The worldscale points between routes change daily with market supply and demand. They are only locked down when a charter party is agreed. There are however, fixed costs in each route which do not change with the market each day, such as canal fees, which are also included in the worldscale book, but which are usually fixed for each vessel size throughout a year. These fixed costs must be added to the total cost of the voyage, in addition to the previously mentioned worldscale points x worldscale flat rate.

Although worldscale is the shipping industry standard reference for negotiating freight rates for spot charters and COAs, the worldscale reference is usually not used for period charters, which simply use lump-sum dayrates.

Role of the Baltic Exchange
The Baltic Exchange in London is a market for matching physical ships with cargoes, in addition to the buying and selling of actual vessels. The Exchange is not a party to the agreed freight transactions and is more of an information and reporting exchange than a traditional financial exchange. The Baltic Exchange, founded in 1744, is so called because it originally matched ships with cargoes of tallow from the Baltic region.

The Exchange publishes settlements for a number of key benchmark routes as part of the Baltic International Tanker Rate Assessment (BITRA) at 4pm

London time each day. These BITRA settlements indicate where the spot freight market is trading against the worldscale index. For example, BITRA may show that hypothetical route Z today may have settled at WS125, and yesterday may have been WS120.

Independent of the Baltic Exchange, the McGraw-Hill publications 'Platts Clean Tankerwire' and 'Platts Dirty Tankerwire' perform a similar price discovery role with daily settlements for key routes.

Along with being a general reference to indicate where the market for each route is trading, the BITRA and Platts numbers are used as references by shipping companies, trade houses and others for hedging or speculation on tanker freight markets. Hedging or speculation on freight rates involves financially settled contracts called freight derivatives or freight forward agreements (FFAs). Certain routes are actively traded in the FFA market. Commonly traded FFAs, traded directly between counterparties with no exchange involved, are:

Table 11-3

FFA clean tanker routes:
1. BITRA Continent/USAC 37kt (Route: **TC2**): Rotterdam (Netherlands) -New York
2. Platts Singapore-Jap 30kt (Route: **TC4**): Singapore-Chiba (Japan)
3. Platts AG-Jap 55kt (Route: **TC5**): Ras Tanura (Saudi Arabia) -Yokohama (Japan)

FFA dirty tanker routes:
1. BITRA AG-Jap 260kt (Route: **TD3**): Ras Tanura (Saudi Arabia) - Chiba (Japan)
2. BITRA WAF-USAC 130kt (Route: **TD5**): Off Shore Bonny (Nigeria) - LOOP (US)
3. BITRA N.Sea-Continent 80kt (Route:**TD7**): Sullom Voe (UK)–Wilhelmshaven (Germany)

(where: TC = clean tanker; TD = dirty tanker)

Typically, American and European ship-owners and charterers reference BITRA settlements in FFAs, while Asian counterparties prefer to reference Platts Tankerwire settlements.

The Baltic Exchange and Platts do not get involved in the forward markets themselves; they merely report settlement prices for spot market activity.

Incoterms
In international trade it is important that commonly used terms are standardized so that all parties have a clear understanding of their obligation when entering contracts. Standard trade terms used in shipping are contained in a book called Incoterms published by the International Chamber of Commerce. Some of the most commonly referred to Incoterms in tanker freight are five methods of delivery pricing: FOB, CFR, CIF, landed cost and DDP.

FOB (Free on Board) means the seller is no longer responsible for the oil once it passes the ship's rail in the port of loading and the buyer then assumes all risks and costs of ownership from that point, including clearing oil for export.

CFR (Cost and Freight, or C+F) indicates that the seller pays for all costs, such as clearing oil for export and freight to bring the oil to delivery port. However, the buyer is responsible for insuring the cargo while in transit. Once the cargo passes over the ship's rail at the discharge port, any further costs are the responsibility of the buyer.

CIF (Cost, Insurance and Freight) is used when the seller pays for all costs, such as clearing oil for export and freight to bring the oil to delivery port, and is also responsible for obtaining marine insurance on the cargo while in transit. Once the cargo passes over the ship's rail at the discharge port it is the responsibility of the buyer.

Landed cost is the same as CIF plus the seller pays the import duties at port of delivery.

DDP (Delivered Duty Paid) requires the seller to deliver the oil to the buyers specified location with all insurance, freight costs and duties paid. This is one of the most convenient methods of delivery if one is a buyer. The DDP price includes all costs.

Single hulls and double hulls
Many nations and individual ports have enacted legislation requiring tankers to have double-hulls, which are basically two layers of steel on a hull, such that if the external hull is breached, oil will be contained by the internal hull. A variation of the double hull structure is a single hull with double bottom.

In response to the grounding of the Exxon Valdez in Prince William Sound in 1989, which resulted in the largest oil tanker spill in US history of approximately 260,000 barrels of crude oil, the US congress enacted legislation to ban any tanker vessel of 5,000 DWT or larger without a double hull from US waters from 2010. Single hull vessels will be allowed to call on a few US deepwater ports and lightering areas until 2015. The legislation is contained within the Oil Pollution Act of 1990, which is referred to in the shipping industry as OPA 90.

The International Maritime Organization (IMO) is the agency of the United Nations which issues regulations governing international shipping. The IMO's MARPOL (marine pollution) convention in particular affects oil tankers. The full convention name is the International Convention for the Prevention of Pollution from Ships.

Following the US OPA 90 legislation, the MARPOL convention Part 13G was put in place internationally by the IMO which requires the phasing out of single hulls globally by certain dates. The single hull tanker phase-out applies only to tankers with a cargo capacity of 5,000 deadweight tons (DWT) and above. Of the current global fleet of 8,771 chemical and oil tankers, 5,469 tankers (62%) are 5,000 DWT and above. The final phase out for single hull tankers for these large vessels is Dec 31, 2010.

Largest oil tanker spills since 1967

Table 11-4

Vessel	Date of Spill	Location	Oil Spilt (mt)	Hull type
Atlantic Empress	1979	Off Tobago, West Indies	287,000	single
ABT Summer	1991	700 nautical miles off Angola	260,000	single
Castillo de Bellver	1983	Off Saldanha Bay, South Africa	252,000	single
Amoco Cadiz	1978	Off Brittany, France	223,000	single
Haven	1991	Genoa, Italy	144,000	single
Odyssey	1988	Off Nova Scotia, Canada	132,000	single
Torrey Canyon	1967	Scilly Isles, UK	119,000	single
Sea Star	1972	Gulf of Oman	115,000	single
Irenes Serenade	1980	Navarino Bay, Greece	100,000	single
Urquiola	1976	La Coruna, Spain	100000	single
Hawaiian Patriot	1977	300 nautical miles off Honolulu	95000	single
Independenta	1979	Bosphorus, Turkey	95000	single
Jakob Maersk	1975	Oporto, Portugal	88000	single
Braer	1993	Shetland Islands, UK	85000	single
Khark 5	1989	Off Atlantic coast of Morocco	80000	single
Aegean Sea	1992	La Coruna, Spain	74000	double (OBO)
Sea Empress	1996	Milford Haven, UK	72000	single
Katina P.	1992	Off Maputo, Mozambique	72000	single
Prestige	2002	Off the Spanish coast	63000	single
Exxon Valdez	1989	Prince William Sound, Alaska	37000	single
Erika	1999	Bay of Biscay France	22000	single

Source: European Maritime Safety Agency

An interesting point to note is that clean oil tanker spills are often not seen as being as dangerous as black oil spills for birds and other shore dwelling animals. This is because clean oils often evaporate into the atmosphere relatively quickly.

The Jones Act
The private merchant marine is often referred to as the fourth arm of defense in the US as it is called into national service during military sealifts of troops and equipment at times of war. The increase in the use of flags of convenience, particularly ships registering in Panama and Liberia, due to less restrictions and lower costs compared with registering under the US flag has in the past led to worries of possible shortages of sealift equipment during times of war.

Following World War I, the US government decided it needed to encourage more vessels to register under the US flag and be manned by US citizens. The Merchant Marine Act of 1920, more commonly known as the Jones Act, requires that any vessel moving cargo between two US ports must be registered in the US. Foreign registered vessels can load and discharge at US ports, but they cannot move cargo between US ports. The need for a large US-flagged merchant marine was proven during WWII when a very large portion of the civilian shipping industry became an essential component of supply lines for the Allies.

The Jones Act flag requirement was temporarily waived for almost three weeks during hurricane Katrina in 2005.

Tanker ship construction, lifespan, and scrapping
Shipbuilding is a labor intensive industry. Over the past hundred years, shipyards have followed inexpensive skilled labor from Belfast in Northern Ireland to Japan, then South Korea and now, most recently, mainland China. Most tankers can last up to 40 years in service. Every year a percentage of old tankers are decommissioned and then scrapped. Scrapping occurs because of mandatory phase outs required by governments and increasing maintenance and repair costs as vessels become more worn out due to years in service.

A tanker contains a vast amount of steel, and the process of disassembling such a large vessel is profitable if one has inexpensive labor. Ship breaking primarily takes place on beaches in India, Bangladesh, Pakistan and China. The process of scrapping begins by running the vessel onto a beach at high tide. When the tide goes out, laborers go to work disassembling the vessel on the beach.

This continual scrapping creates demand for new vessels, which are referred to as newbuilds.

Pipeline

All oil passes through a pipeline at some stage on its journey to a consumer. Pipelines are more efficient, reliable and safe than the land-based alternatives, tanker truck and rail tanker car.

Oil pipelines falls into two basic categories, black oil pipelines and clean product pipelines. Black oil pipelines transport crude oil to refineries. Clean, or white oil, product pipelines transport refined products such as gasoline, diesel, and jet fuel from refineries to distribution centers near end users.

Long-haul lines are also known as mainlines, or trunk lines. Mainlines are large diameter pipes with fewer delivery points than short-haul pipes. Mainlines usually operate in so called fungible mode, such that the shipper will receive the same quality of oil it placed in the pipeline for shipment, but not necessarily the same molecules they put in.

Short-haul lines are also known as spur lines, stub lines or delivery lines. Short-haul lines have smaller diameter pipes with more delivery points than long-haul pipes. Short-haul product pipelines usually operate on batch mode, not fungible mode, in that the shipper will receive the same molecules it tendered for shipment.

Locations where pipelines meet are called hubs. If a hub is at a port, it is known as a marine terminal. If a hub is servicing a number of crude oil wells then it is referred to as a gathering station. There is generally a large storage capacity at pipeline hubs, marine terminals and gathering stations.

As oil frequently changes hands at hubs and marine terminals, they are locations where price discovery occurs in the oil market.

A major hub for crude oil pipelines in the US is the Cushing Hub in Oklahoma, just north of the US Gulf Coast, which is where the NYMEX WTI crude oil futures contract is physically deliverable. The primary hubs and marine terminals for oil price discovery in the US are:

Table 11-5

- New York Harbor	- Group Three (Tulsa, Oklahoma)
- Gulf Coast (Texas-Louisiana coast)	- Chicago
- Cushing, Oklahoma	- Los Angeles

Internationally, the following locations are important price discovery pipeline hubs and/or marine terminals for oil:

Table 11-6

> - ARA (Amsterdam, Rotterdam, and Antwerp) Netherlands/Belgium
> - Genoa/Lavera on the Italian/French Mediterranean coast
> - Singapore
> - Arab Gulf, especially Ras Tanura in Saudi Arabia and Dubai in the UAE

Oil is propelled through pipelines using electrically-powered centrifugal pumps located every 20 to 100 miles, depending on the terrain. Most pipelines move oil at between 3 and 8 miles per hour. Chemicals known as drag reducing agents (DRAs) are sometimes added to oil in order to speed its flow through a pipeline. Pipelines operate 24 hours a day and are monitored constantly. Often, when pipelines pass through urban areas they are buried.

Pipelines are usually built of steel. Apart from damage due to someone accidentally digging into lines, the major cause of wear is corrosion. Cathodic protection is a process where pipelines in use are subjected to a low voltage electric charge in order to counteract corrosion.

Pipelines can become clogged or constricted by wax hydrocarbon molecules, asphaltene hydrocarbon molecules, gas hydrates, diamondoids (a type of diamond occurring in hydrocarbon deposits) and organic scale produced by bacteria.

Devices called pigs are periodically sent through pipelines to clean the lines. Smart pigs, which have sensors, can detect areas which are damaged and may need to be repaired. Pigs can also be used to segregate batches of oil. Pigs are propelled through the pipe by the fluids in the pipe.

The US pipeline market
Many pipelines in the US are independently owned, in that they are not owned by the companies moving the oil through them. Pipelines crossing US state lines are defined as common carriers, regulated by the Federal Energy Regulatory Commission (FERC). Common carriers must publish publicly and apply the same tariffs to all those wishing to transport oil on the pipeline.

Clean product pipelines in the US tend to move products from the large number of US Gulf Coast refineries to the large consuming areas along the Northeast via pipelines such as the Colonial products pipeline from Houston, Texas to Linden, New Jersey.

An unusual feature of the US Northeast New England market is that there are relatively few pipelines within this region because of its large populations living close to the coast. Therefore, a large volume of products from New York Harbor have to be moved by barge along the coast of New England. Similarly, Florida has few pipelines, and barges deliver fuel to terminals around the coast.

Another feature of the US pipeline network is that there isn't a pipeline network crossing the Rocky Mountains, apart from relatively small exceptions, such as the Kinder Morgan Pipeline serving Phoenix and the Longhorn products pipeline. This leaves California and other western states relatively isolated from the large international trade in finished products and crude oil which occurs east of the Rockies. This increases the probability of occasional supply squeezes west of the Rockies.

Significant US clean product pipelines
Table 11-7

- Colonial Pipeline (Houston to New York)
- Plantation Pipeline (US Gulf Coast to Washington, D.C. Area)
- Explorer Pipeline (US Gulf Coast to Chicago)
- TEPPCO (US Gulf Coast to Chicago)
- Buckeye Pipe Line (New York Harbor to Northeast and Midwest)
- Williams Pipeline (Tulsa, Oklahoma to Minnesota and Wisconsin). The Tulsa location is an important product price discovery point called 'Group 3' - as prices were originally set by three railroad companies in Tulsa, Oklahoma.

Significant US crude oil pipelines
Table 11-8

- Seaway Pipeline (US Gulf Coast to Cushing, Oklahoma)
- Capline Pipeline (St. James, Louisiana to Patoka hub in southern Illinois)
- Trans Alaska Pipeline System (TAPS) (Alaskan North Slope (ANS) Prudhoe Bay field to the southern coast of Alaska where it is loaded onto tankers for shipping to Californian refineries)
- Enbridge Pipeline System (Alberta, Canada to Chicago and the Great Lakes)
- Spearhead Pipeline (US portion of Enbridge Pipeline system from Chicago to Cushing, Oklahoma)

US pipeline focus: The Colonial Pipeline (CPL)
One of the more important US pipelines for clean products is the 1,500-mile Colonial products pipeline which runs from the US Gulf Coast to Linden, New Jersey near New York Harbor (Fig. 11-4). Products move along the Colonial mainline at between three to five miles an hour such that it takes approximately eighteen to twenty days for products from the US Gulf Coast to reach Linden. The cost per gallon is currently approximately 2 to 3 cents to move a product along the length of the Colonial Pipeline.

Colonial Pipeline   Fig. 11-4

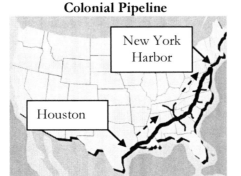

The Colonial Pipeline pumps the same sequence of products approximately every 5 days, which it calls a cycle. There are typically six 5-day cycles per month for a total of 72 cycles per year. The cycles throughout the year are named 1 through 72 (before January 2006, the CPL cycles were 10 days long, there were 36 cycles in a year and each cycle was split into two – called the 'front phase' or the 'front-half' and 'back phase' or the 'back-half'). One may hear a trader say that they will take delivery in the 5th cycle which indicates that he/she will take delivery of product in the 5th 5-day period of the year, January 21 through 25. Similarly if a trader is taking delivery on the 35th cycle he/she is taking delivery June 21st through 25th. The CPL usually schedules shipments up to maximum of 30 days in advance.

The process of requesting space on the pipeline is called nominating, and the request is handled by individuals called schedulers.

Products move along the Colonial Pipeline via either segregated batches or fungible batches.

Segregated batches use pigs or other simple mechanical devices to isolate specialty products from other products within the pipeline. The shipper has to pay an additional fee for this service and it is used when the shipper is moving a product which cannot tolerate even minor contamination by other products. Segregated batches are fairly rare compared to fungible batches.

Fungible batches allow each shipper to get the same specification product at the other end of the pipeline, but not necessarily the exact molecules they put in. Fungible batching allows for more efficient use of a pipeline. The CPL permits a number of defined fungible products. Fungible products are those meeting a set of CPL specifications which can be commingled with other quantities of petroleum product from different producers meeting the exact same one of the CPL specifications. This allows for a certain grade of gasoline, for example, from one refinery to be mixed with the same grade from a different shipper so long as both meet the CPL specifications for that grade.

Each of the fungible products moved on the CPL from the US Gulf Coast along the Eastern States to New Jersey are assigned a two character code which indicates which specification that product meets (Table 11-9). Many of these codes will be familiar to those involved in the US oil market.

Table 11-9

> **CPL Grade 54:** Aviation Kerosene:
> Referred to as: <u>Gulf Coast Jet 54</u>
> **CPL Grade 61:** Ultra Low Sulfur Diesel-#2D 40 Cetane - 15ppmwt sulfur
> Referred to as: <u>Gulf Coast Ultra Low Sulfur Diesel</u>
> **CPL Grade 74:** Low Sulfur Diesel-#2D 40 Cetane – 500ppmwt sulfur
> Referred to as: <u>Gulf Coast Low Sulfur Diesel</u>
> **CPL Grade 88:** Heating Oil - 2000 ppmwt sulfur
> Referred to as: <u>Gulf Coast Heating Oil</u>
> **CPL Grade M1:** Conventional Gasoline-87 Octane–Non-Oxygenated-7.8 psi RVP
> Referred to as: <u>Gulf Coast Unleaded 87 Gasoline</u>

Mainline batch sizes on the CPL vary from 75,000 to 3,200,000 barrels. The smallest mainline batch of 75,000 barrels may be made up of three 25,000-barrel fungible batches. Segregated batches on the CPL must be a minimum of 75,000 barrels. Minimum delivery quantities are 5,000 barrels delivered from mainline itself or 2,500 barrels delivered on stub lines off the mainline.

In order to further maximize the efficiency of batch processing in the pipeline, the CPL utilizes a process called sequencing. Sequencing allows for multiple products to be sent along the same pipeline without any physical barrier in between (Fig. 11-5). Pipeline operators monitor the density of the oil flowing through the pipeline to determine when the product changes to the next in the sequence. As a result of hydrodynamic properties of the products, very little of the oils actually mix. For example, for a 25,000-barrel batch of products, which occupies almost 50 miles of a 10-inch-diameter pipeline, the amount of mixing would be approximately 75 barrels. If a pipeline is shut down temporarily, say due to a power outage, the volumes of both compatible interface and transmix increase as hydraulic pressure decreases. The small amount of product that mixes is known as either compatible interface or transmix.

If the interfacing products are similar, such as kerosene and low sulfur diesel, the mixture is known as compatible interface. Due to their molecular similarity, the products can tolerate a very small amount of the adjacent product material and still meet end user product specifications for the lower grade product. The small amount of interface material is mixed, or downgraded, with the lower quality product.

Transmix is a mixture between two products that are not similar enough to be considered compatible, such as gasoline and low sulfur diesel. Transmix must be reprocessed by a refinery at the delivery end of the pipeline to separate the two products.

The sequencing of multiple products changes throughout the year based on seasonal demand.

Example of product pipeline sequencing   Fig. 11-5

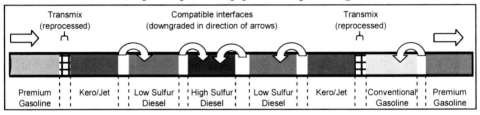

<u>Significant non-US pipelines</u>
Sumed Pipeline, with a capacity of about 2.5 million bpd, links the Ain Sukhna terminal on the Gulf of Suez with Sidi Kerir on the Mediterranean (Fig. 11-6). Sumed is primarily used to move Saudi and other Middle Eastern crudes to Europe across Egypt. A smaller amount of oil, 0.8 million bbl/d, moves through the Suez Canal in suezmax tankers.

Sumed Pipeline   Fig. 11-6

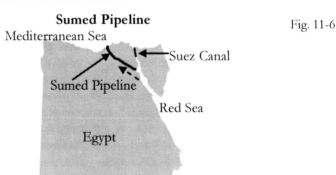

The East-West Crude Oil Pipeline is used to transport crude from the oil fields on the east of Saudi Arabia to the refineries in the west of the country and the Yanbu terminal on the Red Sea for export to European markets (Fig. 11-7). The pipeline, also referred to as the Petroline, has a 4.8-million-bpd capacity. Currently, only half is being used. Most Saudi oil is exported via tanker from the east coast of the country through Straits of Hormuz with the Petroline as a backup in the event that exports through the Straits are not possible.

East-West Pipeline

Fig. 11-7

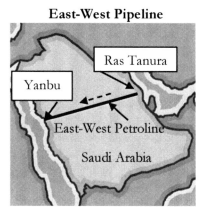

BTC (Baku-Tbilisi-Ceyhan) pipeline, with a capacity of 1 million bpd, brings crude oil from the Caspian Sea to the Mediterranean via Baku (Azerbaijan), Tbilisi (Georgia) and Ceyhan (Turkey.) The pipeline, began deliveries in 2006 (Fig. 11-8).

Fig. 11-8

Baku-Tbilisi-Ceyhan Pipeline

Source: EIA

The Druzhba (friendship) pipeline network brings crude oil from West Siberia, the Urals, and the Caspian Sea to Europe (Fig. 11-9). Druzhba is the longest pipeline network in the world at 2,500 miles. The capacity of the pipe is 1.4 million bpd. Heavy and light oils are mixed in the pipeline to produce Urals export blend.

Druzhba pipeline system

Fig. 11-9

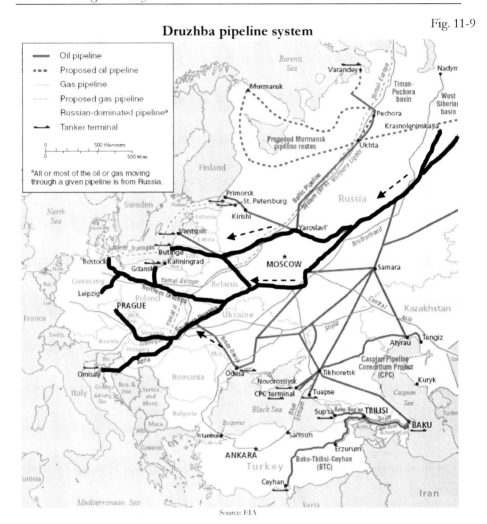

Source: EIA

The Enbridge pipeline network brings over 1.5 million barrels per day of crude oil mostly from the Western Canadian State of Alberta down into the United States. Much of the oil from Alberta is heavy crude produced from oil sands.

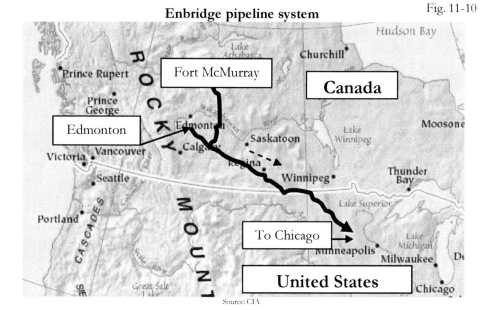

Enbridge pipeline system

Fig. 11-10

Source: CIA

Tank truck

Tank trucks are one of the most expensive methods of transporting oil. Because of the expense, they are usually only used for the last few miles from a hub or terminal to the end user, such as a retail filling station. The main advantage of a tanker truck is flexibility.

A tank truck, or dealer tank wagon (DTW) as it is sometimes called, is loaded with product, such as gasoline or heating oil at a distribution center called a truck rack loading facility. Tank trucks carry between 1,000 and 10,000 gallons of fuel. Tank trucks usually have several compartments to carry several types of fuel on the same journey.

While many large oil companies own their tank trucks, dealer tank wagons are very often owned by independent small companies or individuals unaffiliated with any oil company. These independent local distributors of oil are known as jobbers.

The oil price which the jobber pays at the rack loading distribution center is called the rack price. The price the gas station owner pays the tanker owner is called the dealer tank wagon price and is basically the rack price marked up to cover the jobber's cost.

Gasoline vaporizes quite easily. When the vapor reacts with sunlight, it is a major cause of smog. Many states require gasoline vapor recovery systems which are usually circular plastic vapor boots on the retail pump nozzle. Any gasoline that vaporizes while an automobile is being refueled is captured by the vapor recovery device and pumped back into the underground tank. The vapor is passed through a pipe within the pipe containing the gasoline, or via a second external pipe. Tanks truck operators load this vapor onto their trucks while they are filling the retail station tanks with gasoline. The tank truck owner brings the vapor back to the truck rack, where it is collected and sent for processing in a refinery. Since gasoline vapors contain primarily very light hydrocarbons, they cannot be blended straight back into the gasoline pool as it would raise the RVP of the gasoline above acceptable limits.

Retail stations typically carry one to three day supply of gasoline in their underground tanks. Though many retailers will offer three octane ratings of gasoline, they will only have two separate storage tanks with high octane gasoline and low octane gasoline. The high octane and low octane gasoline can be blended at the pump to produce the middle octane gasoline.

Specialized tank trucks

As propane, which is a product of both oil refining and separation from wet natural gas, is transported under pressure, a specialized truck called a bobtail truck is used (Fig. 11-11). Bobtail trucks are usually much smaller than gasoline or heating oil tank trucks.

A bobtail truck  Fig. 11-11

Source: EIA

Other specialized low profile tank trucks are used for refueling aircraft at airports.

Railcar tanker

Railcar tankers, or tank cars, are a little less expensive than truck tankers, but not as flexible. Railcar tankers are used to transport crude, finished products, and petrochemicals that may be potentially dangerous to ship on roads. Railcars are also used to transport lube oils, bitumen, and ethanol, which can be difficult to move via pipelines. Railcar tankers are particularly common in Russia.

In the past, tank cars used have cylindrical domes on their roof to allow for expansion of oil in transit. Modern tank cars do not have such domes and instead a small volume of the tank is left empty to allow for expansion.

Aircraft tanker
Civilian tankering usually occurs on short haul flights whereby airlines purchase more jet fuel than necessary at one airport and forgo purchasing fuel at other airports. The tankering of jet fuel is not a very efficient use of energy as one expends large amounts of energy lifting the additional fuel. Tankering typically only occurs when some of the airports an airline is flying to have unusually high jet fuel costs or have limited stocks of fuel available. Tankering may also occur if the opportunity cost spent refueling an aircraft every time it lands is higher than the cost of carrying excess fuel. In other words, an airline may be able to squeeze a few extra flights per day from an aircraft if it refuels less often. Tankering of jet fuel is sometimes temporarily restricted by airport authorities when jet fuel stocks at a particular airport are running low.

Dedicated aircraft oil tankers specifically designed to haul jet fuel are normally only used for military mid-air refueling.

CHAPTER TWELVE
STORAGE

The many reasons to store oil include seasonal demand for certain oil products; to hedge against unforeseen supply disruptions; a trader may be able to sell oil in the future at a price greater than the price of oil today plus the cost of storage; for national and strategic defense; and to provide supplies to those unable to afford oil during short price spikes.

Storage structures
In order of increasing expense, hydrocarbons are stored in underground spaces, above ground tanks, and tanker ships.

Underground spaces
Depleted reservoirs, aquifers and salt caverns are employed for underground storage.

Underground spaces are predominantly used to store natural gas. In order to create a buildup in underground pressure so that the natural gas can be removed, a certain amount of base gas, also called cushion gas, is required to remain constantly in an underground storage facility. Base gas is never removed while the storage facility is operating. The quantity of natural gas which is injected and withdrawn each year leading up to and during winter heating demand is known as working gas. It is this level which the market watches very closely.

Fig. 12-1

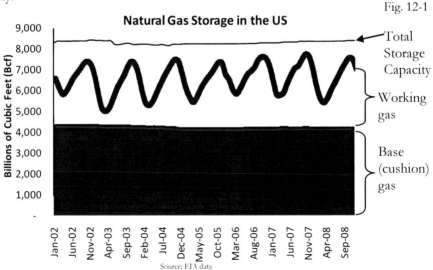

Source: EIA data

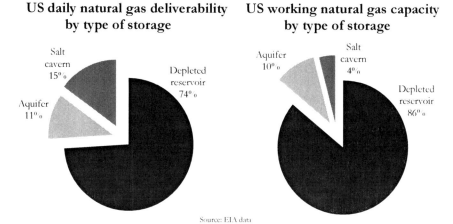

US daily natural gas deliverability by type of storage

US working natural gas capacity by type of storage

Source: EIA data

Depleted reservoirs are old natural gas reservoirs reused as storage facilities. These reservoirs are used for natural gas storage and cannot be used for liquids such as crude oil. At peak storage in a depleted reservoir, approximately 50% of the gas is base gas.

Aquifers are also used to store natural gas, but not for liquid petroleum. This storage method involves pumping natural gas underground to displace the water table in an aquifer. Aquifers are not as well sealed as depleted reservoirs. When methane leaks into the atmosphere, it is a more powerful greenhouse gas than carbon dioxide. Natural gas has to be dehydrated as it is removed from an aquifer, which adds to the storage cost. Because of these environmental and operational reasons, aquifer storage is only used in areas where a depleted reservoirs or salt caverns are unavailable. Base gas requirements can account for as much as 80% of the peak gas stored in an aquifer in order to keep pressures sufficiently high.

Salt caverns are used to store natural gas, propane and crude oil. Salt caverns are formed by hollowing out tall salt domes or flat salt beds by pumping water into the formation to dissolve the salt in a process known as salt cavern leaching. The brine is then pumped out to leave a hollow cavern. As salt is an impermeable rock, base natural gas requirements for salt caverns can be relatively low, at around 30% of the total peak gas stored in the cavern.

The geothermal gradient ensures that any crude oil stored in an underground cavern is slowly mixed and prevents heavy components of the oil from settling to the bottom. Finished oil products such as gasoline and diesel cannot be stored in salt caverns or any underground natural space as they have to meet a tight set of specifications which means they have to be stored in clean steel tanks.

The largest propane storage facility in the US is comprised of underground leached salt dome caverns located at Mont Belvieu, Texas. The facility has a capacity of over 70 million barrels. Mont Belvieu is located near the Houston Ship channel, where NGLs are produced in refineries and fractionated from wet natural gas. A second large storage area with a 28 million barrel capacity for NGLs is located in the Bushton/Conway/Hutchinson region of Kansas. Texas Eastern Transmission (TET) pipeline moves propane used for winter heating from South Texas where Mont Belvieu salt domes are located northward to New York and Massachusetts. The pipeline is now named Texas Eastern Products Pipeline, owned by TEPPCO Partners, L.P., although NGL traders continue to refer to it as TET.

Above ground tanks
Tanks can be used to hold crude oil and refined intermediate or finished products. Clusters of tanks, called tank farms, are located at crude oil production sites, refineries, major pipeline hubs, and marine terminals. A tank farm used to store crude oil from producing wells is known as a stock tank battery.

Above ground tanks are most often constructed from carbon steel painted white to reflect the suns heat and reduce evaporation. A tank's fill pipe is usually very close to the bottom of a tank to avoid a potentially dangerous static buildup from oil movement and also to reduce vapor from splashing.

Tanks are located underground at retail filling stations for safety reasons. The primary challenge with such a situation is that leaks from underground tanks are much more difficult to detect and repair.

Tank farms typically have secondary containment structures such as raised earthen dikes surrounding the compound and impounding basins into which leaks can flow. Another important safety concern during any major storm is that tanks need to contain a minimum amount of oil to withstand high winds and to prevent the tank from floating due to flooding, both of which may break connecting pipes and cause leaks.

Tanks are classified into three types depending on the pressure under which contents are stored: atmospheric tanks, low pressure tanks and pressure vessels.

Atmospheric tanks have fixed roofs and are used to store crude oil, bitumen, residual fuel oil and other non-volatile products at atmospheric pressure.

Atmospheric tanks with peaked, or cone, roofs often have one weld on the roof which is weaker than those on the sides of the tank. The weak roof seam is designed to break before any other welds in the tank, allowing emergency venting of dangerous vapor buildups into the atmosphere before the integrity of the tank sides are threatened.

Low pressure tanks often have floating roofs to prevent evaporation. Contact between oil and the floating roof also reduces growth of organisms in the oil as contact with oxygen and moisture in the atmosphere is reduced. Volatile products such as gasoline, diesel and jet fuel are stored in both atmospheric and low pressure tanks.

The floating roof is supported by pontoons around the edge which prevent it from sinking into the oil. The floating roof is connected to the sides of the tank by what are called shoes. A floating roof may be open to the elements or may be contained under a peaked/cone roof (Fig. 12-3). If the floating roof is under a cone roof then there will be emergency vents at the side of the tank just below the cone roof.

Fig. 12-3

Fixed roof   External floating roof   Internal floating roof

Gasometers are a particular type of low pressure tank used to store gases. A wet gasometer consists of one tank inverted above a second tank which contains water or oil. The inverted tank pushing down into the water or oil seals the gases at a constant and stable low pressure. The pressure comes from the weight of the inverted tank. Dry gasometers are single tanks which have a diaphragm which moves higher and lower with the amount of gas stored in the tank, similar to floating roof tanks. Gasometers used to be a common sight in inner cities when town gas was produced from coal. In addition to storing gas their main purpose was to keep pressure at a constant level within gas pipes. Electrical pumps are more often used these days to maintain pressure in gas pipes and gas is usually stored in depleted reservoirs, aquifers and salt caverns.

A pressure vessel is used to store materials such as butane or propane which may need to be stored at high pressure. Pressure vessels can be spheres, spheroids, or cylindrical horizontal pressure tanks called bullets. Pressure vessels are often located at a remote part of a tank farm and bullets, in particular, are pointed away from buildings and other storage tanks.

Specialized insulated tanks are used to store contents at very low temperatures. These tanks are known as low temperature tanks and cryogenic tanks. Low temperature tanks are used for storing gases which are easily cooled to a liquid at temperatures down to -60°F (-51°C). Cryogenic tanks are used for storing liquefied natural gas (LNG) at temperatures as low as -259°F (-162°C).

Types of above ground tanks   Fig. 12-4

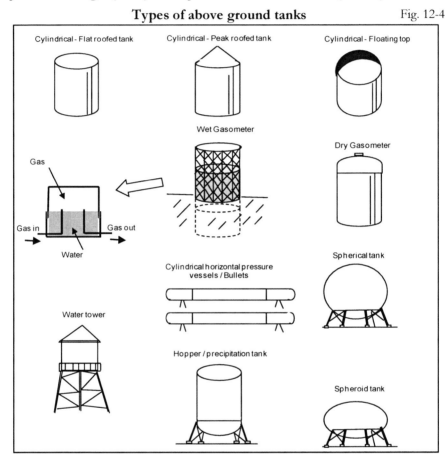

Tanker ships
Oil tankers can be pressed into service as temporary storage facilities. Storing oil in a tanker ship is usually the most expensive form of storage and is used when land based storage is full – a situation described by traders as being at tank tops.

When a converted oil tanker is permanently docked as storage, it is referred to as a Floating Storage and Offloading (FSO) vessel. FSO's are technically unable to resume operations as transport tankers. If a storage vessel is technically able to quickly resume operation in the future as a transport tanker it is known as a Floating Storage Unit (FSU).

Storage categories

The two general categories of storage are commercial and strategic.

Commercial storage

Commercial storage, also known as discretionary storage, can be broken into primary, secondary and tertiary.

<u>Primary storage</u> is oil stored within the oil industry as part of normal day-to-day operations. Primary storage includes oil stored by producers, refineries, natural gas processing plants, oil terminals, pipelines and stocks on tankers.

Unlike many commodities, oil is very expensive to store. Expenses include insurance, maintenance and security staff, the purchase cost of the actual tank, and the opportunity cost of money tied up in stored oil. The high cost of storage has been an incentive to reduce oil stored commercially. For this reason, over most of the past twenty years, oil companies have been attempting to minimize the amount of oil stored by moving toward just in time inventory management.

There is a base level of oil in pipelines, refinery tanks and so on which can never really be removed from storage unless the facility in which it is contained is closed permanently. Pipelines must contain oil over their entire length and this can never be removed unless the pipeline is shut down. This minimum level of stock is referred to as the minimum operating level.

This minimum operating level of oil in the commercial oil industry system is the reason why the weekly US DOE crude oil commercial stock inventory always tends to be above 260 million barrels (Fig. 12-5). The US oil industry is not idly sitting on 260 million barrels of crude oil. This is an amount of crude oil which cannot be removed from the system without causing operational difficulties.

Fig. 12-5

US Industry Crude Oil Storage

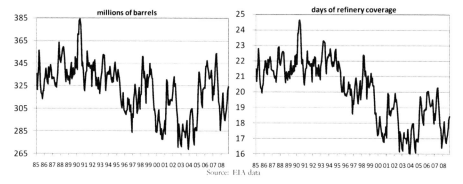

Source: EIA data

Although it is commonly cited and used as a reference (such as on the right-hand axis of the above chart), it is not completely accurate to say that if we have 320 million barrels of crude in commercial storage, and as US refineries process roughly 17 million barrels per day, we have 19 days of a crude oil supply cushion. Unless one shuts down the oil industry, to extract the minimum operating crude, there are in fact roughly three and a half days of useable crude oil supply [(320-260) / 17]. This shows how the oil market is truly a continuous flow process. Apart from the US government, there are very few individual organizations sitting on vast quantities of stored oil.

This should not alarm anyone, as to shut down the entire global oil supply system would be quite a challenge. Pipelines and storage facilities have been continuously bombed, sabotaged, vandalized, stolen from and hit by natural disasters such as hurricanes since the oil industry began. The security of the system is its diversity. The oil industry is dispersed enough to cope such that two or three days supply has proven to be quite sufficient.

Secondary storage is oil stored by professional or semi-professional oil consumers such as oil stored at power stations or in the tanks of distributors.

Tertiary storage is oil stored by end users, such as heating oil stored in individual homes, or storage by industrial consumers excluding power stations.

Strategic storage
Most governments around the world do not store oil themselves. Instead, most governments mandate that oil industry participants, such as crude producers and refineries, carry larger storage than they would if they had discretion.

The US and China are among the few countries in which the government itself stores large volumes of oil. The US has no mandated minimum storage level for private industry. The Chinese approach is known as a dual oil reserve strategy as both the government stores oil in addition to requiring large Chinese oil companies to hold minimum amounts of oil. The government of China began filling its 102 million barrel capacity crude oil SPR in 2006.

US Strategic Petroleum Reserve (SPR): The US federal government decided to store large quantities of crude oil beginning in 1977 following the Arab embargo of 1973-1974 and the resulting oil price shock. Crude oil in the SPR is contained in a number of salt caverns near the coast of Texas and Louisiana.

The US government is the largest single oil consumer in the world and the SPR is the largest single storage facility in the world. The SPR has the capacity to hold up to 727 million barrels of crude oil for an indefinite period of time. At the beginning of 2009, the SPR held approximately 700 million barrels of crude. The SPR can be emptied at a maximum rate of 4.4 million barrels per day.

The US Energy Policy Act of 2005 requires the SPR storage capacity to be raised to 1 billion barrels. In the January 2007 State of the Union address, President Bush announced the intention to double of the SPR capacity to 1.5 billion barrels by 2027.

Fig. 12-6

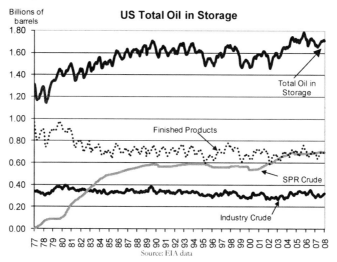

The terms under which crude can be released from the SPR are relatively vague such that any situation which the President or Congress defines as an emergency can be used.

When the US SPR was launched, government and industry amount of oil in storage rose steadily from 75 days of consumption to 97 days of total US consumption. However, since 1980 the more oil the government stored the less the US private oil industry stored relative to demand (Fig. 12-6 and 12-7).

Fig. 12-7

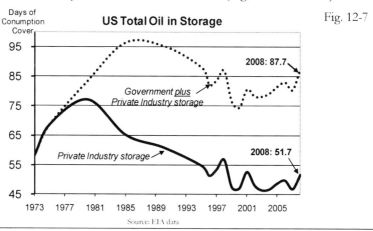

One reason for lower private storage was a more efficient oil industry seeking to keep a reign in expensive storage costs and new technology permitting just in time inventory management.

The SPR has been tapped by private industry several times since 1985 (Table 12-1). In addition to sales, exchanges have occurred whereby one grade of oil was swapped for another, or oil loaned to a company to be replenished later. The frequency of such loans and sales has increased in the past few years.

Table 12-1

SPR Crude Oil Sales and Exchanges	
Crude Oil Sales	
	Volume of Barrels
Dec 1985 Test Sale	1 million
Aug 1990 Test Sale	4 million
Jan 1991 Desert Storm Sale	17 million
Mar 1996 - Jan 1997 Federal Budget Reduction*	28.1 million
Non-Emergency Sale	
Sep-05 Hurricane Katrina Sale	11 million
Crude Oil Exchanges	
Apr 1996 Arco pipeline	0.9 million
Aug 1998 Maya quality	11 million
Jun 2000 Citgo/Conoco Channel Block	0.5 million
Jul 2000 Heating oil reserve swap	2.8 million
Sep 2000 Time swap	30 million
Oct 2002 Hurricane Lili loan	0.3 million
Sep 2004 Hurricane Ivan loan	5.4 million
Sep 2005 Hurricane Katrina loan	9.8 million

Source: DOE

Approximately two thirds of the SPR crude is sour. Sour crude is less expensive than sweet crude. It is often questioned in the trade why the SPR contains any sweet crude at all. If there is a cut in crude oil supplies to the US, there will likely be a surplus of US refinery capacity capable of handling an all sour SPR crude. Apart from a Maya crude exchange in 1998, almost all of the crude released from the SPR has been sweet – which lends further credence to the fact that private industry is simply leaning on the government as a storage provider of last resort.

US heating oil reserve: Unlike crude oil which can be stored for many years, finished products such as gasoline and diesel cannot be stored for very long at all. As light molecules in finished products evaporate over time, some of the product will be oxidized, and bacteria will metabolize other parts of the oil. For these reasons, finished products in storage are turned over every few months to keep them within useable specifications.

The 2 million barrel Northeast Home Heating Oil Reserve was created by the US government in July 2000 following perceived industry shortages during the winter months of December 1996 and Jan-Feb 2000. Two million barrels is a very small amount of product relative to US winter heating oil demand of 1.6 million barrels per day. No heating oil has ever been released from this reserve.

US naval petroleum and oil shale reserve: When people discuss strategic reserves in the US, they never mention, for good reason, the US Naval Petroleum and Oil Shale Reserves (NPOSR). The oil contained in oil shale is not liquid, is prohibitively expensive to produce, and is a clear net waste of energy.

Inventory/stock reporting
There are several major inventory numbers which are reported publicly and are referenced by the market. Barrel counting economists called supply and demand (S&D) analysts use these inventory numbers to try to predict future oil prices from these fundamental reports. Seven commonly cited inventory reports capture primary and secondary, but not tertiary, stocks.

Table 12-2

US DOE EIA– weekly, monthly and annually
US API – weekly and monthly
EU plus Norway Euroilstock – monthly
OECD IEA - monthly (but with a significant time lag)
Japan METI – monthly
Korean KNOC - monthly
IE Singapore - weekly

US Department of Energy (DOE) Energy Information Administration (EIA)
The petroleum division of the EIA collects information which comprises the three components of the Petroleum Supply Reporting System (PSRS): the Weekly Petroleum Status Report (WPSR); the Petroleum Supply Monthly (PSM); and the Petroleum Supply Annual (PSA).

US Weekly Petroleum Status Report
Submission of data to the EIA is mandatory under the US Federal Energy Administration Act of 1974 for those asked to provide data. Part of the report uses a survey method and estimations are made for the non-surveyed portion of the market. For example, data are collected from a sample of 62 US companies which carry or store 1,000 barrels or more of crude oil. Other parts of the weekly canvas include all of the operating refineries and blending terminals in the US.

Data must be submitted by 5pm EST on the Monday following the 7am stock level of the previous Friday.

Those attempting to predict the weekly EIA numbers have a very difficult task because there is a lot of noise in the data. There is the possibility of input error, or oil in transit or in floating storage not being included. A supertanker or two of crude (2 to 4 million barrels) not being counted can swing the weekly numbers and move global market prices. These errors and other faults of the system notwithstanding, the EIA weekly report is an excellent source of data.

As the US government and industry produces the most reliable source of timely and accurate information on oil storage, the market places a large emphasis on weekly US stock numbers from the DOE which are released at 10:30am EST every Wednesday, or Thursday if there was a holiday since the prior week's number. The US consumes approximately 25% of the world's oil and, therefore, US stock numbers are seen as a barometer of global supply and demand. The US numbers are also important as they also reflect the inventory where deliveries of NYMEX oil futures contracts occur.

The US reports its stock numbers by Petroleum Administration for Defense Districts (PADDs). PADDs were defined in World War II for the purpose of petroleum allocation when rationing for the war effort was implemented. Oil storage numbers continue to broken down into these regions for reporting and analysis purposes by the DOE. PADD1 is broken into three sub-districts for analysis purposes as it is a major oil consuming region, particularly the US Northeast.

US PADD map   Fig. 12-8

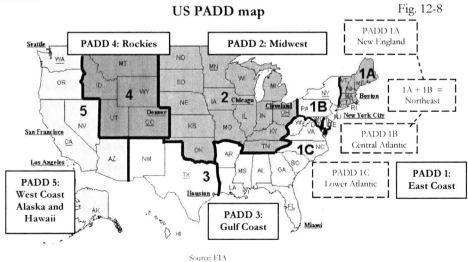

Source: EIA

EIA Petroleum Supply Monthly

The EIA publishes a Monthly Energy Review which provides more commentary and historical data than the weekly report. The EIA also reconciles, via revisions, its weekly data to the monthly data, which is carried out by another survey of the oil market.

EIA Petroleum Supply Annual

Every January the EIA publishes an annual Energy Outlook which reviews recent trends and looks 20 years into the future.

American Petroleum Institute (API) weekly statistical bulletin

The API is a private US oil industry organization. The API began publishing oil market statistics in the 1920s. It only surveys its members and these members are not mandated by law to submit numbers. API stats published on Tuesdays at 4:30pm are often discounted by the market in favor of the EIA numbers which are released at 10:30am EST each Wednesday. Although they are supposed to reflect the same data, the API numbers are frequently quite different to the EIA numbers.

Many oil industry players, such as Valero, the fourth largest refining company in the US, are not members of the API and do not submit data to the weekly report.

Similar to the EIA, the API publishes a monthly report which reconciles to the weekly reports.

International Energy Agency (IEA) Monthly Inventory Report

OECD nations consume just over 55% of daily global oil production. The OECD IEA publishes a Monthly Oil Market Report (OMR) which contains the Monthly Oil Statistics (MOS) aggregating demand, supply and stock numbers submitted by OECD member countries.

The IEA also gathers data from non-OECD countries and consultants where it is available to enhance the report.

Actual product delivery numbers are used when statistic collection agencies have such demand data. Where data on deliveries of finished products are not available, such as in the former Soviet Union and mainland China, supply and demand analysts at the IEA, will calculate what is referred to as apparent demand, or implied demand. Apparent demand is production less net exports for that product.

The IEA forecasts demand a year ahead and creates what is termed a call on OPEC. The call on OPEC is the required supply from OPEC to keep global oil stock levels unchanged. The IEA forecasts demand using economic models which take into account economic forecasts of the OECD.

While the IEA charges for most of its data, some of its information the IEA provides is free to the public after a period of time.

Euroilstock monthly report
Euroilstock was originally founded by private oil industry members and the European Commission, which is part of the EU government, to collect data on European oil stocks.

Since 1986 the report has been prepared solely by private oil industry agency Stichting Euroilstock, which is Dutch for Euroilstock Foundation, located in Amsterdam and not affiliated to the EU government.

Two separate Euroilstock reports are published each month. The data refers to the first 15 joining European Union member countries plus non-EU Norway as of the end of the previous month. Data submission to Stichting Euroilstock is voluntary. Market participants do not pay a lot amount of attention to the Euroilstock reports because they are voluntary, are not timely, and are subject to significant and frequent revisions.

Euroilstock Inventory Reports are published at 1400GMT on the 7th business day of each month. The report provides a month-end country by country breakdown of stocks of crude, finished motor gasoline, middle distillates, fuel oils, and naphtha plus refinery crude intakes. The data is submitted to Euroilstock on the fifth business day of each month.

Euroilstock Refinery Reports are published at 1400GMT on the 15th business day of each month. This report provides data on refinery throughput and production figures as well as refinery output of gasoline, middle distillates, fuel oils and naphtha. The data for the report is submitted on the thirteenth business day of the month.

Japan's METI
Japan's Ministry of Economy, Trade and Industry (METI) produces monthly oil demand, supply and storage numbers. Japan is also a member of the IEA.

South Korean storage

In March 2002, Korea became the second Asian nation, following Japan, to become a member of the IEA. Korea National Oil Corporation (KNOC) publishes oil statistics monthly.

IE Singapore weekly inventory report

The Government of Singapore's International Enterprise (IE) agency publishes a weekly report on Thursday's at 4pm local Singapore time with inventories of various products and crude on the island state. The market discounts these numbers as, even though Singapore is a large trading hub and a benchmark location for oil prices, there is quite a lot of storage in countries surrounding Singapore, such as Malaysia and Indonesia, which can make stock changes in Singapore somewhat misleading in terms of regional balances.

CHAPTER THIRTEEN
SEASONALITY

There is an element of seasonality in the oil market as it involves events which occur at a similar time each year and affects demand, storage, transportation, and supply.

Demand seasonality
Measures of temperature dependent demand

Energy markets have defined seasons which depend on the need for heating and cooling. The two seasons are the winter heating season and the summer cooling season (Table 13-1). The low energy demand transitions between the two seasons are known as the shoulder periods. Heating Degree Days (HDD) and Cooling Degree Days (CDD) are measures of heating and cooling energy required.

Table 13-1

HDD < 65°F...................Winter: Nov-Mar	
CDD > 65°F...................Summer: Apr-Oct	

Buildings in winter require heating to keep temperatures inside above a comfortable 65°F. HDD is measured by comparing the simple average of high and low temperatures against 65°F. For example, if the average of the high and low temperatures for a day is 35°F then the HDD value for that day is 30.

CDD measures air conditioning cooling demand. CDD is also measured by using the simple average of the high and low temperature compared against 65°F. For example if the average of the high and low temperatures for a day is 85°F then the CDD for that day is 20.

Heating demand – Western Europe is the largest consumer of oil products for heating during winter, followed by the US Northeast and Midwest. Japan, Canada and South Korea are smaller consumers of heating oil.

Table 13-2

Heating oil demand

	Typical summer demand millions of barrels per day		Typical winter demand millions of barrels per day
OECD Europe	1.9	→	2.9
US	0.9	→	1.6
Canada	0.3	→	0.4
Japan	0.5	→	0.7
South Korea	0.12	→	0.17

Source: IEA data 2006/2007

Though most demand for heating oil is logically in the winter season, there is heating oil demand during the summer. Many consumers fill their tanks during the summer to take advantage of seasonally lower prices. More significantly, local heating oil distributors fill their storage tanks during the summer after having run them down during winter.

Of the just over 111 million households in the US, most use natural gas or electricity for winter heating (Table 13-3). Natural gas, if it is available via pipeline is usually cheaper than heating oil. Only 8 million US households use heating oil as their primary source of heat. Most of the households using heating oil in the US are in the Northeast (PADD 1A and 1B). The Midwest (PADD 2) relies more on natural gas and propane than heating oil for space heating. While many consumers fill their tanks before the heating season, most consumer tanks are not large enough to provide sufficient storage for an entire winter and many have to refill four or five times during heating season. Heating oil demand in the US usually peaks in January at 1.6 million barrels per day and then falls to around 900,000 bpd during the summer months.

US Household Installed Primary Heat Source   Table 13-3
(millions of households)

	Total US		Northeast		Midwest	
Nat Gas	58.2	52%	11.4	55%	18.4	75%
Electricity	33.7	30%	1.6	8%	3.5	14%
Heating Oil	7.7	7%	6.2	30%	0.7	3%
Propane	6.0	5%	0.4	2%	1.9	8%
Other	5.5	5%	1	5%	1.1	4%
Total Housholds	111.1		20.6		25.6	

Numbers of particular interest to the oil market

Source: EIA data

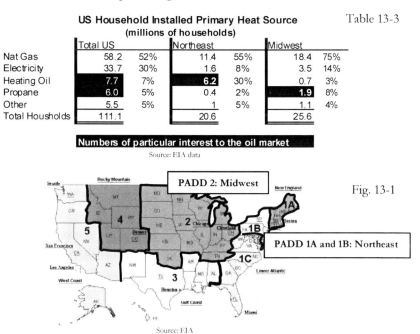

Fig. 13-1

Source: EIA

Propane demand – Propane is used by 9.4 million households in the US for cooking and heating. Of the households using propane, just over half, or 4.9 million households, use propane as their primary source of winter heating. Propane heating demand is especially concentrated in the Midwest. Propane demand in the US which runs at about 0.9 mmbpd during the summer increases to roughly 1.7 mmbpd during the winter.

Agricultural use of propane for crop drying in the US, mostly corn, which accounts for about 8% of total US propane use, is also a seasonal factor. Propane is used for crop drying during the US fall season, September through November, and is dependent on crop size and moisture content of the crop.

Kerosene demand – Japan and Korea use more kerosene than heating oil for winter home heating. In Japan, kerosene demand rises to approximately 1.2 million barrels per day in winter, which then falls to around 500,000 bpd during the summer. In Korea, kerosene demand rises to around 400,000 bpd during the winter from just over 150,000 bpd during the summer.

Jet fuel traders keep an eye on Northern Asian temperatures as most of the molecules which make up kerosene are the same as those used in jet fuel.

A further seasonal draw on kerosene is its use as a blending stock in the US to create winterized diesel, which is referred to as arctic diesel in Europe. Winterized diesel is diesel mixed with a small amount of kerosene range molecules, and although it contains slightly less energy per gallon, it flows more easily in cold weather situations.

Natural gas heating and cooling demand - The two primary seasonal components to natural gas demand are winter heating demand and summer cooling demand throughout the US. Natural gas is frequently used in boiler rooms in winter and at power generation facilities to produce additional power for air conditioning during the summer.

Gasoline demand –Gasoline demand increases during the summer as people drive to more outdoor activities and schools close, allowing families go on driving vacations (Fig. 13-2). Traditionally, summer driving season in the US runs from Memorial Day, which is the last Monday of May, to Labor Day, which is the first Monday in September. During summer gasoline season, refineries switch to what is referred to as max gasoline mode from winter max distillate mode. This switch can move approximately 2% of refinery production molecules from gasoline to distillate and back depending on the season. Refineries change their crude feedstock slate in addition to changing refinery processes to make the switch.

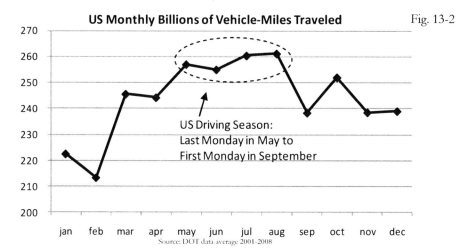

US Monthly Billions of Vehicle-Miles Traveled — Fig. 13-2

US Driving Season:
Last Monday in May to
First Monday in September

Source: DOT data average 2001-2008

Another seasonal factor affecting gasoline demand occurs during particularly heavy snow storms in the US Mid-west and Northeast, when gasoline demand can fall for a couple of days as schools are closed until roads are cleared.

Jet fuel demand - Similar to gasoline demand, jet fuel demand increases during the Northern Hemisphere summer as people take vacations. However, contrary to what one would initially expect, jet fuel prices tend to be lowest during the summer and highest during winter because of heating oil demand. The reason is that both jet fuel and heating oil come from the same part of the barrel – middle distillates. Jet fuel must compete with heating oil within a refinery system for production. A refinery can decide to produce more heating oil at the expense of jet fuel production and vice versa.

The premium of jet fuel prices above heating oil prices, called the jet diff, or regrade, widens during the winter to encourage refineries to divert more middle distillates molecules to making jet fuel. The jet diff falls during the spring and summer as refineries do not need such an incentive.

Jet fuel demand seasonality is usually well correlated with Revenue Passenger Miles (RPM), or Revenue Passenger Kilometers (RPK), as it is known outside of the US. RPM, or RPK, is the distance traveled by paying passengers.

In the US, the Air Transport Association (ATA) in addition to the FAA produce monthly RPM statistics which are frequently referred to by traders to gauge jet fuel demand. In Europe, the Association of European Airlines (AEA) produces RPK numbers on a weekly basis. Globally, the International Air Transport Association (IATA) produces monthly RPK numbers covering most international airlines.

Flu season: Flu season runs from November through March in the northern hemisphere with a peak in February. The flu virus is contagious and epidemics usually peak after 3 weeks.

H5N1, a lethal avian flu which emerged in Hong Kong in 1997, has killed farmers and others coming in contact with sick birds. H5N1 has the potential, if it mutates into a more rapidly spreading virus which can move from human to human instead of from birds to humans, to become the next pandemic. While only 2% of infected Americans died having been infected by the 1918 Spanish flu, known as the French Flu in Spain, H5N1 appears to kill a much higher percentage of those becoming infected. Many who die from H5N1 are young healthy adults. H5N1 re-emerged in Southeast Asia in 2003 killing over 50 people, most of whom had contact with birds.

In 2002-2003, a viral outbreak in Asia named SARS (Severe Acute Respiratory Syndrome) killed more than 800 people. The virus had emerged in October 2002 in Southern China. The SARS outbreak effectively shut down international economic activity in Asia in the second quarter of 2003. Luckily the spread of the virus was stemmed by quarantining those infected and it appeared to be contagious only when an infected person had a fever, which made it more difficult to spread.

Monitoring flu season is important in oil markets as demand can swiftly collapse, as evidenced by SARS, until a pandemic passes.

Residual fuel oil demand: Residual fuel oil is used for shipping and electrical power generation. Electrical power generation peaks during the summer as individuals turn on air conditioning. Residual fuel is used for a very small portion of overall US power generation and is therefore only slightly affected by power demand. Individual generators may have switching capability to burn oil rather than natural gas, but in the overall demand situation, this only has a slight seasonal effect. In Asia, where residual fuel accounts for a much higher portion of refinery production, utilities burn more residual fuel for electricity generation.

Bitumen demand: Bitumen is used to make asphalt. Asphalt requires heat to prevent it from setting and therefore can only be transported over relatively short distances. During the summer, asphalt can be transported over longer distances before it begins to set. Bitumen demand, therefore, increases during summer road surfacing season of May through September.

Chinese Lunar New Year: China has recently become a major factor in the global oil market. There is a sharp drop off in Chinese oil demand prior to and during Chinese New Year celebrations, which is usually in late January of February. Chinese demand tends to pick up just after the Chinese New Year.

Storage seasonality

Storage of heating oil, gasoline, propane, and natural gas are all highly seasonal. For all of these products, inventories build during low demand periods of the year and draw during high demand periods.

Transportation seasonality

Rhine water levels: The Rhine is a major transportation mechanism for delivering finished products to Germany and other large European consuming nations from the ARA (Amsterdam-Rotterdam-Antwerp) product hub. Low water levels during the European Summer can make it difficult for product barge traffic to navigate the Rhine. Low water levels thus have a temporary depressing effect on prices in ARA, and cause temporary higher prices in Germany and other major inland, Rhine-dependent, nations.

Baltic icing: Russia exports about 1.5 million barrels per day of crude oil and residual fuel oil through four ports on the Baltic Sea: Primorsk (Russia), Tallinn (Estonia), Ventspils (Latvia) and Butinge (Lithuania). Freezing seas limit passage to ice-class vessels from late December until April. Ice-class tankers are specialized tankers capable of navigating through frozen waters during winter months. Late or early freezing in the Baltic, combined with the availability of ice-class tankers, can affect marginal supplies from this region.

Supply seasonality

Refinery turnarounds: Refineries carry out scheduled maintenance and upgrades in the shoulder demand periods between heating oil season and summer gasoline seasons.

During the summer, increased atmospheric temperature increases the likelihood of a refinery overheating. This leads to unscheduled shutdown of equipment which is why some traders factor temperatures at refinery locations into their supply and demand models.

North Sea winter storms: Storms in the North Sea can partially shut down offshore platform production for several days and also can affect loading of tankers. If the storm is severe enough, damage to production platforms could reduce production for several months.

Hurricane season in the Gulf of Mexico (Jun-Nov): Atlantic hurricane season runs from June 1 to November 30 with a peak in September (Table 13-4 and Fig 13-4).

Fig. 13-4

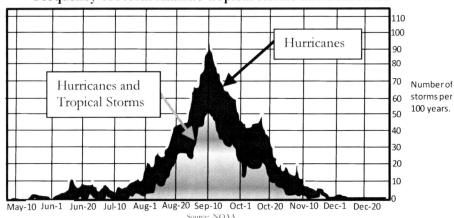

Frequency of North Atlantic tropical storms and hurricanes

Source: NOAA

Tropical storms and hurricanes by month, for the period 1851-2007 (North Atlantic region)		
Month	Total	Average
January-April	5	<0.1
May	19	0.1
June	80	0.5
July	102	0.6
August	347	2.2
September	466	3.0
October	281	1.8
November	61	0.4
December	11	0.1
Total	*1,372*	*8.7*

Table 13-4

Source: NOAA

Hurricanes are usually stronger and longer lasting than usual if water temperatures are warmer than usual, even by a couple of degrees. In recent years the number and intensity of hurricanes has been increasing as tropical water temperatures have been rising.

There is also a correlation between hurricanes and the warmth of the El Niño and cooler La Niña Pacific temperature variations, which increase and reduce wind shear, respectively, in Atlantic hurricane regions. Increased wind shear reduces hurricane development.

Employees are typically evacuated from offshore production rigs if they are in the path of a hurricane and production is shut in. Damage to an offshore crude production facility can take from several months up to a year to repair. Hurricane Ivan in September 2004 and Hurricane Katrina in August 2005 both caused significant damage to offshore US gulf coast oil production and onshore refineries.

There is also the temporary disruption to imports of crude oil from tankers, which will not offload until the hurricane has passed. Refineries onshore may have to shut down due to power outages or direct hit damage and flooding.

Stages in hurricane development
Tropical depression - winds up to 38mph: Tropical depressions are low pressure systems with maximum sustained winds of 38mph. Sustained is defined as a 10 minute average. Tropical depressions do not have clouds in swirling patterns or eyes.

Tropical storm have sustained winds of between 39 and 73mph. The swirling storm cloud pattern emerges, but the storm does not have an eye yet. The storm is given a name at this stage.

The anti-clockwise swirling pattern of a hurricane in the northern hemisphere is due to the coriolis effect, which is caused by the earth spinning the atmosphere through space relatively faster at the equator than at the poles. Therefore, parts of the storm further from the equator are held back by the relatively slower movement of the atmosphere further from the equator. Tropical storms and hurricanes spin clockwise in the Southern hemisphere.

Hurricane have sustained winds of 74mph or greater. The hurricane will have an area of the lowest pressure within a defined eye at the center of a swirling cloud pattern. Hurricanes carry the same name as the tropical storm from which it developed.

Once a hurricane has formed, its intensity is categorized on the Saffir/Simpson hurricane scale (Table 13-5).

Saffir/Simpson hurricane scale   Table 13-5

	Category	Sustained Winds	Potential Damage
	1	74-95mph	No real damage to building structures. Damage primarily to unanchored mobile homes, shrubbery, and trees. Also, some coastal flooding and minor pier damage.
	2	96-110mph	Some roofing material, door, and window damage. Considerable damage to vegetation, mobile homes, etc. Flooding damages piers and small craft in unprotected moorings may break their moorings.
'Major Hurricanes' as defined by the US National Hurricane Center.	3	111-130mph	Some structural damage to small residences and utility buildings, with a minor amount of curtainwall failures. Mobile homes are destroyed. Flooding near the coast destroys smaller structures with larger structures damaged by floating debris. Terrain may be flooded well inland.
	4	131-155mph	More extensive curtainwall failures with some complete roof structure failure on small residences. Major erosion of beach areas. Terrain may be flooded well inland.
	5	155+ mph	Complete roof failure on many residences and industrial buildings. Some complete building failures with small utility buildings blown over or away. Flooding causes major damage to lower floors of all structures near the shoreline. Massive evacuation of residential areas may be required.

Source: NOAA

Naming conventions

Names for tropical storms are created by the World Meteorological Organization (WMO), an agency of the United Nations, located in Geneva. The WMO cycles through six lists for Northern Atlantic hurricanes with each list consisting of 21 names. Countries which have experienced severe damage as a result of a hurricane can request a name to be retired from use for ten years.

There are no names beginning with Q, U, X, Y or Z. Once there have been 21 tropical storms or hurricanes, then the naming convention switches to the Greek alphabet, such that storm number 22 is alpha. The first time the Greek alphabet had to be used was in 2005.

<u>Monitoring hurricane status</u>

The National Oceanic and Atmospheric Administration's (NOAA) national Hurricane Center has a very good web site for monitoring hurricanes. Also during hurricane season, the US Minerals Management Service (MMS) publishes daily updates on its web site detailing any disruption to supply as a large amount of crude is produced from offshore federal lands. The MMS divides US offshore crude oil production into three regions, Western, Central and Eastern (Fig. 13-5).

Gulf of Mexico MMS Outer Continental Shelf Regions   Fig. 13-5

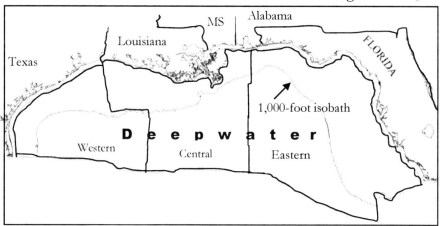

Source: MMS

CHAPTER FOURTEEN
RESERVES

There are no universally accepted definitions of oil reserves. Industry groups, regulators, and governments each have their own set of definitions. Many OPEC producer nations publish reserves figures which are not reserve estimates at all, but appear to be a total of past production plus future estimated production.

Within oil companies there are often a range of reserves estimate best illustrated by an old oil industry story:

> "[an oil company executive was] interviewing as potential employees a geologist, a geophysicist, and a petroleum engineer (the kind that estimates reserves). One question asked was, "What is two times two?" The geologist answered that it was probably more than three and less than five, but the issue could use some more research. The geophysicist punched it into his palmtop computer and announced that it was 3.999999. The petroleum engineer jumped up and locked the door, closed the window blinds, unplugged the phone, and asked quietly, "What do you want it to be? (Deffeyes, K.S. 2005)

Although there are no universally accepted definitions of reserves, there is a general framework described in this chapter within which most definitions exist.

Reserves terminology

The total discovered and undiscovered crude oil in a region is referred to as Original Oil In Place (OOIP), also known as Petroleum Initially In Place (PIIP).

Resources are the total amount of discovered and undiscovered oil which has been produced in the past and which are estimated to be technically feasible to recover in the future. Resources are also known as the Ultimate Recoverable Resource (URR).

Reserves are part of resources estimated to be both technically and economically producible in the future.

Resources and reserves are assigned a probability by geologists, geophysicists and petroleum engineers. These estimates can and do often change.

Historical production plus reserves is known as Estimated Ultimately Recoverable (EUR), or Ultimate Estimated Recovery (UER).

Reserves acronyms

Table 14-1

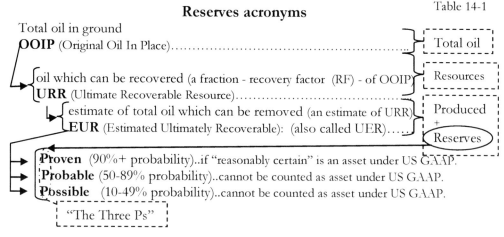

Total oil in ground
OOIP (Original Oil In Place).. Total oil

oil which can be recovered (a fraction - recovery factor (RF) - of OOIP) Resources
URR (Ultimate Recoverable Resource)...................................

estimate of total oil which can be removed (an estimate of URR) Produced
EUR (Estimated Ultimately Recoverable): (also called UER)..... +
Reserves

Proven (90%+ probability)..if "reasonably certain" is an asset under US GAAP.
Probable (50-89% probability)..cannot be counted as asset under US GAAP.
Possible (10-49% probability)..cannot be counted as asset under US GAAP.

"The Three Ps"

Reserves are often categorized into groups referred to as the three Ps depending on their probability (Table 14-2):

Table 14-2

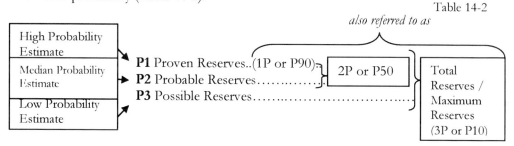

also referred to as

High Probability Estimate		
Median Probability Estimate	**P1** Proven Reserves..(1P or P90)	2P or P50	Total Reserves / Maximum Reserves (3P or P10)
	P2 Probable Reserves.............		
Low Probability Estimate	**P3** Possible Reserves...............		

P1, P2 and P3 are the individual components of reserves and 1P, 2P and 3P is the terminology used to describe the cumulative components. 1P, 2P and 3P are also referred to as P90, P50 and P10, respectively.

Proven reserves are generally estimates with a probability greater than 90% under present technical and economic conditions.

Proven reserves are estimated using prevailing prices and costs at the time of the estimate and not forecasted prices. If the price of oil rises, proven reserves may increase to include reserves which could not have been economically produced at lower prices. Likewise, if crude oil prices fall, proven reserves may fall.

To be categorized as proven, detailed seismic analysis, test wells, core sampling, flow tests, well-logging and other studies have to be carried out to define the quantity of oil to a relatively high degree of certainty. Because of the rigorous testing involved, proven reserves are also referred to as measured reserves.

Actual production of crude oil yields more valuable information to enhance the quality of reserves estimates than estimates prior to production. For this reason, proven reserves are often broken out into: Proven Developed Producing (PDP); Proven Developed Non-Producing (PDNP); and Proven Undeveloped (PUD).

Probable reserves are those from which production is more likely than not. This is an estimate with a probability equal to or above 50% and less than 90%. Probable reserves are also called inferred reserves. Classification of probable reserves involves a less rigorous assessment than proven reserves, and relies on inferring reservoir size from much fewer tests and appraisal wells than used for proven reserves, or even just 3-D seismic analysis and no drilling at all.

If the price of oil rises, it may become economical to drill test and appraisal wells and construct expensive production facilities which may move reserves from probable to proven.

Possible reserves are those from which production is less likely than probable. Possible reserves have a chance equal to or above 10% and less than 50% of being economically and technically feasible. Reserves are classified as possible when a general knowledge of the geology points to the presence of oil, but expensive on the ground tests have not been carried out, or if the price of oil would have to be much higher in order to pay for production completions.

The sum of proven, probable and possible reserves is called maximum reserves.

The three Ps probabilistic disclosure method is recommended by the Society of Petroleum Engineers (SPE) and the World Petroleum Council (WPC). Variations of the three Ps probabilistic disclosure method are used in financial reporting in Canada, the U.K., Norway and Australia.

The alternative method to the three Ps probabilistic method is the deterministic method (Table 14-3). Until 2009, the US SEC had been one of the main proponents of the deterministic method by requiring it for US-listed public companies. The goal of the deterministic method is to arrive at a single reasonably certain conservative figure such that reserves numbers are only published if a producer is very close to 100% certain that those reserves can be produced. The deterministic method does not provide as much information.

Deterministic method	Probabilistic method
Reasonable certainty	Different outcomes assigned probabilities
Conservative (90%+) single number	Three estimates typically: Conservative (90%+ probability) Median (50%+ „ „ „) Optimistic (10%+ „ „ „)
US SEC filings (until Dec 15, 2009)	- SPE/WPC recommendation - US SEC filings after Dec 15, 2009

US SEC definition of reserves: Under US Generally Accepted Accounting Practices (GAAP), companies can choose to expense finding and development (F&D) costs in one of two ways: successful efforts or full cost.

Successful efforts (SE) accounting permits companies to write off F&D expenses against profits until reserves become proven. Once reserves become proven, F&D costs associated with them can be capitalized. Capitalizing costs on an oil company's balance sheet enables the company to defer expenses. These deferred expenses are offset against future production revenues.

Full cost (FC) accounting allows companies to capitalize all F&D costs, not just those associated with successful wells. Deferred F&D costs are then expensed against future revenue. As FC accounting tends to be less conservative than SE, the US SEC limits the amount of F&D expense companies can defer. An SEC 'ceiling test' linked to proven reserves places a limit on the total costs which may be capitalized for each individual country the company operates in.

Whether a company chooses to use SE or FC, the ability to defer expenses and increase in the appearance of short term profitability provides an incentive to classify reserves as proven as early as possible.

Another major incentive to book reserves as proven early is that investors look at proven reserves as a key figure in valuing an oil company.

A further incentive is that management of most oil producers is judged, and paid, by how well it replaces reserves with new proven reserves.

In order to prevent the over-optimistic booking of proven reserves, the US SEC, which regulates US publicly traded companies, issued a ruling in 1978 permitting proven reserves to be booked only if production is "reasonably certain." The test of reasonable certainty was if "economic producibility is supported by either actual production or conclusive formation test."

The SEC "conclusive formation test" clause was interpreted to means that contact with the reservoir through drilling and well-flow testing takes place.

By requiring reasonable certainty, the SEC did not permit probable or possible reserves to be included in public financial statements.

The SEC also did not permit 3-D seismic analysis alone to define proven reserve size, except in the deep offshore USGC. One reason for the exclusion was that in 1978, 3-D seismic analysis was almost nonexistent.

The US SEC also did not permit the inclusion of oil sands reserves as proven oil reserves, until a method to extract the heavy oil had been constructed. This is because discovery of oil sands is not an issue. Everyone knows the oil is there. However, the gigantic capital investment required for oil sands production is the major impediment to production.

The US SEC definition of proven reserves was based on prices prevailing on the most recent December 31. The use of a single day's price is unusual in a market as volatile as oil, and was also throwback to 1978 when spot oil markets were not as transparent as today.

The SEC announced in late 2008 that it would be revising its reserves disclosure requirements for financial reports filed after Dec 15, 2009. The new ruling permits proven reserves to include oil sands and seismic 3-D analysis alone to define reserves rather than having to drill into the reservoir. The ruling also permits disclosure of probable and possible reserves and requires companies to use an average oil price of oil at each month end over the 12 months prior to the filing date as the reference price to determine if reserves are economic.

How reserve estimates are arrived at
There are five commonly used reserve estimation techniques. The methods are listed in increasing order of accuracy, expense, and time implemented:

Table 14-4

Nominal, Volumetric	Used prior to production. Inference based on geological and engineering survey data. Not as accurate as estimates using actual production data.
Decline curve analysis, Material balance, Reservoir simulation	Used once production has begun. Based on production performance data

<u>Nominal methods</u> used before production begins compare a field to another known field using rule-of-thumb analogies. This method may be used when the boundaries of a continuous productive rock formation have been observed around an exploratory well, but no other tests have taken place. The nominal method is not very accurate, but better than nothing.

<u>Volumetrics</u> is also known as the geologist's method. It is useful for estimating reservoir volumes before actual production begins. Volumetrics involves using geological and engineering data on porosity, hydrocarbon saturation of those pores, net pay thickness, and areal extent to estimate the amount of oil or gas. Because volumetrics does not use flow testing or actual production flow data, it is not as accurate as other methods.

<u>Material balance</u> uses measurements over time of reservoir pressures to estimate the amount of oil remaining. Material balance is more accurate than volumetrics as early production data is used.

<u>Decline curve analysis (DCA)</u> is used once production rates have begun to naturally decline. Following a early plateau almost all oilfields settle into a steady inexorable decline. Production rates and pressure declines are plotted and extrapolated to give an estimate of oil volumes.

<u>Reservoir simulation</u> involves using many sources of sample and test data to construct a complex computer model of the reservoir and an estimate of the oil it contains.

Global reserves: How much and where?

Estimating global reserves is difficult because data used to answer this question are somewhat unreliable. In addition, as reserves from different organizations are created on dissimilar bases, they can be difficult to compare and cannot be easily summed to give a precise global reserves number.

The infinite reserves approach simply has reserves growing indefinitely to meet growing demand. This method assumes that discoveries are correlated with prices, but ignores the fact that large discoveries have, in fact, tended to occur somewhat randomly throughout the entire history of the oil industry and are not highly correlated with prices. The method also assumes that there is unlimited oil to be found, such that oil prices will always revert to a long term mean. The method appears to be used by the OECD International Energy Agency (IEA).

The macro approach involves carrying out a big picture analysis of the entire planet using a combination of remote sensing equipment such as satellite imagery in addition to some rough physical geological data. This approach is used by the US Department of the Interior Geological Survey (USGS) to generate estimates of undiscovered reserves. The USGS last published a macro survey in the year 2000.

The USGS macro approach, while an interesting approach, should be treated with caution as it may create a false impression of accuracy. No oil company would ever drill based on USGS year 2000 macro reserves estimates in any region. Instead the USGS approach is useful for a very general and highly inaccurate sense of how much oil is undiscovered.

The field-by-field method involves independently going field-by-field through the oil world and obtaining geological/engineering estimates along with details of the methodologies used to generate those estimates. The sheer magnitude of manpower required for such a survey, as well as the lack of access to field test data, limits the accuracy of any possible estimate using this method. A few consultants, such as IHS (parent of oil analysts Cambridge Energy Research Associates - CERA) and Wood Mackenzie, try to back into field by field estimates, but it is incredibly difficult to be accurate as there are so many variables and a lack of honest data.

The decline curve analysis approach applies DCA methods on a global scale. This method is a fundamentally sound approach, but suffers from some proponents trying to be overly accurate by specifying an exact day when a peak in oil production will occur. The data used only supports a 10-15 year range.

One of the reasons regional and global oil production follows bell shaped curves is that if one sums the lognormal probability distribution of production rates from a large number of individual reservoirs, the sum will tend to form a normal distribution, or bell shaped curve. This is based on a mathematical concept known as the central limit theorem.

DCA on a global scale has its roots in 1956 when a geophysicist named Marian King Hubbert working at Shell carried out on an analysis of oil production in the US lower 48 states. He predicted production in this region would peak in 1970 and then begin to fall steadily and irreversibly. Although he was a respected oilman, the industry didn't put much stock in his forecast. US oil production did in fact peak in 1970 and, apart from a brief respite in the early 1980s from Alaskan production, it has been in decline ever since (Fig. 14-1). Hubbert later went on to work at the USGS.

Fig. 14-1

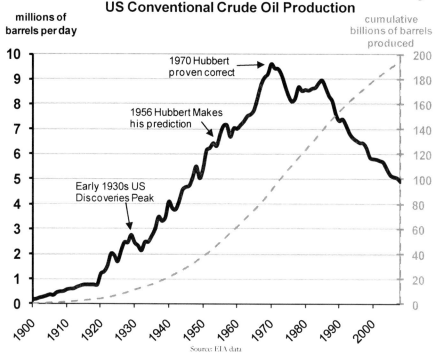

US Conventional Crude Oil Production

millions of barrels per day

cumulative billions of barrels produced

1970 Hubbert proven correct

1956 Hubbert Makes his prediction

Early 1930s US Discoveries Peak

Source: EIA data

Hubbert's peak oil prediction was based on a relatively simple observation. He noticed that US oil discoveries peaked and began to decline in the early 1930s. Hubbert deduced that US oil production rates may also follow a similar bell shaped curve after a time lag.

The length of the time lag between the peak in discoveries and production is dependent on how much oil was originally discovered and the annual rate of production since discovery. The level of discoveries and the rate of production in the US permitted the time lag to be 40 years.

Although available global discovery data is not precise, there is a general acceptance that global discoveries of crude oil peaked and began a steady decline in the early 1960s despite advances in technology (Fig. 14-2). If depletion occurs globally at the same rate as it did in the lower US then world production should peak and begin to decline between 2005 and 2015.

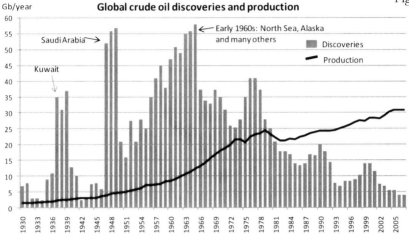

Fig. 14-2

Source: C.J.Campbell, ASPO. Revisions backdated. Rounded with 3-year moving average. Gb=billion barrels

Few in the oil market doubt that global production rates will form a similar bell-shaped curve to that of the US, and the list of producing countries witnessing the same peak and steady decline pattern increases each year (Table 14-5).

Table 14-5

Peaking oil production by country

country	year of peak	country	year of peak
USA	1970	Colombia	1999
Venezuela	1970	Australia	1999
Indonesia	1977	Uzbekistan	1999
Romania	1977	Norway	2001
Tunisia	1980	Oman	2001
Peru	1982	Yemen	2001
Cameroon	1985	Canada (conventional)	2002
Egypt	1993	Vietnam	2003
Syria	1995	Turkmenistan	2003
Gabon	1996	Mexico	2004
Argentina	1998	Malaysia	2004
UK	1999	Denmark	2004
		Italy	2005

Source: BP data

A traditional argument used against oil production peaking is that many believe discoveries increase when oil prices rise. Unfortunately, since 1859, discoveries have had an unfortunate tendency to be random and uncorrelated to high oil prices. Furthermore, although oil prices have rallied many times in the past forty years and technology has improved, no major change to the long-term decline in discovery rates has emerged since the 1960s.

A second argument used against peak oil theory is that so many false predictions have been made in the past. This is not a valid argument against the theory; it is more of a comment on human nature and large groups in that there is always someone somewhere making a prediction of some event.

The trust but don't verify approach involves contacting each oil producer and asking them for a proven reserves number. If producers don't provide a number then the previous year's number is used. This method is used in the publications Oil & Gas Journal, World Oil, and the BP Annual Statistical Review for most of the reserves they list.

The trust but don't verify figures are most commonly cited in general media because they are simple single numbers published by large reputable organizations. However, the numbers have flaws which render them unreliable.

Firstly, there is no universal consensus on a single definition of crude oil. Does it include condensates, NGLs and heavy oil?

Secondly, with most of the world's reserves estimates, one does not know the probability assigned to the proven reserves number or the test methods used to classify reserves as proven. The differences in volume between one definition and another can be colossal.

Thirdly, between 1982 and 1988, six OPEC members added 322 billion barrels to reserve estimates, almost doubling OPEC reserves overnight – although no major oil field had been discovered during this time (Table 14-6 and Fig. 14-3). The dramatic increase in proven reserves was as a result of OPEC quota shares being linked to reserves estimates.

Finally, since 1988 six OPEC members have produced over 225 billion barrels of crude, without any major new discoveries, but their stated reserves have hardly changed. It appears that these OPEC producers simply stopped estimating reserves.

The six producers which increased reserves during the 1980s and ceased accounting for production from these reserves account for two thirds of the global reserves numbers in the trust but don't verify approach.

**Six OPEC members highly unusual reserves increases
and failure to revise since for production**

Table 14-6

	Venezuela	Kuwait	Iran	Iraq	UAE	Saudi Arabia	Total
1980	20	68	58	30	30	168	374
1981	20	68	57	32	32	168	377
1982	25	67	56	**59**	32	165	**405**
1983	26	67	55	65	32	169	**414**
1984	28	**93**	59	65	32	172	449
1985	**54**	92	59	65	33	171	475
1986	56	95	**93**	72	**97**	170	582
1987	58	95	93	**100**	98	170	613
1988	59	95	93	100	98	**255**	699
1989	59	97	93	100	98	260	707
1990	60	97	93	100	98	260	708
1991	63	97	93	100	98	261	711
1992	63	97	93	100	98	261	712
1993	64	97	93	100	98	261	713
1994	65	97	94	100	98	261	715
1995	66	97	94	100	98	261	716
1996	73	97	93	**112**	98	261	733
1997	75	97	93	113	98	262	736
1998	76	97	94	113	98	262	738
1999	77	97	93	113	98	263	740
2000	77	97	100	113	98	263	746
2001	78	97	99	115	98	263	749
2002	77	97	**131**	115	98	263	780
2003	77	99	133	115	98	263	785
2004	80	102	133	115	98	264	791
2005	80	102	137	115	98	264	796
2006	80	102	137	115	98	264	796

1980s reserves inflation

■ = Unusual Increases

Fig. 14-3

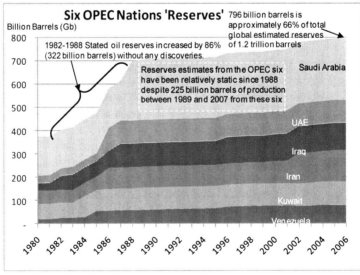

Six OPEC Nations 'Reserves'

Billion Barrels (Gb)

796 billion barrels is approximately 66% of total global estimated reserves of 1.2 trillion barrels

1982-1988 Stated oil reserves increased by 86% (322 billion barrels) without any discoveries.

Reserves estimates from the OPEC six have been relatively static since 1988 despite 225 billion barrels of production between 1989 and 2007 from these six

Source: BP data

Bearing the flaws of the trust but don't verify numbers, they place total global proven reserves between 1.12 and 1.32 trillion barrels (Table 14-7). To put the estimates in perspective, since the beginning of the oil industry in 1859, globally we have consumed just over 1.3 trillion (1300 billion) barrels, of which over 90% has been consumed since 1960.

To put global reserves into a more user friendly format, a frequently mentioned statistic is the Reserves to Production (R/P) ratio. Production is currently around 30 billion barrels (Gb) per year. The R/P ratio is the ratio of estimated reserves to current annual production rates and has been around 40 over the past decade based on the trust but don't verify reserves estimates.

Crude Oil Reserves Estimates

Table 14-7

	BP statistical Review		Oil & Gas Journal	World Oil
	billions of barrels	% of total	billions of barrels	billions of barrels
1. Saudi Arabia*	264	21%	262	262
2. Iran*	138	11%	136	132
3. Iraq*	115	9%	115	115
4. Kuwait*	102	8%	102	101
5. UAE*	98	8%	98	70
6. Venezuela*[1]	87	7%	80	53
7. Russia	79	6%	60	74
8. Libya*	42	3%	41	34
9. Kazakhstan	40	3%	30	not reported separately
10. Nigeria*	36	3%	36	37
11. US	29	2%	22	22
12. Canada[2]	28	2%	179	12
13. Qatar*	27	2%	15	20
14. China	16	1%	16	16
15. Brazil	13	1%	12	12
16. Mexico	12	1%	12	12
17. Algeria*	12	1%	12	11
18. Angola*	9	1%	8	9
19. Norway	8	1%	8	8
20. Azerbaijan	7	1%	7	not reported separately
Top 20	1,162	94%	1,252	1,001
other nations	76	6%	65	119
Total	**1,238**	**100%**	**1,317**	**1,120**
Subtotals:				
OPEC[3]	**935**	76%	**910**	**849**
Non-OPEC	**303**	24%	**407**	**270**
Middle East[4]	**756**	61%	**739**	**711**

Notes on above table: * = OPEC countries   OPEC reserves include 1980s reserve inflation.
[1] World Oil does not include NGLs or Venezuelan heavy oil.
[2] Oil & Gas Journal includes Canadian tar sands estimate much larger than World Oil and BP.
  BP and World Oil only Canadian tar sands currently under development.
[3] Ecuador, which is a member of OPEC, is not in the top 20 reserves, but is added here.
[4] Middle East = Saudi Arabia, Iran, Iraq, Kuwait, UAE, Qatar, plus non-OPEC: Oman, Syria, and Yemen
Dates of data: BP figures for Dec 2007. Oil & Gas Journal data as at Jan 2007. World Oil data is for Dec 2005

CHAPTER FIFTEEN
ENVIRONMENTAL REGULATIONS

Environmental regulations affecting the oil market fall into three general categories. There are petroleum-specific fuel regulations designed to protect health and property from the byproducts of burning oil. Storage and transportation of oil is also regulated to prevent toxic spills and dangerous explosions. Additionally, greenhouse gas regulations are intended to stem the growth of gases in the atmosphere, particularly carbon dioxide (CO_2).

Petroleum-specific fuel regulations
International trade in oil relies on industry standards for fuels which are minimum specifications products should meet. These standards are set by organizations such as ASTM International or the ISO. With a few exceptions, international standards do not focus on pollutants unless they affect the energy-related performance of a fuel.

Within individual countries, another layer of standards in addition to those of the ASTM and ISO are often imposed which seek to restrict petroleum pollution.

Creating unique standards for each country would isolate small economies from international oil trade and result in higher prices and potential shortages. To avoid this situation, nations with relatively small populations such as Australia, or developing economies such as China, have standards of their own which are often harmonized to standards of either the US, Europe or Japan.

US regulations tend to be followed in the Americas. European regulations set by the EU are followed by many Asian countries in addition to Australia and Brazil. Only a few nations outside of Japan itself follow Japanese petroleum regulations. Singapore follows regulations from all three regions, depending on the fuel.

This book will focus on environmental regulations in the US which has tended to lead the world in setting petroleum-related regulations over the past forty years.

US environmental regulation
The US Clean Air Act (CAA) of 1963 was the first major piece of legislation the federal government enacted in order to reduce air pollution at a national level. The CAA was subsequently amended in 1967, 1970, 1977 and 1990.

The US Environmental Protection Agency (EPA) was created in the 1970 amendment to the CAA to be responsible for creating and enforcing National Ambient Air Quality Standards (NAAQS), which are maximum acceptable pollutant levels in the atmosphere.

The EPA produces criteria documents which outline the adverse affects of what are known as criteria pollutants (Tables 15-1 and 15-2). Criteria pollutants are those which have measurable negative effect on humans, animals, vegetation and property. Maximum acceptable levels for each airborne criteria pollutant are outlined in NAAQS. There are two types of NAAQS: primary and secondary. Primary standards are those which protect human health. Secondary standards are those which protect animals, vegetation and property.

In addition to the EPA, the state of California Area Resource Board (CARB) sets its own air quality standards. Due to severe automobile pollution the CARB was established by the state of California in 1967 to set and enforce standards for that state. California is the only state allowed to set its own more restrictive air quality standards and emission controls than the EPA.

The EPA NAAQS and CARB air quality standards set limits on pollutants 1 through 6 shown in Table 15-1. The CARB has a seventh standard, for sulfates, which the EPA does not regulate.

Criteria pollutants
Table 15-1

1. Ozone (a.k.a. O_3 or smog)   = VOCs + NOx + sunlight
2. Particulate matter (PM) (a.k.a. soot) = SO_2 + NOx + ammonia + VOCs + particle emissions
3. Nitrogen dioxide (NO_2)
4. Carbon monoxide (CO)
5. Sulfur oxides (SOx)
6. Lead (Pb)
7. Sulfates: CARB standards only, and not regulated by the EPA.

   where: VOCs = Volatile Organic Compounds are organic chemicals, such as benzene, which have a high vapor pressure and easily evaporate into the atmosphere.

      NOx = Nitrogen Oxides forming at high temperatures when some nitrogen in air is oxidized.

Another frequently used rhyming way of listing the pollutants is:
Table 15-2

TOX (toxins):	Particulate matter, carbon monoxide; lead and sulfates
NOx:	Nitrogen Oxides
VOCs:	Volatile Organic Compounds, such as benzene
SOx:	Sulfur Oxides

Ozone is present in the upper atmosphere, where it filters some of the suns ultra-violet light. Ground-level ozone, which forms as a result of VOCs and NOx reacting in sunlight, is commonly referred to as smog. Smog causes and aggravates many respiratory ailments. Particulate matter, commonly referred to as soot, also causes respiratory issues in addition to an unsightly haze.

Diesel engines and gasoline engines pollute in slightly different ways, partially because of differences in the ways the fuels are burned and partially because of the unique properties of the individual fuels. Diesel engines are large emitters of particulate matter and NOx, small emitters of CO and VOCs, and emit 40% lower CO_2 than gasoline. Gasoline engines are small emitters of particulate matter and large emitters of CO, VOCs and NOx.

While the EPA and CARB set air quality standards, individual states are delegated the task of enforcing the standards with regulations, pollution monitoring and fines.

If the level of a criteria pollutant is frequently above the maximum set in NAAQS over several consecutive years, as many urban areas of the US are, then the area is referred to as a nonattainment area for that criteria pollutant. The EPA provides funding, research and guidance to states with nonattainment areas. The EPA can also issue sanctions.

Each state is required to produce, and submit to the EPA for approval, a State Implementation Plan (SIP) which outlines the actions it will take to meet the NAAQS set by the EPA.

Although the EPA sets minimum standards for vehicle fuels, the state plans can mandate standards more restrictive than EPA national standards. Allowing individual states to set unique fuel standards has resulted in a plethora of boutique fuels, with many slight variations. Boutique fuels resulted in the unintended consequence of an inefficient higher priced gasoline market. To prevent further fragmentation of the gasoline market the US 2005 Energy Policy Act (EPACT) capped the number of boutiques fuels such that no area within the US can create a new type of gasoline unless it completely replaces an existing type of gasoline or unless it is a gasoline which limits volatility to 7.0psi RVP.

In addition to fuel standards, the EPA and CARB set vehicle emission standards. These emission standards require various vehicle hardware components. Apart from California, states cannot require vehicle emission standards more restrictive than the EPA national standards. EPA vehicle emission standards are tailored for several classifications of vehicle, depending on the fuel burned, gasoline or diesel, the weight of the vehicle, such as light duty-heavy duty, and the environment it is used, such as on-road or off-road, marine or locomotive. There are four EPA regulated vehicle emissions:

US EPA regulated emissions

Table 15-3

1. Carbon monoxide (CO)
2. Hydrocarbons (HC) – fuel molecules evaporating from the fuel system before combustion or not being completely oxidized (burned) and escaping through the tailpipe as VOCs.
3. Nitrogen oxides (NOx)
4. Particulate matter (PM) – only applies to diesel engines and not gasoline vehicles.

The reason the Clean Air Act forbids individual states from enacting their own more restrictive vehicle emission standards than the EPA is so automobile manufacturers do not have to manufacture different vehicles for each state which would quickly become a very expensive and unwieldy process. The vehicle emission standards set by the CARB are more restrictive than those set by the EPA. In some limited circumstances, such as the California Low Emission Vehicle program discussed later in this chapter, states other than California are permitted to enact legislation to require new vehicles to meet strict Californian emission standards rather than EPA standards.

In an effort to make the criteria pollutant air monitoring service more real-time and user friendly to the general public, the EPA publishes a color coded Air Quality Index (AQI) to allow those with respiratory ailments determine if they should carry out any outdoor activity (Table 15-4). This index is published daily on the EPA web site in addition to newspapers and on television.

Table 15-4

Air Quality Index (AQI) Values	Levels of Health Concern
0 to 50	Good
51 to 100	Moderate
101 to 150	Unhealthy for Sensitive Groups
151 to 200	Unhealthy
201 to 300	Very Unhealthy
301 to 500	Hazardous

Source: airnow.gov

A summary of the US environmental regulatory process is shown in Table 15-5.

US regulatory process summary   Table 15-5

EPA sets National Ambient Air Quality Standards (NAAQS) against six (seven for California) atmospheric criteria pollutants which are measured.

Individual states put together and implement a State Implementation Plan (SIP) to meet NAAQS.

In order to meet NAAQS, the EPA and CARB set limits for vehicle emissions.

Emission limits are achieved by:
 Fuel standards – any state can create a restrictive boutique fuel
 Vehicle hardware – only EPA and CARB can require

Areas which continuously fail to meet NAAQS for a criteria pollutant are deemed non-attainment areas against which more severe restrictions on fuels and/or hardware are applied.

Fuel Standards
Gasoline and distillate fuel content regulations have been introduced since the 1960s in California and since the early 1970s in the entire US. In relation to gasoline, most of the EPA, CARB and individual state fuel content regulations cover sulfur, aromatics, benzene, vapor pressure, olefins, metals (lead, manganese, phosphorous), distillation profile, and oxygenates. In addition, fuels such as gasoline cannot be any more polluting when burned than a baseline for VOCs, NOx and toxic air pollutants (TAP). The EPA, CARB and individual states use fuel content restrictions to define conventional gasoline, reformulated gasoline, low RVP gasoline, and winter oxyfuels gasoline, among others.

Distillates, such as diesel and heating oil, have also come under tighter environmental regulations in recent years. Sulfur content limits for distillates have been the primary area addressed.

Vehicle hardware
Vehicle hardware combats pollution by preventing evaporation while refueling, storing fuel in a vehicle, or moving the fuel to the engine; burning petroleum more thoroughly so that there are fewer hydrocarbons emitted into the atmosphere; and capturing and neutralizing any remaining pollutants from the exhaust stream (Fig. 15-1).

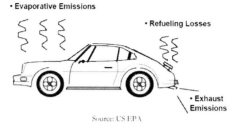

• Evaporative Emissions   Fig. 15-1

• Refueling Losses

• Exhaust
  Emissions

Source: US EPA

The EPA's regulations covering vehicle emissions require the use of catalytic converters, on board diagnostics (OBD) and evaporative control systems, among other hardware devices (Table 15-6). The EPA requires most states to implement regular inspections of vehicles to ensure that emissions control devices are operating correctly.

Emissions reducing vehicle hardware
Table 15-6

- Catalytic converters on gasoline exhausts and newer diesel exhausts.
- Catalytic diesel particulate filter (CDPF) on newer diesel exhausts.
- On-board diagnostics (OBD) computer system. OBDs, which are required on all new vehicles in the US, notify the driver via a check engine light of a failure in emission control hardware. The fault condition is stored on a vehicles computer memory which can be accessed at a service station.
- Evaporative emission control systems ensure fuel doesn't evaporate on its way to the engine from fuel lines in the vehicle. Any evaporating fuel is trapped and returned to the engine.
- Standardized fuel pump nozzles and vehicle fuel tank ports prevent evaporation during vehicle refueling at filling stations.
- Exhaust gas recirculation (EGR) returns unburned hydrocarbons to the combustion chamber.
- Fuel injection to optimize combustion ensures the fuel is burned thoroughly.
- Computer-controlled fuel systems also optimize combustion.

California Low Emission Vehicle (LEV) program

The State of California is the only state permitted to set vehicle emission standards in addition to the federal EPA. California adopted a Low Emission Vehicle (LEV) program in 1990. The fist phase was called LEV I. A second stricter phase began in 1998, called LEV II. The goal of California's LEV program has been to set emission standards stricter than federal requirements, for all vehicles; and, secondly, to put a minimum percentage of Zero Emission Vehicles (ZEV) on the road. Under California's LEV II program, car manufactures must classify their vehicles under one of the following categories:

LEV: Low Emission Vehicles are the lowest, most polluting, standard for 2004 model-year onward Californian vehicles, but still stricter than Tier 2 federal emission standards (see below) used by 45 States and District of Columbia.

ULEV: Ultra Low Emission Vehicles produce 50% fewer emissions than the average 2003 model-year new vehicles sold in California.

SULEV: Super Ultra Low Emission Vehicles generate 90% fewer emissions than average 2003 model-year new vehicles sold in California.

PZEV: Partial Zero Emission Vehicles meets SULEV emission standard, but also have zero evaporative emissions and a 15 year/150,000-mile warranty.

AT-PZEV: Advanced Technology Partial Zero Emission Vehicles meet SULEV tailpipe emission standard, but also have zero evaporative emissions and a 15 year/150,000-mile warranty. AT-PZEV have the capability to use novel zero emission technologies such as hybrid batteries.

ZEV: Zero Emission Vehicles have zero polluting evaporative or tailpipe emissions. These vehicles are 98 percent less polluting than 2003 model-year vehicles. Examples of ZEVs are electric and hydrogen fuel cell vehicles.

California vehicle emissions ratings   Fig. 15-2

LEV → ULEV → SULEV → PZEV → AT-PZEV → ZEV

Car manufacturers selling cars in California must comply with fleet average reductions in non-methane organic gas (NMOG) emissions from 2004 through 2010. The standard becomes more stringent over time.

Manufacturers also must have at least 10% percent of ZEVs in their light-duty vehicles (LDV) fleet for sale in California from model year 2005 onwards. A system of credits exists such that PZEV and AT-PZEV can be used to offset a part of the 10% ZEV requirement.

While a zero emission vehicle (ZEV) uses electricity or hydrogen fuel cells, partial zero emission (PZEV) status can be achieved with relatively simple technology in conjunction with a gasoline engine. Most polluting emissions occur when automobiles are started as the catalytic converter is cold. PZEV automobiles run very lean, with a low gasoline to air ratio, until the engine cylinders heat up. The catalytic converters are located close to the exhaust manifold so that they heat up more quickly. Multiple and larger catalytic converters are often used. In addition to being closer to the exhaust manifold, PZEV catalytic converters heat up quicker as they are made from tubular steel rather than cast iron. Several PZEV automobiles have electronic systems to circulate exhaust gases back to the combustion chamber, so that any remaining hydrocarbon molecules in the exhaust being burned.

In addition to very low tailpipe exhaust emissions, PZEV vehicles must have zero evaporative emissions. Evaporative emissions in PZEVs are trapped by canisters filled with carbon. The vapors are sucked out of the canisters by the engine and burned. Stainless steel gasoline tanks rather than plastic and thick fuel lines further reduce evaporative emissions.

Apart from California, individual states are not allowed by the EPA to depart from federal vehicle emissions standards. An exception to this restriction has been permitted for the States of New York, Maine, Massachusetts, and Vermont, which have been permitted by the EPA to adopt California's LEV program, although their implementation dates are different.

Unlike the Californian LEV I program, the EPA's initial federal emissions program consisted of a single standard, a so called Tier 1 emission standard - in other words, no LEV, ULEV, and SULEV etc. – just a single standard. The federal EPA Tier 1 emission standard was outlined in 1991 and includes vehicle emission limits for non-methane hydrocarbons (NMHC), nitrogen oxides, carbon monoxide and particulate matter, all measured using Federal Test Procedures (FTP) and applicable for the full useful life of the vehicle. From model-year 1996 and thereafter, all light-duty vehicles (LDVs) and light-duty trucks (LDTs) had to comply with the federal Tier 1 emission standards. The EPA's federal Tier 1 single standard was equivalent to the lowest, most polluting, of the Californian LEV I standards.

In 2002, the EPA implemented the federal Tier 2 emissions program which is similar, but not exactly the same as California's LEV II which had been applicable from model year 2004 vehicles onwards in California. In the federal Tier 2 program, vehicles are classified using terminology of Bins, such as Bin I, Bin II etc., rather than California's LEV, ULEV, SULEV etc., acronyms. Federal Tier 2 emissions standards became effective with model-year 2007 vehicles and manufacturers have had to make available for sale certain percentages of vehicles in each Bin.

US vehicle fuel efficiency regulations
There are three regulations in the US which have attempted to improve the efficiency of the entire US vehicle fleet: CAFE standards; the luxury car tax, and the gas guzzler tax.

CAFE standards: The US EPA can regulate fuels and emissions but cannot set fuel efficiency regulations. States cannot issue fuel efficiency regulations either. The only organization in the US permitted to set fuel efficiency regulations is the Department of Transportation's National Highway Traffic Safety Administration (NHTSA.)

As a result of the high oil prices of the 1970s, the US federal government enacted the Energy Policy and Conservation Act of 1975. The goal of the Act was to increase efficiency of the US automobile fleet. The law directs the NHTSA to establish Corporate Average Fuel Economy (CAFE) standards.

As at the beginning of 2009 the sales-weighted CAFE standards are 27.5mpg for cars and 22.5mpg for light trucks (under 8,500lbs), minivans, and SUVs. The light truck standard changes to 23.1mpg for 2009 model year and 23.5mpg for 2010 model year.

Although the EPA does not set or enforce CAFE standards, the EPA does administer the engines tests to determine fleet fuel efficiency on behalf of the NHTSA in laboratory conditions designed to simulate road performance. In US new car dealer showrooms, a large sticker on the vehicle window is required to clearly show the EPA mpg test rating.

Automakers are fined for not achieving the CAFE standards if their fleet of vehicles sold does not meet these standards. Between 1983 and 2006 model years, a total of $735 million in CAFE fines were levied. No US or Asian auto manufacturer has ever been fined.

European vehicles account for fewer than 8% of US auto sales, but 100% of CAFE fines. European automobile makers such as Porsche, BMW, and Mercedes, simply do not sell small cars in the US. As CAFE standards only come to bear on 8% of vehicles sold, the standard has been alleged to be a form of import tariff against EU car manufacturers rather than an efficiency tool.

In the past CAFE standards have been criticized as merely rubber-stamping whatever vehicle efficiency level the US auto industry is currently operating at, instead of aggressively pushing efficiency improvements. The US Energy Independence and Security Act of 2007 tries to address this by requiring car manufacturers to increase fleet efficiency for new cars to 35mpg by 2020.

Luxury car tax: This federal tax which was first imposed in 1990 was phased out on Jan 1, 2003. However, we are now living with the results. Between 1990 and 2002 a tax of up to 10% was levied on the cost of a car above $38,000 (the value changed slightly each year.) The tax only applied to cars weighing less than 6,000 pounds. This ill-conceived tax incentivized consumers, if they were going to purchase a vehicle for more than $38,000, to buy a large vehicle and save paying the luxury tax. Luxury SUV sales rocketed, and car sales fell. Most of these heavy vehicles are of course still on the road.

Gas guzzler tax: The gas guzzler tax was created in the Energy Tax Act of 1978 and was designed to discourage the purchase of inefficient vehicles. The tax is paid by the consumer and not by the manufacturer. The tax is posted on the same window sticker as the EPA mpg rating at a new vehicle dealership. The gas guzzler tax only applies to cars. SUVs and light trucks are not affected, which is another incentive to purchase a heavy vehicle.

Gas guzzler taxes apply to cars that get less than 22.5mpg, with taxes beginning at $1,000 and increasing to $7,700 for cars which get 12.5mpg or less.

In addition to the SUV and light truck exemption, which encourages the purchase of these inefficient vehicles, 80% of the gas guzzler tax is once again paid on cars sold by European vehicle manufacturers, which account for a tiny amount of total vehicle sales in the US.

In summary, the current CAFE standards and gas guzzler taxes, in addition to the now repealed luxury car tax, encouraged SUV sales. They affected only a tiny portion of the cars sold in US, primarily those sold by European luxury car manufacturers. The results of these three pieces of legislation are now pretty evident.

Despite the fact that engines over the past 30 years have become vastly more efficient at delivering power and improved acceleration; the number of heavy vehicles, such as light trucks and SUVs, has actually reversed overall fleet fuel efficiency. The high-point for vehicle fleet efficiency in the US over the past 30 years was 1987.

Storage and transportation regulations

The Department of Labor's Occupational Safety and Health Administration (OSHA) sets standards and provides training on worker safety and health. OSHA standards affect many parts of the oil industry such as storage, transportation, refinery operations, and oil rig/production operations.

The U.S. National Fire Protection Association (NFPA) maintains a standard: NFPA 704, also called the fire diamond, which is a placard attached to storage and transportation containers to enable emergency workers to quickly evaluate risks associated with petroleum products in addition to other Hazardous Materials (HAZMAT).

The fire diamond has four sections: Blue = health; Red = Fire; Yellow= reactivity; and white= special hazards such as biohazards or radioactivity. The numbers indicate the scale of the danger 0 = inert and 4 = very high.

Fire diamond example   Fig. 15-3

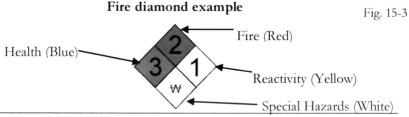

313

The US Department of Transportation in addition to the United Nations also have codes for each type of HAZMAT which are displayed on storage tanks.

Greenhouse gas regulations

Greenhouse gases are those which trap the sun's heat or reduce the planet's ability to reflect the sun's heat and thus raise the planets temperature.

Of the six greenhouse gases - carbon dioxide, methane, nitrous oxide, sulfur hexafluoride, HFCs, and PFCs – most scientists believe that carbon dioxide is the primary culprit in raising temperatures. The cause of carbon dioxide buildup in the atmosphere has been the rapid burning of fossil fuels such as coal, oil, and natural gas over the past 150 years.

While the burning of coal for power generation and industrial process is the largest generator (40%) of additional carbon dioxide in the atmosphere, the burning of oil products for transportation is a close second (35%).

Kyoto treaty

The United Nations Framework Convention on Climate Change was amended in 1997 to become the Kyoto Protocol. The protocol was agreed in Kyoto, Japan. A protocol is a treaty awaiting ratification by the parties to it. The treaty came into effect for signatories in February, 2005. The treaty requires the top 35 industrialized signatories to reduce emissions of greenhouse gases by 5.2% compared to 1990 levels by 2012.

Almost all of the signatories to the Kyoto Protocol ratified the protocol except for the US. The primary reason the US refused to ratify the protocol was the fact that energy consumption is closely linked to GDP growth. In addition, two very large developing countries, China and India, were exempted from the emission reduction framework.

Until 2007, the EPA had argued that it had no jurisdiction over carbon dioxide emissions. This changed with the decision of the US Supreme court that the EPA was required to regulate carbon dioxide as an air pollutant under the Clean Air Act. A change in US national policy concerning greenhouse gases now appears increasingly likely.

Although the US federal government has not yet ratified the Kyoto Protocol, various states, including a group of ten Northwestern US states, formed an alliance called the Regional Greenhouse Gas Initiative (RGGI) to develop methods of capping carbon dioxide emissions and trading emissions rights.

The State of California has also announced legislation requiring manufacturers selling automobiles in that state to reduce vehicle carbon emissions.

CHAPTER SIXTEEN
NEW ENGINE TECHNOLOGIES

Two areas of technology which are significant for the oil industry are hybrid engines, and alternative, non-gasoline or non-diesel, fuels

Hybrid engines

Hybrid automobiles are those which derive energy from two or more sources of power. The most common type of hybrids are gasoline-electric vehicles. Natural gas-electric vehicles are rarely found. Hybrids which use electricity as one of their sources of power are referred to as hybrid electric vehicles (HEV).

HEVs carry batteries. The batteries usually never need to be plugged into an outlet and derive all of their power from a gasoline engine. These batteries enable a gasoline powered vehicle to have a smaller horsepower engine than an equivalently sized vehicle would normally have. The batteries provide extra power when needed, such as when the driver is accelerating from a stop. A smaller engine is more efficient as it weighs less and also uses less fuel to overcome its own inertia. Hybrid automobiles use the energy produced from braking to charge the batteries – which is called regenerative braking. When one steps on the brakes, a generator rather than brake pads slows the car. Hybrid automobiles also are often designed so that the gasoline engine shuts off when idling, to save fuel. This feature is referred to as an idle start/stop mechanism. Some hybrid vehicles use the batteries to maneuver at slow speed, only activating the gasoline engine when high speeds are required.

Battery technology has not advanced as quickly as many would have hoped in recent years. Three primary challenges with batteries are cost, weight and the long time they require to charge.

Until recently most hybrid automobiles used nickel metal-hydride (NiMH) batteries. Alternative lithium-ion (Li-ion) batteries occupy roughly one third of the space and weight of the equivalently charged NiMH batteries. Li-ion batteries are commonly used in mobile phones and laptops. Although Li-ion batteries are currently more expensive than NiMH batteries, they are expected to become the predominant hybrid battery within a few years.

HEVs produce better mileage in the city than on the highway, which is the opposite of regular gasoline engines. This is because of the regenerative braking and the way the gasoline engine shuts off at stop lights or stalled traffic. On a highway, hybrids are less efficient because they have to carry the heavy batteries which do little to aid the engine at highway speeds.

In addition to HEVs, there are bi-fuel, or dual-fuel vehicles which are also considered hybrids because they use more than one source of power. Dual fuel vehicles often use diesel and compressed natural gas (CNG) or diesel and propane.

Dual-fuel hybrids are mostly used on city buses because of their relatively low emissions and the lack of CNG and propane filling stations outside of major cities. In addition to environmental benefits, dual-fuel vehicles can take advantage of the cheaper of two fuels. Despite their ability to switch fuels to the cheapest source, dual-fuel systems are expensive to maintain and will likely continue to be a niche environment-driven market.

Hybrid electric vehicles make sense if one drives predominantly in stop-go city driving, but require high gasoline prices to pay for the additional cost of the batteries, electrical motor and regenerative brakes.

Alternative, non-gasoline and non-diesel, transportation fuels

Four common alternative fuels are natural gas, propane, ethanol and biodiesel. Compressed natural gas (CNG) is natural gas stored under compression as a gas in a special tank in an automobile. Propane is commonly referred to as liquefied petroleum gas (LPG) or autogas by consumers. Ethanol is an alcohol produced by fermenting and distilling corn or sugar cane. Ethanol can be blended with gasoline or can be used as a replacement for gasoline. Biodiesel is a diesel engine fuel produced from plant oils, such as soybean oil, or animal fats. Biodiesel can be used by itself or blended with hydrocarbon diesel. Biodiesel is burned in a diesel engine and cannot be used in a gasoline engine.

All of the above fuels are niche transportation fuels because crude oil-based gasoline and diesel are more easily carried on a vehicle, more easily transported to end users, and most importantly, gasoline and diesel are less expensive.

A fifth alternative to gasoline is electric power stored in batteries. Electric powered cars typically use lead-acid batteries to store power. The cost to produce NiMH batteries is over ten times that of lead-acid batteries at present. The range on electric vehicles is fairly short, usually less than 150 miles, whereas most gasoline and diesel tanks permit 300 miles or more. Recharging batteries can take hours. Electric powered vehicles are typically used in situations, such as in cities or golf courses, where a short range limitation is not an issue. As the electricity to recharge the batteries typically comes from coal and nuclear power-generation stations, using electric powered vehicles does not necessarily reduce pollution; it simply changes the location where pollution occurs.

The final alternative to gasoline is hydrogen fuel cell technology. Hydrogen fuel cells produce electricity by reacting hydrogen (H) and oxygen (O) to produce water (H_2O). The hurdles facing hydrogen fuel cell technology are hydrogen storage, production and delivery.

Hydrogen storage: Hydrogen contains more energy potential by weight than hydrocarbon fuels, which makes it very useful for space vehicles where weight is a primary issue. However it is very difficult to compress hydrogen such that it contains the same energy by volume of hydrocarbons. This makes it difficult to fit a large amount of hydrogen into a small automobile gas tank.

A vehicle powered by hydrogen needs very large tanks to have a similar range as a typical gasoline vehicle. Larger tanks are more vulnerable in car crashes as hydrogen is highly flammable.

In order to squeeze a large weight of hydrogen into a small space, high pressures and/or extreme cold are one possible, but awkward, solution. Another alternative, with more promise, is the process of storing hydrogen by chemically binding it to solids such as powders or liquids. Hydrogen readily combines with some metals, to form metal hydrides. Hydrogen also readily binds to carbon, such as in carbon nanotubes. If the combination process is easily reversible, such that the hydrogen can be simply unbound from the metal or carbon, the result would be a greater and more inexpensive storage capacity compared with storing hydrogen as a compressed gas or super-cooled liquid.

Hydrogen production: Over 95% of hydrogen produced today is from steam methane reforming (SMR), a process in which high-pressure steam is applied to methane, the primary component of natural gas. Natural gas is not cheap, and one has to wonder why one would bother expending energy converting the natural gas to hydrogen, when one can simply burn the natural gas in an automobile using existing technology. Natural gas as a source of hydrogen is not the solution.

The ultimate goal in hydrogen production is to split water. Splitting water requires large amounts of electricity. The additional electricity to split water comes from burning coal or nuclear power, which are realistically the only two sources large enough to produce sufficient electrical power for large scale hydrogen production.

In short, the so called hydrogen economy is mere hype unless one builds a large number of nuclear and coal power stations to fuel the hydrogen economy.

Hydrogen delivery: The infrastructure for delivering hydrogen in large volumes to consumers will have to be built from scratch. This is the least of the challenges associated with hydrogen fuel cell technology.

CHAPTER SEVENTEEN
OIL PRICES

Wholesale trade in physical oil occurs via either term supply contracts or spot supply contracts.

Term supply contracts, also referred to as off take agreements, are contracts in which a seller of oil agrees to supply a buyer with specific quantities of oil at scheduled dates in the future. The oil is known as term barrels. The majority of oil traded physically is by means of term supply contracts.

The opposite of term supply contracts are spot supply contracts. Spot supply contracts are contracts for delivery of a quantity of oil at a specific location as soon as is operationally possible, often within a day or two. Spot supply contracts only account for a small portion of international oil trade; however, as they are barrels sold to the highest bidder willing to take immediate delivery, the prices at which spot supply transactions occur will reflect up to the minute oil market news and any short term changes to supply and demand.

Benchmark pricing

Until the 1980s, there was very little spot oil traded. Oil prices in term supply contracts were set at fixed prices by major oil companies unilaterally or in direct negotiation. The pricing was called contract pricing and a price agreed in a contract would not change for weeks, months or even years at a time, depending on the contract term.

Although fixed contract pricing is still used in parts of the oil industry, since the mid-1980s almost all pricing for term supply contracts has changed to benchmark pricing which references spot supply prices. Benchmark prices used in this manner are also called price markers

How spot benchmarks are set

At the end of each trading day, a snapshot is taken of spot supply trading in oil prices of several benchmark grades of oil. The daily snapshot of oil prices comes from two sources: futures exchanges and trade journals.

Futures exchanges: Oil trades on two main futures exchanges, the NYMEX (New York Mercantile Exchange) and the ICE (InterContinental Exchange) based in London. A small amount of oil is traded on other futures exchanges in Tokyo, Shanghai and Dubai.

<u>Trade journals:</u> Oil also trades in what is called the OTC (Over-The-Counter) market. The OTC market is regulated and well-defined by common law and industry convention. The OTC market is not physically based anywhere. Traders from all over the world deal directly with each other over the phone, using a broker, internet instant messaging, or electronic trading platforms. Contracts typically reference either New York or London contract law. In order to trade in the OTC market one must be relatively experienced in oil trading and well capitalized as one may have to establish short term credit lines with counterparties. Trade journals monitor trading which occurs in OTC markets, and produce daily assessments of each benchmark oil price. The daily price assessment often occurs in what are known as pricing windows, which is the period of time the trade journal will monitor spot-market trading very closely, such as for 30 minutes or so at the end of the trading day, and arrive at an assessment price based on the transactions during this time period.

The most widely used trade journal for OTC benchmarks in oil markets is Platts Oilgram, which is owned by McGraw-Hill. Platts publish a high and low price for several grades of oil at the end of every business day and traders will use the MOP, or Mean of Platts high and low, prices as their benchmarks. Other oil market trade journals which are used for specific benchmarks are Petroleum Argus, APPI, OPIS, and CMAI among many others.

Spot market oil is traded by oil buyers and sellers 24 hours a day in the open international free market, although by convention almost no oil trading occurs on Saturday and Sunday. Spot market oil traders tend to be physically based in Singapore, London and New York, all of which have stable legal systems and a tradition of free markets. Geneva and Connecticut are also popular bases for oil trading operations.

Some NYMEX and ICE futures and OTC trade journal grades of oil have become spot market benchmarks, or markers, as a result of market convention. Non-benchmark grades of oil, such as the price in your local area or country are set at a premium or discount to these benchmark grades. The premium/discount is determined by transportation costs, taxes and quality differences between the grade of oil in your local area and the benchmark grade.

The following chart shows only the most commonly referenced benchmark oil prices (Fig 17-1). There are over a hundred benchmark prices covering most grades of petroleum around the world.

Main global oil market benchmarks

Fig. 17-1

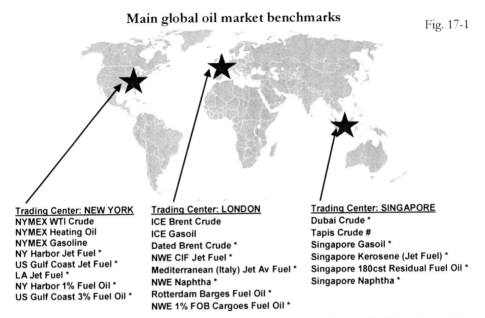

Trading Center: NEW YORK
NYMEX WTI Crude
NYMEX Heating Oil
NYMEX Gasoline
NY Harbor Jet Fuel *
US Gulf Coast Jet Fuel *
LA Jet Fuel *
NY Harbor 1% Fuel Oil *
US Gulf Coast 3% Fuel Oil *

Trading Center: LONDON
ICE Brent Crude
ICE Gasoil
Dated Brent Crude *
NWE CIF Jet Fuel *
Mediterranean (Italy) Jet Av Fuel *
NWE Naphtha *
Rotterdam Barges Fuel Oil *
NWE 1% FOB Cargoes Fuel Oil *

Trading Center: SINGAPORE
Dubai Crude *
Tapis Crude #
Singapore Gasoil *
Singapore Kerosene (Jet Fuel) *
Singapore 180cst Residual Fuel Oil *
Singapore Naphtha *

NYMEX = New York Mercantile Exchange (Settlement Prices come from daily close of 1st Nearby Futures Contract)
ICE – InterContinental Exchange (London) (Settlement Prices come from daily close of 1st Nearby Futures Contract)

* = Indicates that the Settlement Prices come from 'Platts Oilgram' Daily journal
# = Indicates that the Settlement Prices come from the 'APPI' Twice Weekly

Using benchmarks in formula pricing

The exact calculation by which benchmark prices are used is specified in oil supply contracts. The use of calculations in a supply contract is known as formula pricing, as one or several benchmarks will be referenced in a formula. For example, a supply contract formula for a retail gas station in New Jersey may say that the station has to pay the previous business days NYMEX gasoline price (the benchmark), plus a transportation cost, when it receives a delivery. As another example, the price of jet fuel at an airport in Northern Spain may be linked to the price of a benchmark jet fuel price which is assessed by Platts trade journal based on OTC trading for delivery at a major oil hub in NWE (North-West-Europe.) Jet fuel at an airport in Ireland or Portugal may also be linked to the same benchmark price of jet fuel at NWE. As yet another example, diesel prices at a tanker truck depot in Columbus, Ohio may be linked to the benchmark price of heating oil, which is similar to diesel, on the NYMEX futures exchange.

Formulae sometimes reference more than one benchmark, but the general concept remains the same.

Even OPEC crude oil producers use benchmarks and formula pricing in sales contracts for their crude. Individual OPEC producers use this formula to arrive at an Official Selling Price, or OSP, for each type of crude. The OSP premium/discount to the benchmark(s) in each producer's formula is usually revised at the end of each month to apply to crude deliveries taking place the following month in order to keep the crude oil competitively marketed to refineries. Again, the premium or discount of the OSP to the benchmark(s) will take into account the quality difference between the benchmark crude and the crude grade being priced in addition to relative demand for the crude being priced.

For example, Nigeria, an OPEC producer, may sell its Bonny Light crude at Platts (the trade journal) Dated Brent crude (a European crude oil benchmark) plus or minus an official premium or discount. This OSP formula method is sometimes known as related pricing. In the example of Nigerian Bonny Light, the crude is being sold on a Dated Brent-related basis.

Most non-OPEC crude oil producers also use formulae linking the price of the crude they are selling to a benchmark. The formula price is again adjusted on a daily or monthly basis by adding or subtracting a premium or discount, often called a constant, although it changes periodically. Mexico, for example, a non-OPEC crude oil producer, sells three of its crude oils on a formula basis which use several benchmarks, and which are available publicly:

Isthmus crude = 0.4(WTS + LLS) + 0.2(Dated Brent) – constant

Maya crude = 0.4(WTS + US Gulf No.6 oil 3%) + 0.10 (LLS + Dated Brent) – constant

Olmeca crude = 0.333 (WTS + LLS + Dated Brent) + constant

Where: Isthmus, Maya and Olmeca are three grades of Mexican crude oil and WTS (West Texas Sour crude), LLS (Light Louisiana Sweet crude), Dated Brent crude and US Gulf No.6 oil 3% (a grade of residual fuel) are four different benchmarks assessed by Platts daily.

Variation and alternatives to formula pricing
A variation of benchmark formula pricing is called freight netback pricing, whereby the price for a grade of oil at one location is linked to a benchmark oil price at another location, adjusted for a benchmark freight cost published by the Baltic Exchange, Platts, or some other trade journal. This method is used to determine prices in obscure locations where little trading in physical oil occurs.

Another alternative is called posted pricing. Posted prices are bids from refineries and others seeking to buy crude. Posted prices are named literally for sheets of prices which, in years past, were posted by potential buyers at refinery gates. Posted prices are not an obligation to buy, but are starting points for negotiating a market price. Very often premia are paid above posted prices, known as posting-plus pricing.

In the US, some crude oils are traded based on posted prices. There have been issues over the years with regard to posted prices being low-balled in order to avoid paying federal government royalties. Most posted prices these days appear to be closely correlated to benchmark prices, though posted prices themselves are not used as benchmark prices.

Oil price terminology at various delivery points

When physical oil prices are quoted, they are in relation to a certain point in the delivery chain. Crude oil accounts for most of the price. However, each step in the refining and transportation chain adds a little onto the cost of the oil. Not all grades of oil follow the following pricing sequence, or they may follow, but in a different order:

Wellhead price is the price of crude oil at the mouth of the well.

Cargo price is the price of oil (crude or product) on a large ocean-going tanker.

Refinery gate price is the price of crude oil going into a refinery or finished products as they leave a refinery.

Barge price is the price of oil, on a barge vessel usually on rivers or close to shore with a capacity typically between 8,000 and 50,000 bbls. Some barges have a capacity of up to and over 100,000bbls. Barge prices, because there is an additional handling cost, are higher than cargo prices.

Pipeline price is the price of oil at a specified point in a pipeline.

Rack price is the price of oil at a wholesale truck tanker terminal where delivery trucks load oil to bring to filling stations, homes and offices.

Dealer tank wagon (DTW) price is the wholesale price of oil delivered to retail outlets such as a gasoline stations. This is the price the gas station owner pays.

Retail price is the price of oil to end user, which typically includes a retail tax and a small markup from the retail station owner.

The effect of government taxation on retail prices

The main reason for differences between retail oil prices in various parts of the world is government taxation. In particular, developed Western European nations have very high population densities and this allows the use of mass transportation and higher taxes on automobile fuels. Developed countries with low population densities such as the US, Canada and Australia are unable to

Fig. 17.2

implement the same transportation mechanisms and thus the gover
these countries find it more difficult to tax automobile fuels. Anot
for differences between retail prices is that, due to the lack of eco
scale, smaller nations tend to have less efficient distribution and
mechanisms (Fig. 17-2).

Example of retail price difference between US and UK

January 2009 **Gasoline** Price in U.K.	
Retail Price	US $**4.80**/gallon (GBP 0.87/liter)
Govt Tax	3.40
Distribution and marketing	0.40
Refining	0.04
Crude Oil	0.96

January 2009 Gasoline Price in U	
Retail Price	US $**1.66**/gallon
Govt Tax	0.39
Distribution and marketing	0.27
Refining	0.04
Crude Oil	0.96

Oil prices and the US dollar

Oil is traded globally in US dollars. It is often asked why oil is not m
traded in Euro, Yen or some other non-US dollar currency. This st
comes up any time the US dollar is weakening against other major curt
for anti-US political reasons.

The international oil market uses the US dollar for very simple reas
most important being that oil itself is a currency. As a currency, oil is
in value to the US dollar. If the US dollar weakens against oil cons
currencies then, all other things being equal, oil prices will rally in L
terms. If the US dollar strengthens against oil consumption currencies
other things being equal, oil prices will fall in US dollar terms.

For several additional reasons, it is preferable for oil be traded in
currency and the US dollar in particular. Firstly, the US dollar is the mo
convertible and liquid currency with the lowest transaction cost, which
lower costs for consumers. Secondly, trading oil in a single currency i
easier to compare oil prices internationally. More efficient arbitrage er
more fair, transparent and competitive market for oil, and lower pr
consumers. Thirdly, if one trades oil in multiple currencies before it reac
end consumer then each of the currency transaction costs have to be ar
the price of oil. By trading oil internationally in a single currency, such
US dollar, consumers only have to pay one currency transaction cost.
oil producers not wishing to hold US dollars can convert their dolla
whatever currency they wish immediately.

summary, nobody tells the international free market that it must use US
ollars when trading oil. The oil industry chooses to use the US dollar because
makes practical economic sense for everyone involved.

il price units

lthough oil at a retail level is sold in either gallons or liters, depending on the
ountry it is sold, oil in the wholesale market only trades in barrels (bbls),
allons (gals), and metric tonnes (mt) in different regions around the world.
he conversion between volumes is straight forward. There are approximately
79 liters in a US gallon and there are 42 US gallons in a barrel. Conversion
tes from metric tonnes to barrels, and the units in which different grades of
l are traded in wholesale markets worldwide are as follows:

Commonly used oil market units   Table 17-1

	Region	Traded unit	Commonly used conversion (bbls/mt)
Crude oil	Worldwide	bbls	7.58 bbls*
Gasoil/Diesel/Heating oil	Asia	bbls	7.45
	Europe	mt	7.45
	US	gals	n/a
Jet fuel/Kerosene	Asia	bbls	7.88
	Europe	mt	7.88
	US	gals	n/a
Gasoline	Asia	bbls	8.33
	Europe	mt	8.33
	US	gals	n/a
Naphtha/Natural gasoline	Asia	mt	9
	Europe	mt	8.9
	US	gals	n/a
Residual Fuel Oil	Asia	mt	6.5
	Europe	mt	6.35
	US	bbls	n/a

*Used only for **Brent** crude in Europe.

CHAPTER EIGHTEEN
FORWARD OIL MARKETS:
FUTURES and SWAPS

The price of oil for immediate delivery is called the spot price or cash price. The price of oil for delivery at a specified date in the future is called a forward price. Oil for delivery in the future is most commonly traded using exchange traded futures contracts and Over-the-Counter (OTC) swap contracts.

A useful feature of futures and swaps is that, if one chooses, one never actually has to take or make delivery of physical oil. As one doesn't actually have to get involved in the physical oil market such contacts are referred to as paper barrels as opposed to real physical wet barrels. Paper barrels are also called derivative instruments as their value is derived from physical oil prices but may not necessarily result in ownership of physical oil.

In certain circumstances, such as holding a futures contract past its expiry date, one can take or make delivery of physical oil. In this way, paper contract values ultimately converge to physical oil prices. Less than 1% of paper barrel contracts such as futures or swaps are converted into physical oil, but it is still important that the link between paper and physical exists as it ensures that paper contracts have real, and not just theoretical, underpinnings.

Charting the various dates and forward prices of either futures or swaps creates a forward curve of prices going out into the future (Fig 18-1). A forward curve is also called a forward strip or term structure. Each benchmark grade of oil has its own forward curve.

Fig. 18-1

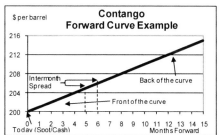
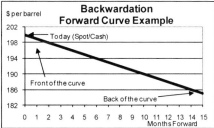

Parts of the forward curve closer to expiry are referred to as being at the front-end of the curve and further along the curve into the distant future is referred to as the back-end of the curve. Price differentials between the front of the curve and further out parts of the curve are referred to as front-to-backs, or time spreads. The price spread between a one month and the next is called an intermonth spread or a switch.

Forward curves can be in contango, which is upward sloping as one goes further into the future, or in backwardation, which is downward sloping. Contango usually occurs when there is too much oil around today relative to today's demand and implies that there may be money to be made in storing oil as one can sell oil on the forward curve at a higher price than today's low price. Backwardation usually occurs when there is a relative shortage of oil today and the forward curve discourages storing oil as one can sell oil today at a better price than in the future.

As time progresses oil forward curves flip from backwardation to contango and vice versa. In addition to the relative availability and scarcity of physical oil today, contango and backwardation at the front of the curve are caused by shortages or gluts of storage facilities, respectively. Contango and backwardation can also be caused by imbalances in the volume of oil producer and consumer transactions along the forward curve.

Contango intermonth spreads are limited by the cost of building and operating storage and the finance charge to temporarily buy oil, all of which is termed the cost of carry. When the marginal monthly cost of storage is the same as an intermonth contango spread, the market is referred to as being at full carry. Backwardation is unlimited by any such constraint and intermonth price spreads in backwardation have no theoretical limit.

Backwardation is far more common than contango. The economist John Maynard Keynes in his 1930 'Treatise on Money' stated that backwardation was more "normal" because of the tendency in the 1920s of commodity producers to hedge more than commodity consumers. While such overweight producer selling situations do occur, there are many cases of oil forward curves where there is persistently more consumer buying than producer selling (in jet fuel, and residual fuel, for example). A more robust theory as to why backwardation is more predominant than contango is that traders in most commodity markets can easily arbitrage a contango curve by building more storage and selling forward to finance the storage. As the carry trade is such a popular and simple trade, contango only exists for brief periods before new storage facilities are constructed and forward curves are sold. However, backwardation is not a market condition which a commodity market can arbitrage out of existence.

Unlike interest rate forward curves, oil forward curves usually show actual prices in dollars per barrel, metric tonne, or gallon, and not interest points. One can back out the yield from borrowing or owning oil. A forward curve in contango implies that oil as an asset has a negative yield for those owning oil (owner of oil has to pay storage costs to continually own oil), and backwardation implies that oil has a positive yield (owner of oil can collect money by selling oil today and buying it back cheaper in the future).

The very back end of the oil forward curve is the current market estimate of finding and development costs for marginal crude oil barrels (Fig. 18-2). In other words, if one were to go out today and raise capital, explore for oil, hopefully find oil, build production facilities and get the oil to market – how much would that oil have to be priced at to provide a decent profit by the time one had all of that completed? During the 1990s the long-term marginal F&D cost was estimated by the market to be around $20 per barrel as relatively easy to find land-based fields were being exploited. However, these forward curve market estimates were proven very wrong. Forward curves historically have not provided reliable forecasts. Since 2003 the marginal F&D cost has risen to above $80/bbl as conventional crude discovery continues its long-term decline and high cost oil, such as that found deep offshore and oil sands, is now the source of marginal barrels.

NYMEX WTI Crude Oil Term Structure on first trading day of each year

Fig. 18-2

Crude oil production is not very seasonal and so crude forward curves tend to be fairly smooth. Demand for some refined finished products is seasonal and is reflected in forward curves (Fig 18-3). The seasonality of individual products nets out which is why crude oil is not very seasonal. Heating oil, propane, gasoline, jet fuel, and bitumen have the most pronounced seasonality of demand. Relatively constant production from refineries means that seasonal products which are refined in low demand periods must be stored until high demand periods. This results in seasonal forward curves with contango and backwardation each year. The contango in the forward curve encourages storage operators to buy oil products in the low demand seasons and store them until the high demand season.

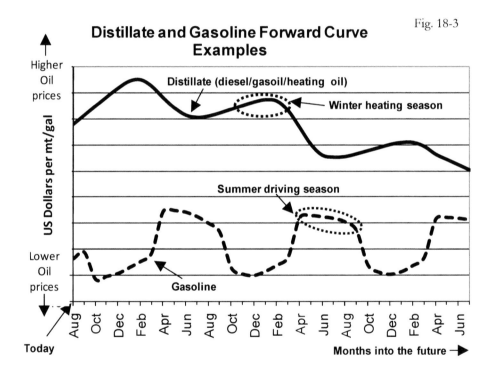

Fig. 18-3

In addition to forward curves for various grades of oil, there exist quite liquid forward curves for petrochemicals, plastics and oil tanker freight rates. In this chapter, however, we will concentrate on forward curves for oil, as similar principles apply to the other markets.

Spreads between oil forward curves

As there has always tended to be spare capacity globally in oil storage, refining, and transportation, changes in spreads due to fundamental economic and seasonal differences have typically not been as significant as changes in absolute prices of crude oil up or down. This means that oil spot prices and forward curves are relatively highly positively correlated (Fig. 18-4).

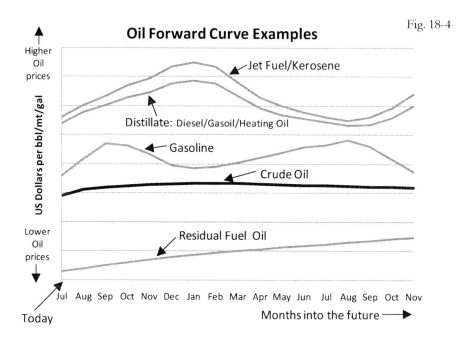

Fig. 18-4

Oil traders will talk about having either flat price positions or spread positions. A flat price position is also known as an outright position. Having a flat price position means that a trader is exposed to the simple directional movement of oil prices up or down, for example, from \$101/bbl to \$102/bbl or \$100/bbl. A spread position means a trader is exposed to the price differences between, for example, two grades of oil, or the same grade of oil over two different time periods, or the same grade of oil in two locations. The market risk associated with any spread is also known as basis risk.

Although there are exceptions, spreads are more often than not quoted as positive numbers, i.e. the higher price minus the lower. If one is buying a positive spread, ones market value increases in value if the spread widens in absolute terms, and if one is selling a positive spread ones market value increases in value if the spread narrows in absolute terms. Four common oil market spreads are crack, arbitrage, relative value and time spreads.

Crack spreads, also known as a refinery margin spreads, are finished product prices, such as gasoline, minus crude oil prices. This is the refinery's margin for refining crude oil. Frequently, a refinery will look to hedge the spread between several of its finished products against crude oil in what is known as a basket crack spread. Some commonly referred to basket crack spreads in the US are the 321 spread (if one is selling the spread: buy 3 crude, sell 2 gasoline, and sell 1 heating oil futures/swaps) and the 211 spread (if one is selling the spread: buy 2 crude, sell 1 gasoline, and sell 1 heating oil futures/swaps.)

Arbitrage spreads, or arbs, are oil prices in one region minus prices for the same, or a very similar, grade of oil in another region. For example, jet fuel in New York versus Rotterdam. Arbs are primarily a function of transportation rates between regions. Although there are thousands of arbs, the most well known and commonly traded arb in the oil market is the TI-Brent crude oil arb, sometimes referred to as the Atlantic arb.

Relative value spreads, also called diffs, are the spreads between two different but closely related oil products, such as gasoline and diesel, from the same region. An example of a commonly traded relative value spread is the diff between naphtha and gasoline prices in Rotterdam. A particular type of relative value spread is known as a regrade, which is the relative value spread between different grades of gasoline in the US, or between kerosene and gasoil in Asia.

Time spreads, also known as a calendar spread, are differences between the price of oil in one time period versus the price for the same grade of oil in another time period. Time spreads are used to trade seasonality and storage economics among other reasons. An example of a time spread is the price difference between December 2012 and December 2013 crude oil. A variation on the time spread is the box spread, which is a time spread involving two crack, arbitrage or relative value spreads. For example, one could buy a crack spread for September and sell a crack spread for December.

Several spreads between grades of oil are more frequently traded than others. These spreads result in a family tree of spreads between benchmark grades of oil around the world as shown on the following diagram (Fig. 18-5).

Fig. 18-5

Commonly traded spreads between benchmark grades of oil

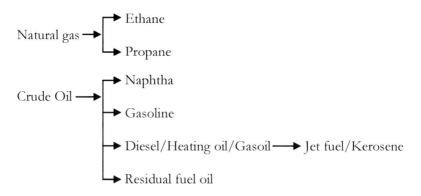

Exchange traded futures and OTC Swaps

The two basic types of forward oil derivative contracts are exchange traded futures contracts and Over-the-Counter (OTC) swap contracts.

Exchange traded futures contracts

A futures contract gives one the right to buy a standardized quantity of oil for delivery at a particular date in the future.

If one buys futures then one is said to be going long futures and the trade will make money if market prices rise. If one sells futures then one is said to be going short and the trade will make money if market prices fall. If one has a long or short position and one adds a trade in the opposite direction then one is unwinding, or closing out, a market position. If one had a long or short market position but then bought or sold an equivalent amount to unwind the position completely then one is said to have squared out a position.

To trade futures one opens an account with a futures broker and posts an initial cash margin deposit. The amount of initial margin depends on how many contracts one wishes to buy or sell. The initial margin one posts is passed from the broker to an exchange clearinghouse which manages the credit of the entire exchange. If the market subsequently moves against a trader, he/she will be asked to immediately post additional margin, called variation margin, to cover the loss. The notice to post variation margin is called a margin call. If one does not post the variation margin then the position will be unwound by the exchange. The exchange will use any margin it has received to cover any costs on closing out the position. The variation margining process occurs at the end of each trading day based on end of day settlement prices.

There are only five actively traded, or liquid, futures contracts for petroleum globally, three listed on the NYMEX and two on ICE. The following are the five oil futures contracts (Table 18-1):

Table 18-1

Main oil futures contracts

NYMEX (New York) Futures	Ticker root	ICE Futures (London)	Ticker root
WTI Crude Oil (contract size: 1,000 bbls)	CL	**Brent Crude** (contract size: 1,000 bbls)	LCO (Reuters) CO (Bloomberg)
Heating Oil (contract size: 42,000 gals)	HO	**Gasoil**** (contract size: 100 metric tonnes)	LGO (Reuters) QS (Bloomberg)
Gasoline* (contract size: 42,000 gals)	RB		

Notes: *42,000 gallons = 1,000 barrels; **Gasoil is a European product similar to heating oil in the US;
1 metric tonne of gasoil = 7.45 barrels.

The NYMEX is regulated by the US Commodity Futures Trading Commission (CFTC) and is a public company. The ICE Futures exchange in London is also a public company and is regulated by the UK Financial Services Authority (FSA). There are other small oil exchanges in Japan, China, and the UAE but for a variety of reasons they are not widely used outside of those countries (Table 18-2).

Table 18-2

Global oil futures contracts

Oil Futures Contract	Exchange	Annual Contract Volume	Contract Size	Average number of bbls traded per day
WTI Crude Oil	New York Mercantile Exchange	121,525,967	1,000bbls	467,407,565
Brent Crude Oil	Intercontinental Exchange	59,728,941	1,000bbls	229,726,696
WTI Crude Oil	Intercontinental Exchange	51,388,362	1,000bbls	197,647,546
Gasoil	Intercontinental Exchange	24,509,884	100mt	70,230,245
Gasoline	New York Mercantile Exchange	19,794,439	42,000gals	76,132,458
Heating Oil	New York Mercantile Exchange	18,078,976	42,000gals	69,534,523
Gasoline	Tokyo Commodity Exchange	7,529,706	50KL	9,107,772
Fuel Oil	Shanghai Metals Exchange	12,005,094	10mt	2,932,013
Kerosene	Tokyo Commodity Exchange	2,350,819	50KL	2,843,501
Brent Crude Oil	New York Mercantile Exchange	548,128	1,000bbls	2,108,185
Crude Oil	Tokyo Commodity Exchange	1,489,018	50KL	1,801,085
Gasoline	Central Japan Commodity Exchange	3,635,329	10KL	879,443
Heating Oil (US)	Intercontinental Exchange	205,072	42,000gals	788,738
Oman Crude Oil	Dubai Mercantile Exchange	200,892	1,000bbls	772,662
Kerosene	Central Japan Commodity Exchange	2,685,345	10KL	649,627
Gasoline (US)	Intercontinental Exchange	14,805	42,000gals	56,942

Source: The CRB Commodity Yearbook 2008

Futures contracts prices are looked up using a ticker symbol when using professional market data software, such as Reuters or Bloomberg. The futures ticker, similar to a stock ticker uniquely identifies a futures contract. A futures ticker code follows the following format:

	Ticker Root	+	Month Code	+	Last Digit of Year
Example:	**CL**		**Z**		**5**
	(NYMEX crude oil)		December		2015

Therefore, **CLZ5** is the ticker for NYMEX WTI crude oil, December 2015. The month codes are follows:

Table 18-3

	Code		Code
January	F	July	N
February	G	August	Q
March	H	September	U
April	J	October	V
May	K	November	X
June	M	December	Z

The futures contract in the month which is closest to expiry is known as the first nearby contract, front month contract, or prompt contract. The monthly contract second closest to expiry is known as the second nearby contract.

The terms red, blue, and green are sometimes to describe contracts one, two and three years ahead. For example, if today is October 12, 2011, the front month contract will be the November 2011 contract; the second nearby contract will be the December 2011 contract; red dec will be the December 2012 contract; blue dec will be the December 2013 contract; and green dec will be the December 2014 contract.

If one holds onto a futures contract past its expiry date, which very few do, one may have to take or make delivery of physical oil at the delivery location outlined in the futures contract specifications. Most, but not all, traders square out front contract positions before they expire so that they don't have to deal with physical delivery.

At the end of trading each day, settlement prices for each futures contract are posted by the exchange. These settlement prices are determined by trading over the last few minutes of the trading day. Daily settlement prices are used for mark-to-markets, which show the financial state of positions at the end of each trading day and are used for calculating variation margin payments required.

Because of the small initial margin required by an exchange clearinghouse, a trader can get exposure to $70,000 worth of oil for a deposit of around $5,000, which is leverage of 14 to 1. With such high leverage, small market movements can result in large profits or losses relative to the initial margin deposit. This available leverage is why futures are frequently viewed with fear by the general public. Those unable to manage the high leverage are known as weak hands and are often stopped out of good positions because of very small adverse market movements. One could reduce the leverage and become a strong hand by posting the full value of the oil ($70,000 in our example above) with ones broker, or simply keep it in one's own bank account ready to pay margin calls at the end of each day if the market moves. Such a strategy of reducing leverage of a long oil position to zero is called fully collateralizing an oil position.

Over-The-Counter (OTC) swap contracts
An OTC contract is similar to a futures contract in that one can buy and sell oil derivative contracts for dates out into the future. The primary differences between futures contracts and an OTC contracts are that futures contracts are standardized and traded on a futures exchange, whereas OTC contracts are completely customizable and most often traded directly between two parties referred to as counterparties. The OTC market primarily caters to large organizations and not individuals.

There are industry standards for the most common OTC transaction types. An industry funded entity called the International Swaps and Derivatives Association (ISDA) produces standardized contracts and individual contract clauses which OTC oil traders use.

One can create any financial instrument one wishes in the OTC market. The most common type of OTC transactions are swaps. They referred to as plain vanilla structures as they are the most standardized and common. More complicated structures are called exotics.

The term swap comes from the fact that the transaction allows oil price movement risk to be swapped from one trader to another. For example, an airline naturally faces the risk that jet fuel prices will rise. The airline can buy jet fuel on paper via a swap transaction which swaps the risk of prices rising to a bank. If the price of jet fuel rises, the bank will pay the airline the price increase multiplied by the number of barrels, or metric tonnes agreed in the swap. If the price of jet fuel falls, then the airline pays away the benefit of the lower prices to the bank.

In OTC swap terminology, buying oil, whereby a contract increases in value as prices rise and loses value as prices fall, such as an airline hedging against jet fuel price increases, is referred to as paying fixed and receiving floating, or going long a swap.

Selling oil, whereby the contract increases in value as prices fall and loses value as prices rise, such as an oil producer hedging against a fall in crude oil prices, is referred to as receiving fixed and paying floating, or going short a swap.

In addition to trading or hedging the simple direction of oil via a swap, a trader can buy or sell the differential, or spread, between two grades of oil, or the same oil over different time periods, in what is referred to as a floating-for-floating swap. For example, an oil refinery may sell a refinery margin spread, which involves buying crude oil swap and simultaneously selling a finished product, such as gasoline, swap.

Whereas exchange-traded futures contracts expire and can become physical oil (although almost all traders square out of their position before a contract expires to avoid such a physical situation), the OTC market is almost always cash settled with no physical component.

OTC swap contracts are usually cash settled on expiry against the average oil price over a calendar month, or other long period of time. Such average settlement is called Asian settlement, average-rate settlement, or par forward settlement. For example, an OTC swap for December would be cash settled against the simple average of the twenty or so daily settlement prices for the grade of oil over the month of December.

The OTC market is flexible, and one can create an OTC swap that settles against any price, even a single day's price, a so called bullet settlement.

The daily prices used to settle monthly average OTC swaps are typically benchmark oil prices published by a futures exchange or a trade journal.

An important difference between exchange-traded futures contracts and non-exchange traded OTC contracts are that oil futures contract cash flows are not discounted for time-value, whereas OTC swaps are discounted. The reason discounting is important is that those hedging futures positions with swaps need to do slightly more volume in a trade with swaps to achieve the same present value profit and loss impact.

OTC documentation
Non-legally binding term sheets outline the terms of an OTC trade before it has been transacted and is simply to ensure that everyone clearly understands the potential trade. Term sheets are usually only used if the trade is somewhat unusual or non-standard or if the counterparty in the trade is new to OTC transactions.

Trade advices are issued within minutes or a few hours of an OTC transaction being completed. Trade advices are non-legally binding documents used because the formal legal documentation from a bank or trade house back office can take up to 24 hours to be prepared and sent to a counterparty and the market may have moved significantly during that time. A trade advice allows counterparties to spot any errors quickly before a formal trade confirmation is sent out.

Trade confirmations are the legal documents which outline the terms of an OTC trade. Whereas term sheets and trade advices are optional, a trade confirmation is required for every OTC trade. Most counterparties agree to predefined International Swap and Derivative Association (ISDA) terms which are set by the ISDA organization. The use of predefined terms allows an OTC document to typically only be two or three pages.

Table 18-4

Differences between exchange-traded futures and OTC contracts

Exchange trade futures contracts	Over-The-Counter (OTC) contracts
Traded on exchanges only. Two main oil exchanges, NYMEX and ICE, with five liquid petroleum contracts between them.	Usually no exchange involved. Over 100 oil benchmark prices can be used. Some of the benchmarks may reference NYMEX and ICE futures prices, and many others refer to Platts trade journal prices
Standardized with fixed volumes	Customizable for any volume or expiry.
Most common contracts called futures	Most common contract called a swap
Upfront initial margin deposit required. Further variation margin required if trade starts to lose money. Margins are managed by the exchange clearinghouse	Counterparties are effectively lending each other the margin and extending credit lines to each other. Alternatively, bilateral margining between parties.
Exchanges regulations apply.	Industry (ISDA) standards and common law used
Contracts expire on single day, called European or bullet expiry	Contracts most commonly expire against average, called Asian or average-rate expiry, of prices over a month – but can be made to settle against average of any length period from a single day to several years
Contracts expire into physical oil although most traders net out positions before expiry to avoid physical aspect.	Contracts usually cash settled
Few traders hold contracts at expiry	Contracts settle into cash so almost all traders hold contracts into expiry
Anyone with enough cash to post margin can trade	Only relatively sophisticated counterparties, such as large corporations, are involved
Futures brokers execute trade on exchange and earn a brokerage fee	OTC brokers are used or direct dealing occurs between counterparties which have a large volume of trading to execute.
Cost of transaction is futures brokerage cost plus bid/offer spread	Cost of transaction is bid/offer spread plus, if used, OTC broker cost.
No discounting of quantity	Discounting for time value of quantity
Buying (going long) futures	Paying fixed price and receiving floating
Selling (going short) futures	Receiving fixed price and paying floating

The three most common types of OTC Swaps are:

1. Fixed price payer swap (contract increases in value as price rises.)
2. Fixed price receiver swap (contract increases in value as price falls.)
3. Spread swap (contract changes in value as spread changes.)

1. Fixed price payer swap example:

> Scenario: An airline wishes to hedge against rising oil prices.
> Fixed price payer (e.g. Generic Airline Company)
> Floating price payer (e.g. Best Bank)
> Fixed Price (e.g. $100.47 per barrel)
> Underlying benchmark oil (e.g. NYMEX WTI 1st Nearby Futures)
> Tenor (e.g. Calendar year 2012)
> Determination period (e.g. monthly average of daily settlements)
> Settlement payment (e.g. 5 business days following each month end)

2. Fixed price receiver swap example:

> Scenario: An African crude oil producer wishes to hedge against falling oil prices.
> Fixed price payer (e.g. Best Bank)
> Floating price payer (e.g. Exploration and Production Oil Company)
> Fixed price (e.g. $108.34 per barrel)
> Underlying benchmark oil (e.g. Platts Dated Brent)
> Tenor (e.g. fourth quarter of 2012)
> Determination period (e.g. monthly average of daily settlements)
> Settlement payment (e.g. 5 business days following each month end)

3. Spread swap example:

> Scenario: A US airline wishes to hedge against jet fuel prices rallying versus crude.
> Fixed price payer (e.g. Generic Airline Company)
> Floating price payer (e.g. Best Bank)
> Fixed price (e.g. $17 per barrel)
> Underlying benchmark oil (e.g. Platts USGC Jet Fuel 54 *MINUS* NYMEX WTI 1st Nearby Futures)
> Tenor (e.g. June 2012)
> Determination period (e.g. monthly average of daily settlements)
> Settlement payment (e.g. 5 business days following each month end)

Real world example of an OTC swap
Trader A at an airline has term supply contracts to buy physical jet fuel at various airports all over Europe. The price of all of his jet fuel purchases at various locations may be linked to the price of the main European jet fuel benchmark, Platts NWE (North West Europe) jet fuel which is published at the end of each day by Platts. This benchmark price moves up and down every day with supply and demand.

Trader A's management have set their jet fuel expense budget for the year and Trader A is concerned that a military buildup on the border of two large oil producing nations appears to be the precursor to a war he believes may begin in March. The war is not expected to last very long as the defending nation's military is poorly equipped. Oil supply to the world is likely only to be disrupted for a month or so. In addition to the potential disruption to global crude oil supply, Trader A is aware that jet fuel prices rally especially hard when there is military action because of the pull from increased military jet sorties.

On January 10th, Trader A at the airline phones Trader B at a bank and tells him he wants to hedge by buying 20,000 metric tonnes of Platts NWE Jet Fuel for the month of March via a swap. Trader A is not asking Trader B to supply physical oil, he merely wants to enter into a paper derivative contract. If this is the first time Trader A has hedged, Trader B's sales team may fax or e-mail Trader A a non-legally binding term sheet before they transact so that they both can see in paper exactly what they are potentially agreeing to.

Trader B shows Trader A a fixed price offer of US$1,510/metric tonne (jet fuel is traded in US$/mt in Europe) and they agree to transact at that price. The trade has now been 'done' verbally, i.e., once a trader actually says "you are done" the trade is a verbal contract and cannot be backed out of. The bank sales team may, within minutes, fax or e-mail a non-legally binding trade advice to Trader A with the details of the trade, again to ensure that everyone is clear about what has now been transacted.

Later that same day, Trader A and Traders B each have their back office staff fax each other legally binding ISDA-style (market standard) confirmation contracts, which are usually 3-4 pages outlining the transaction in more detail. The confirmations will contain or refer to full definitions of legal terms. Each counterparty signs the ISDA confirmations and fax them back to the other counterparty. Sometimes only one of the counterparties, usually the bank, prepares the confirmations and the other party simply checks it, signs it and faxes it back.

Following are the cash flows, or payments, that will occur at the end of March as a result of the example swap transaction:

	Trader A buying oil	**Trader B selling oil**
Fixed Leg:	Pays Fixed Price (e.g. $1,510/mt)→	Receives Fixed Price
Floating Leg:	Receives Floating Price	← Pays Floating Price

During the month of March the swap will 'price out' each business day. Cash settlement actually occur at the beginning of the following month, usually on the 5th business day of the following month, which in the example case, say is April 6th. On April 6th, Trader A and Trader B back office staff will calculate how much to pay, if anything. Suppose that over the month of March the floating price, which is the average of the 20 or so business day settlement prices published by Platts over the month for NWE Jet Fuel, is $1,550.25.

Trader A would have to pay and Trader B would receive the fixed leg of the swap:

$$\$1,510/mt \ x \ 20,000mt = \$30,200,000$$

Trader B would have to pay and Trader A would receive the floating leg of the swap:

$$\$1,550.25/mt \ x \ 20,000mt = \$31,005,000$$

Payments are actually netted so that in this case the back office of Trader A would simply issue an invoice for $805,000 and the back office of Trader B would wire the funds to Trader A on April 6th. Note that if the price of oil settled on average below $1,510 during the month of March then Trader A would have to pay Trader B.

CHAPTER NINETEEN
FORWARD OIL MARKETS: OPTIONS

Options can be viewed as insurance against the price of oil moving up or down and, just as with insurance, an upfront premium payment is required from the option buyer to the seller. The formal definition of an oil option is that an option is the right but not the obligation to buy or sell oil at a set price in the future. In other words, you have an option whether or not to buy or sell oil at a set price in the future.

For example, say you have bought an option which pays off if oil prices go above $105 over a certain period of time and oil prices during this time do not go above $105, then you simply walk away, having lost only your premium. Compare this to an OTC swap contract or an exchange traded futures contract which is an obligation to buy or sell oil at a set price in the future. With a swap or futures contract you have no choice but to buy or sell oil at the price you transacted. If you entered into a swap to buy oil at $105 over a certain period then if oil prices fall to $75 over that period then you are still required to pay $105 – you have no option.

The buyer of an option typically wires the option premium upfront to the seller within 2 days of the trade date. The seller, also known as a writer, of an option does not have to pay the buyer any money until after the option is exercised. The premium is never returned to the option buyer and therefore the buyer must hope that the payoff from the option contract is greater than the premium paid.

The most an option buyer can lose is the premium paid. A seller of an option, on the other hand, keeps the premium but has potentially unlimited losses.

The seller of an option usually posts collateral, or margin, with the option buyer or an exchange on the buyers behalf. This margin changes as the option value changes.

The price at which an option payoff begins is referred to as the strike price. The strike price is often referred to as the exercise price, or simply K.

There are two basic types of options depending on simple market direction: call options, which profit from rising prices, and put options which profit from falling prices.

A buyer of an option only receives a payoff after the option expires if the market price is above (in the case of a call option) or below (in the case of a put option) the strike price. The seller of the option must give the buyer this payoff. Airlines, shipping companies and other oil consumers typically buy call options. Crude oil producers typically buy put options.

Call option example: Suppose we buy a $105 strike call option for a premium of $5 per barrel. If the underlying market is $110 at expiry the option owner will receive a payoff of $5 which will cover the cost of the premium. If the market expires above $105 then we will be making money net of the premium paid. If the market settles at $105 or below we will receive nothing and our net loss on the strategy is equal to the premium paid (Table 19-1).

Put option example: Suppose we buy a $105 strike put option for a premium of $5 per barrel. If the underlying market is $100 at expiry the option owner will receive a payoff of $5 which will cover the cost of the premium. If the market expires below $100 then we will be making money net of the premium paid. If the market settles at $105 or above we will receive nothing and will simply have lost the premium (Table 19-1).

Table 19-1

Market Price at Expiry ($/bbl)	$105 Strike call option ($/bbl) payoff (net of $5 premium)	$105 Strike put option ($/bbl) payoff (net of $5 premium)
↑	↑	-$5
$120	+$10	-$5
$115	+$5	-$5
$110	0	-$5
$105	-$5	-$5
$100	-$5	0
$ 95	-$5	+$5
$ 90	-$5	+$10
↓	-$5	↓

A payoff diagram is often used to show how the payoff of an option changes with the underlying market price. For example, the data from the above table in payoff diagram format would look as follows (Fig. 19-1 and 19-2):

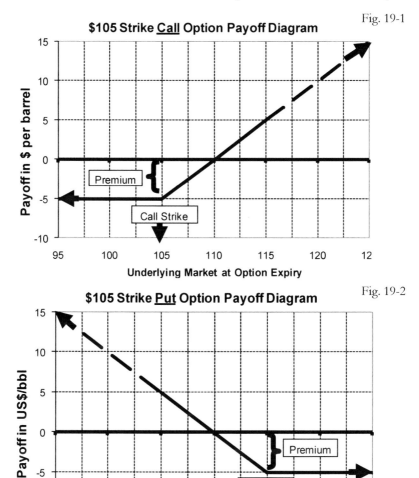

$105 Strike <u>Call</u> Option Payoff Diagram — Fig. 19-1

$105 Strike <u>Put</u> Option Payoff Diagram — Fig. 19-2

Straightforward put and call options, such as those described above, are known as plain vanilla options as they are the most common and standard.

The value of an option changes as the underlying market price moves, or the option gets closer to expiry (a decline in an option's time-value), interest rates change, or market volatility changes. Of course, once a certain premium is locked in on the purchase of an option, a buyer would only be interested in the change in the option's value if he or she were looking to sell the option back into the market before it expires.

Option strike terminology

An option with the strike price set equal to the current market price underlying the option is referred to as an at the money (ATM) option. For example, if the market is currently $105/bbl then a call or put option which is struck at $105 is referred to as an at the money option (Table 19-2).

Intuitively, an option will get more expensive the further it is in the money, all else being equal. For example, a $110 strike put option where the market price of the underlying crude oil is $105 is already $5 in the money and therefore is worth more than an at the money put at $105.

Table 19-2

	In the Money strike	At the Money strike	Out of the Money strike
Put:	Strike over market price	Strike at market price	Strike under market price
Call:	Strike under market price	Strike at market price	Strike over market price
Premium:	More than ATM	--------------	Less expensive than ATM

Option expiry

Most oil options expire in one of four ways, referred to as American, European, Asian and Bermudan style expiries. The names American, European, Asian and Bermudan carry no significance geographically and are simply market shorthand for the expiry type. In the OTC market the most common type of oil option is Asian expiry. The most common types of options on oil futures exchanges are American expiries.

American expiry is where buyer of the option can exercise it at any time from the option purchase date up to and including expiry date. Most exchange-traded oil options are American style.

Asian expiry, also known as average price options (APOs), are options exercised automatically only at expiry if in the money against the arithmetic average price over a period of time, such as a month, quarter or year. Unlike American options, Asian options cannot be exercised at any time and the buyer must wait until the expiry of the option to receive a payoff, if any. Arithmetic averaging only takes into account prices on business days during the period of time. Most OTC oil options are Asian style.

The use of Asian style options is more common than American style options in the OTC world for three reasons: Firstly, oil consumers and producers usually consume and produce oil ratably every day over a period of time and the use of options which match this constant every-day consumption or production pattern is more appropriate than American options which expire against a single day's price; Secondly, by using an average of a number of days prices (Asian options) instead of a single days price (American options), the volatility which is used to price the option is lower and this makes Asian options less expensive than American options. By lengthening the averaging period of an Asian option, say using a quarterly average instead of a monthly average, makes an Asian option cheaper still as the volatility of a quarterly average is lower than the volatility of a monthly average; and thirdly, American options can be exercised at any time until expiry, whereas Asian options are only exercised automatically at the expiry date. The lack of exercise flexibility in Asian options makes them less expensive.

European expiry, also known as bullet expiry, is where the buyer of the option can only exercise against a single price at the end of a period of time. European options are less expensive than American style options as they can only be exercised at the expiry date. European options are more expensive than Asian options as they settle against a single price, which has a higher volatility than the average of a number of prices used with Asian options. European-style options are rarely used in oil markets.

Bermudan expiry, also known as window expiry, are where the option buyer can only exercise the option during a defined time window(s). Bermudan options are rarely used in oil markets.

Table 19-3

Examples of the four most common options expiry methods

American expiry: any time from today until Aug 17, 2013
Asian expiry: monthly average of prices during business days of June 2012
European expiry: single day of Aug 17, 2013
Bermudan expiry: any time during the periods Sep 21-25, or Oct 21-25, 2013

Option payouts

Exchange traded oil options typically turn into futures contracts position at expiry if the buyer exercises in a process known as assignment. In order to realize a cash payout if the option is in the money the option buyer must buy or sell the futures which the options expired into.

OTC oil options are almost always cash settled. In addition to cash settlement, with OTC options there are many different types of more exotic expiry choices. For example, one can purchase an OTC option, called a compound option, which, if exercised, simply gives the buyer another option. As another non-cash expiry example, one can purchase an OTC option, called a swaption, which, if exercised, gives the buyer an OTC swap.

Pricing options

Option are priced using a variety of mathematical models, with the most well-known and commonly used being the Black-Scholes option model. The six Black-Scholes model inputs (Table 19-4) are usually annualized:

Table 19-4

Black-Scholes option model inputs

Volatility	underlying market price
time to expiration	strike price of option
risk-free interest rate	whether the option is a call or put

The Black-Scholes model makes several assumptions about market behavior, such as price changes forming a lognormal distribution and the ability to continuously hedge an option position, which are somewhat unrealistic in the real world.

Despite its flaws, the adjusted Black-Scholes model is still the model most commonly used by oil traders. One of the primary reasons the Black-Scholes model is so popular is that it provides a quick closed-form solution. Popular open-form solutions, such as Monte Carlo simulations, involve testing millions of different scenarios. Carrying out Monte Carlo simulations on a portfolio containing thousands of options can take massive brute force computing power and is usually not the timely solution most traders seek.

Volatility

Most futures and swaps traders simply trade the direction of price movements. Option traders can additionally trade the speed a market moves in any direction. One needs to calculate assign a value to volatility in order to trade it. If one has an option price, say from a futures exchange, and one knows all of the other Black-Scholes option model inputs one can solve for the volatility that the option price implies - so called implied volatility (Table 19-5).

Calculating implied volatility

Table 19-5

✓ Option Price (known)   = *implied volatility* (to solve for)
✓ time to expiration (known)
✓ risk-free interest rate (known)
✓ underlying market price (known)
✓ strike price of option (known)
✓ Call or Put (known)

If one connects the calculated implied volatility of options with various expiry dates in the future, one can generate a term structure, or forward curve of implied volatility (Fig. 19-3)

Fig. 19-3

Volatility Term Structure (forward curve)
Example for a ATM Options

Implied Volatility

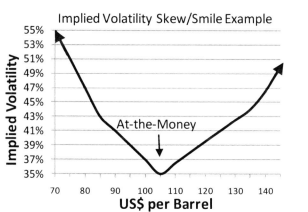

The above chart shows an example of the volatility used for pricing at the money options. However, different volatility numbers are commonly used to price in the money, at the money and out of the money options. Most often, at-the money options have the lowest volatility and all other option strikes have higher volatility, which is called volatility skew or volatility smile because of the way it appears on a chart (Fig 19-4). This smile, although most common, does not always hold and sometimes, rarely, volatility for at the money options can be higher than in or out of the money options:

Fig. 19-4

Implied Volatility Skew/Smile Example

At-the-Money

US$ per Barrel

If one plots the volatilities implied by the market premia for each option strike and for each expiry date one comes up with a more detailed view of the volatility used to price options called a volatility surface which is a 3-dimensional representations of volatility skew (Fig. 19-5).

Implied Volatility Skew/Smile Surface

Fig. 19-5

Option traders will be either long vol, profiting when implied volatility rallies, or short vol, profiting when implied volatility falls. The easiest way to get long volatility is to be net long options, which means having net paid premium. The easiest way to get short volatility is to be net short options, which means having net sold premium.

Option greeks

As the Black-Scholes model inputs change, the value of an option will change. In order to manage the risk involved in the inputs changing and therefore the changing value of a portfolio of options, most option traders will look at an option's 'greeks' which show a trader his risk, or sensitivity, to changes in each of the option model inputs. The greeks are referred to as delta, gamma, vega, theta, and rho. Greeks are especially important for traders managing portfolios, or books, of options containing perhaps hundreds or thousands of different strikes and types of options. For example, if just volatility changes, a trader will want to know how much his option position, or book of options, will increase in value or decrease in value.

Greeks can therefore neatly summarize an option trader's risk in the face of market changes.

By using smaller market movement increments, or granularity, in calculating greeks a trader can get a much better idea of how his option portfolio will behave as various model inputs change. This increase in granularity comes at a price, as the increased number of calculations consumes more computer processing power and time and the output can yield too much information such that it is difficult for a trader to deal with.

In addition to determining the sensitivity to small market movements, it is important to stress test option portfolios for large market movements, such as a sudden 10% or 20% move in prices or volatility. This is particularly important for options on more illiquid grades of oil, the price of which tend to gap, which means a large change in price without incremental trading.

Delta shows the flat price market exposure resulting from an option position. Delta is usually shown in barrels, gallons or metric tonnes. At the money options have a delta of roughly 50% of an option's total notional quantity. For example, if the notional quantity of oil underlying the option is 1 million barrels, an at the money put or call option will have a delta of roughly 500,000 barrels. This implies that the option will make or lose $500,000 if the market moves $1. In general, in the money options have deltas of greater than 50% and out of the money options will have deltas less than 50% of the option notional quantity.

Gamma shows change in delta for a change in underlying market price. In other words, if prices move up or down, how much will a trader's delta change? Gamma is shown in barrels, gallons or metric tonnes per price increment. For example, if gamma of an option or an option portfolio is +10million barrels per +$1/bbl move in prices: if prices move up $1 then the trader goes from being long 75 million barrels to being long 85 million barrels; if prices fall $1/bbl then the trader goes from being long 75 million barrels to being long 65 million barrels.

If a portfolio has several different option strikes then it important for a trader to know the strike locations, option types and sizes of the position. A gamma map, or strike map, achieves this. The strike map is plotted with underlying market prices at differing levels, strikes, and tenor to show where gamma effects are large and where they are not.

Additionally. traders often want to know if the change in delta, i.e. gamma, as the market moves is sufficient to justify an option premium. One method traders use to determine if an option premium is worth paying is called daily breakeven, which shows how much the underlying oil market must move on average each day in order for the change in delta to pay for the premium (Fig. 19-6). This method known as gamma trading assumes that the trader fully hedges his change in delta each day.

Calculating daily option breakeven

Fig. 19-6

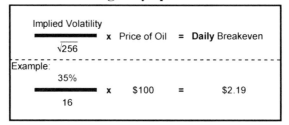

The same method can be used to calculate the weekly breakeven (Fig 19-7).

Calculating weekly option breakeven

Fig. 19-7

Implied volatility, when calculated using the Black-Scholes model, is proportional to the square root of time. Therefore, in the above calculations, the divisor 16 in the daily breakeven formula is the square root of 256, the assumed number of trading days in a year. The divisor 7.2 in the weekly breakeven formula is the square root of 52, the number of weeks in a year.

<u>Vega</u>, which unlike the other derivative greeks is not a Greek letter, shows the sensitivity of an option value to changes in volatility, in other words, if volatility moves up of down, all other things being equal, how much will an option value change. Vega for oil options is usually shown as a dollar amount per 1% move. For example, vega may show that a trader is long volatility such that if volatility moves up 1% he will make $40 million or if volatility moves down 1% he will lose $40 million.

<u>Theta</u> is the change in an option value due to passage of time and is usually measured in dollar amounts per day. The loss of an options value due to the passage of time, as the option approaches expiry, is referred to as time decay – an option buyer loses this time decay and an option seller gains time decay.

<u>Rho</u> is the change in an option value due to changes in interest rates. Rho for oil is usually shown in dollar amount per 50bps (50bps = 50 basis points or half of a percentage) move.

Although the above greeks are the primary ones looked at by traders, there are other second derivative greeks such as the volatility of volatility and the change in gamma for a given change in underlying price.

Option strategies

An option strategy involves creating a portfolio of options to achieve a custom tailored payoff.

The most common option strategy is the collar[5]. An oil consumer hedging against oil prices rising buys a call and sells a put. Conversely, an oil producer hedging against prices falling buys a put and sells a call. In both cases the sold option finances the bought option such that there is no cash premium paid upfront.

Another frequently used strategy is the call spread or put spread. An oil consumer hedging against oil prices rising buys a call at strike $X and sells a call at strike $X+Y. The sold call reduces the premium of the call at strike $X, but also limits the payoff of the option. An oil producer hedging against oil prices falling buys a put at $X strike and sells a put at a strike of $X-Y. Again, the sold put option reduces the premium and limits the payoff of the bought put.

Finally, an option structure most popular with option market makers is the straddle, which involves buying or selling a call option and a put option at the same strike. This trading structure is used to take a position on, or hedge against, rising or falling market volatility.

---

[5] Also known as a fence, tunnel, min-max, risk reversal, or cylinder.

CHAPTER TWENTY
MANAGING OIL PRICE RISK

Commodities are generally the most volatile asset class when compared to equities, currencies, and bonds (Fig. 20-1). This is because commodities are expensive to store and so, apart from precious metals such as gold, large ready to be consumed inventories rarely exist. Low commodity inventories can result in worry about potential shortages. Commodity shortages take months and even years to alleviate, whereas equities, currencies, and bond shortages can quickly be erased by companies issuing more shares, governments printing more currency, and debt issuers borrowing more.

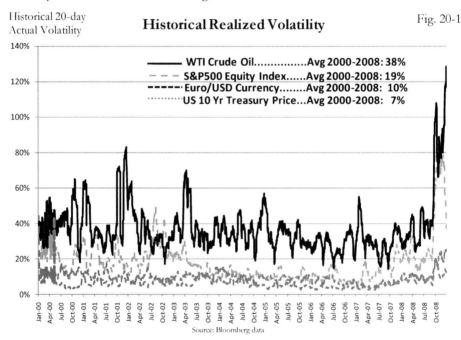

Historical 20-day Actual Volatility

Historical Realized Volatility

Fig. 20-1

WTI Crude Oil.................Avg 2000-2008: 38%
S&P500 Equity Index......Avg 2000-2008: 19%
Euro/USD Currency........Avg 2000-2008: 10%
US 10 Yr Treasury Price...Avg 2000-2008: 7%

Source: Bloomberg data

Hedging involves entering into a transaction to smooth the short term impact of volatile oil price movements on an individual or organization (Fig 20-2). Some organizations will tend to use exchange traded futures contracts and options, such as NYMEX and ICE futures and options contracts, to hedge against oil price movements. Larger organizations will tend to hedge in the OTC market with banks and not on futures exchanges. OTC markets permit access to a much larger range of benchmark oil prices and products such as OTC swaps and OTC options which can be tailored to an individual hedger's needs.

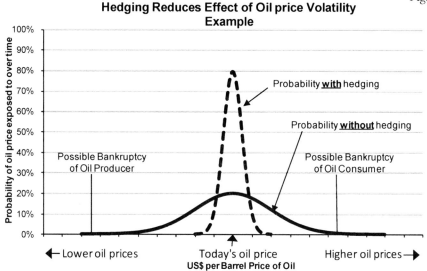

Fig. 20-2

A hedging transaction, such as an airline buying a jet fuel swap or a producer selling crude oil via a crude oil swap, may make money one year and lose the next year. If one looks at the cash flows resulting from hedging over the long term hedging will likely lose a small amount of money as the hedger has to cross a market bid/offer spread. However, there are many advantages to hedging which offset this minor cost.

The two primary reasons to hedge are to reduce short-term cash flow volatility and to maximize return on capital for a target level of risk.

Reducing the volatility of cash flows has many benefits. The risk of bankruptcy is reduced which improves the ability to borrow money and reduces the cost of such borrowing. More accurate planning, budgeting and earnings forecasting is possible. Organizations with more predictable earnings tend to be valued more highly by investors. A competitive advantage is created in being able to withstand short term adverse price movements. Finally, the ability of management to focus on core competencies is enhanced. For example, managers of an airline or a shipping company are not oil market specialists; they are in the business of transportation logistics and may wish to smooth the effects of short-term movements in oil prices in order to concentrate on what they are best at.

The second reason for hedging is based on portfolio theory. Portfolio theory is most often used to optimize an investment portfolio containing equities, commodities, bonds, cash, and other assets, so that the portfolio generates the highest return for the level of risk an investor is comfortable with.

Portfolio theory can also be applied to a corporation. A company is made up of a portfolio of volatile costs and revenues. As part of its portfolio of costs and revenue, an organization's management can raise or lower the impact of oil price volatility by hedging, such that the organization's overall rate of return is maximized for the level of risk which investors are happy assuming. This was referred to by economist Harry Markowitz as placing a company on an optimal efficient frontier.

Why some corporations don't hedge

Management of some corporations with exposure to oil price fluctuations do not hedge because they feel it is a form of gambling. Such managers fail to realize that gambling involves increasing one's risk profile by betting on price movements when one perceives the odds to be in one's favor. In hedging one is generally reducing the risk profile of an organization, and a hedger usually never perceives the odds to be in his favor or against him. A hedger is not out to make money from hedging itself. A hedger is out to maximize an organization's return on capital and reduce short-term cash flow volatility.

Other managers feel that by not hedging, and closing their eyes to the oil market, they are making a safe and low risk choice, but by not hedging they have actually made a passive decision to be subject to the full volatility of the oil market.

Another reason some managers decide not to hedge is that over time there will be periods where hedging transactions result in short term cash outflows. Some investors incorrectly punish management for hedging during these periods even though the long-term results of hedging should be beneficial. Poorly informed media reports can reinforce this incorrect characterization of hedging by labeling hedge related cash outflows as bad bets and hedge related cash inflows as good bets. This can provide a strong disincentive for management not to hedge.

The decision not to hedge should be an active decision. Management should clearly inform investors why they decide to face the full volatility of the oil market when they have an opportunity to manage that risk.

APPENDIX ONE
More on two most important forward markets

WTI is the basis for roughly one third of the world's oil prices, and Brent crude is the basis for the other two thirds.

NYMEX WTI crude oil futures contracts with expiry dates up to 9 years into the future are quoted. One WTI contract is 1,000 barrels (42,000 gallons) and the price is quoted in US dollars per barrel. The minimum tick (price increment) for WTI is 1 cent per barrel and a 1 cent move in a contract price is equal to US$10 (i.e. 1000barrels x .01). The ticker for WTI is CL.

Trading occurs in an open outcry pit on the floor of the NYMEX with brokers (those acting on behalf of clients) and locals (those taking positions on themselves), transact by shouting prices across a circular pit. Floor trading opens at 9am and closes at 2:30pm. The daily settlement price is set based on the last two minutes of trading before 2:30pm (last 30 minutes on contract expiry day).

WTI also trades on a computer system called GLOBEX, which is operated by the Chicago Mercantile Exchange (CME) on behalf of the NYMEX. WTI contracts traded on the floor and electronically are fungible. Although floor trading hours tend to be the most liquid time to execute a trade, most trading volume takes place electronically on the GLOBEX system. Trading on the GLOBEX electronic platform continues almost 24 hours a day from Sunday to Friday evenings. GLOBEX computerized trading starts at 6PM, with a 45-minute break occurring each day from 5:15pm to 6pm.

A WTI futures contract expires on the third business day prior to the 25th calendar day of the delivery month. For example, a June WTI contract will expire on the third business days prior to May 25th (assuming May 25th is a business day). In general, most market participants roll their positions to a contract further along the forward curve before a NYMEX crude contract expires. If one holds a contract past expiry then one must take or make delivery of 1,000 barrels of crude at Cushing, Oklahoma.

Physical delivery of oil at Cushing is normally in 50,000 to 100,000 barrels increments deliverable ratably over a calendar month via pipeline. For example, a June WTI contract which expires on the third business day prior to May 25th is physically deliverable ratably over the month of June. Six domestic grades of crude oil with an API gravity between 37° and 42°, and a sulfur content less than 0.42% by weight are deliverable against a NYMEX obligation at Cushing:

WTI, Low Sweet Mix, New Mexican Sweet, North Texas Sweet, Oklahoma Sweet, and South Texas Sweet. Six non-US grades are also deliverable against TI at a small NYMEX set fixed premium or discount to the TI expiry settlement: Nigerian Bonny Light and Qua Iboe; UK Brent and Forties; Norwegian Oseberg; and Colombian Cusiana. The non-US grades have an API gravity between 34° and 42°.

In addition to trading a NYMEX futures contract at the prevailing market price, one can buy and sell the daily settlement price – called a TAS trade, for Trade at Settlement – during the trading day. TAS is different to MOC (Market on Close), pronounced "M.O.C.", which is where a broker will attempt to buy or sell futures over the settlement period on your behalf.

ICE Brent crude oil is the second most liquid market for crude oil in the world behind WTI. One Brent contract is equal to 1,000 barrels (42,000 gallons). Price is quoted in US$ per barrel. Minimum tick (price increment) is 1 cent per barrel. A 1 cent move in a contract price is equal to US$10 (i.e. 1000barrels x .01). The ticker for Brent crude is LCO (on Reuters) or CO (on Bloomberg).

The Brent futures contract is, just like WTI, physically deliverable, in that if one holds the contract past expiry one has to take or make delivery of 1,000 barrels of Brent blend crude which is produced from more than 20 separate fields in the North Sea, commingled through the Brent and Ninian pipeline systems, and deliverable at a loading terminal in Sullom Voe on the Shetland Islands, off the coast of Scotland.

Brent trades at a discount or premium to WTI in order to encourage or discourage, respectively, arb barrels of Brent price linked crudes moving to the US. This spread between WTI and Brent crude is called the TI-Brent arb, or the Atlantic arb. The spread is the most commonly traded in oil and it determines whether large African and the Middle Eastern crude oil exporters send oil to western or eastern markets.

For example, Atlantic basin crudes, such as the West African (WAF) sweet crude oil Nigerian Bonny Light, are frequently a pivot point in balancing global physical crude oil movements. WAF sweet prices are linked via formula pricing to Brent crude prices. If transatlantic freight rates exceed the premium at which WTI is over Brent, or if sweet refinery margins in Asia are higher than other parts of the world, then more WAF sweets will move to Asia rather than the Americas and Europe until the TI-Brent spread changes sufficiently to restore the flow of some WAF crudes to western mrkets. When actually calculating the true physical Atlantic arb economics, traders most often use OTC traded oil grades Louisiana Light Sweet (LLS) crude and Dated Brent crude, rather than WTI (as it is inland-based delivery) and ICE Brent futures (as it ultimately becomes OTC Dated Brent).

Brent futures contracts expire on the business day preceding the 15th day prior to the start of the delivery month. When a Brent futures contract expires, a trader has a choice between cash or physical settlement. Physical settlement is the default if the cash settlement option is not taken.

If a trader chooses to cash settle an expired Brent futures contract then settlement is made against the ICE Futures Brent Index, published at 12 noon London time the following day, and which is linked to what is known as the 21-day Brent-Forties-Oseberg-Ekofisk price, referred to as BFOE, and discussed in more detail later. A trader must notify the exchange within an hour past the contract expiry time if he wishes to cash settle, and not physically settle, a futures contract.

If a buyer opts to take or make physical delivery, then the Brent contract becomes a 21-day BFOE contract for delivery in the following month. The seller of BFOE is required to give the buyer 21-days notice when nominating the lifting window, which is a short 3-day period in which delivery can take place, in the month following expiry. Shell manages the BFOE loading program at the Sullom Voe Terminal. Although Shell manages the loading program, BP actually operates the terminal.

Unlike oil traded on the NYMEX, Brent futures contracts are only traded electronically on the IntercontinentalExchange (ICE) platform, with no open outcry floor pit, since April 2005. Brent futures contracts with expiry dates up to 9 years into the future are quoted. ICE Brent futures trade Monday to Friday, 1am (11pm Sunday night) to 11pm London local time. Daily settlement is based on the weighted average of traded price over three minutes beginning at 7:27pm London time Monday through Friday, which approximately coincides with the NYMEX WTI settlement window.

In addition to the 3-minute weighted average end of day settlement price for Brent crude, there is a settlement published daily called the BWAVE, or Brent weighted average, which represents the weighted average price of all trades each day. BWAVE is used by some Middle Eastern producers, notably Saudi Arabia, as a benchmark used in calculating their formula Official Selling Price (OSP) as they viewed they view this price as being more representative of the day's trading rather than the end of settlement based on the final few minutes of trading

The ICE Futures Brent market, because of its physical delivery location, is used primarily by European, African, Middle Eastern and Asian traders as a crude oil benchmark.

ICE Brent futures is closely linked to over-the-counter (OTC) Brent contracts which are traded amongst professional oil traders and used as benchmarks in their own right. There are four OTC Brent contracts available to professional oil traders: 21-day BFOE; Dated Brent; CFDs; and DFLs.

<u>21-day BFOE</u>
The 21-day BFOE (Brent-Forties-Oseberg-Ekofisk) market is an OTC forward market, and, although ICE Brent futures can expire into a prompt month 21-day BFOE contract, the BFOE market operates independently of the ICE Futures exchange or any other exchange. The 21-day BFOE market is also known as the cash BFOE or the paper BFOE market.

As trading takes place OTC, 21-day BFOE counterparties, such as large oil majors, trade houses, or banks, have credit lines between one other (i.e. not necessarily any margining) and have to deal with each other directly (no clearinghouse), which is referred to as bilateral trading. A clip size, or parcel size, is 600,000 barrels, referred to as a cargo, instead of the 1,000 barrels with ICE Brent futures. Clips, or parcels, are terms used to refer to minimum tradable volume in all OTC markets and varies with the grade and location of oil.

Not all counterparties want exposure to a 600,000 barrel cargo of BFOE, and so there is an OTC 21-day BFOE partials market where counterparties will trade smaller clip sizes of typically 100,000-200,000 barrels. Partials can often be combined into cargoes.

21-day BFOE is a paper contract traded for several months ahead, although only the first three to six months are liquid. When the month to which a 21-day BFOE contract refers becomes prompt, i.e. it is the next month, a series of events are set in motion. For example, when the prompt 21-day BFOE contract is March, events begin in the first week of February.

In the first week of a month, an equity producer wishing to sell a cargo of Brent, Forties, Oseberg, or Ekofisk, which are four North Sea crudes, and having sold a 21-day BFOE contract for the following calendar month must inform either Shell, BP, StatoilHydro or ConocoPhillips of preferred 3-day loading dates in the following month. Shell, BP, StatoilHydro and ConocoPhillips manage the loading schedules at terminals on the North Sea for Brent, Forties, Oseberg, and Ekofisk respectively. After negotiations, the loading schedules are finalized by these four oil companies a few days later, e.g. the loading schedule for month of March is finalized in early February. Loading dates are also referred to as lifting windows.

Even though the BFOE seller (which usually begins as an equity producer of either Brent, Forties, Oseberg or Ekofisk) knows the 3-day loading dates he has been allocated, the seller of a 21-day BFOE contract is only required to give a buyer 21 days notice prior to the start of the 3-day loading period.

The seller can, if the scheduled lifting window allocated by Shell, BP, StatoilHydro, or ConocoPhillips is late in the delivery month, wait until early in the actual delivery month to notify the buyer of a 21-day BFOE contract of the exact 3-days, so long as he gives the buyer at least 21 days notice.

When giving notice of the loading window to a buyer, the seller must also specify which of the four grades of crude oil is being delivered. Up until 2007, most sellers typically tried to deliver Brent Blend against a BFOE obligation, as it was usually the least expensive, lowest quality, of the four crude oils. However, beginning in 2007, with the addition of heavy higher sulfur Buzzard field production into the Forties Blend, Forties has now become the crude most often delivered by sellers. As Forties Blend sulfur content can vary considerably when some of its component streams are undergoing routine maintenance, a sulphur de-escalator price adjustment is used when Forties Blend crude with greater than 0.6 percent sulfur content is delivered. The price adjustment is in cents per barrel per 0.1 percent weight of sulfur over the 0.6 percent. The adjustment is deducted from the price the buyer must pay. Platts sets the adjustment in consultation with the oil market.

Once a seller has notified a buyer, the 21-day BFOE contract becomes a dated Brent contract, which is a wet barrel physical contract. The crude has been assigned a 3-day lifting timeframe - i.e. it is now dated, hence the name dated Brent. Although the BFOE contract is called dated Brent after the seller has notified the buyer, remember that the oil may in fact be Brent, Forties, Oseberg, or Ekofisk crude, and may, therefore be lifted at one of five different North Sea terminals which the seller has nominated.

Although it is not immediately deliverable (is still 21 days from physical), the dated Brent market is effectively the spot market and the 21-day BFOE market is simply the same market with loading dates yet to be specified.

Until they become dated barrels, the 21-day BFOE contracts are traded amongst a small number, typically well under 15, of oil producers, refineries, banks and trade houses. When two counterparties have offsetting contracts, i.e.

a purchase and a sale with each other, the contracts are sometimes netted off against one another in a process known as a dry book-out. In a dry book-out, parties simply exchange cash for the value difference between the offsetting trades and the contracts never become wet dated barrels.

Once the buyer has been notified of the 3-day laytime and crude oil, he must nominate a specific vessel, which means notifying and requesting approval from the terminal operator to use that vessel, usually at least seven days prior to the 3-day lifting period. Operational tolerance (Op Tol) for delivery volumes on a cargo is plus or minus 1% at the buyer's option. Buyers often load VLCCs capable of carrying over 2 million barrels with several 600,000 barrel parcels.

Just as in futures markets, only a few traders actually want to take physical delivery of crude. More often than not, traders just want to hedge, or speculate using the 21-day BFOE paper contract but to get out of the contract obligation before it expires and becomes a physical wet barrel. If an equity producer of BFOE, or a buyer of a 21-day BFOE contract for a specific month does not want to take physical delivery, he can turn around and sell a 21-day contract to another trader; this trader can in turn sell a 21-day contract for that month to another trader, and so on. This so called daisy-chain of buyers and sellers goes on until a trader wishing to take physical delivery holds the contract, or the month becomes prompt and the 21-day notice period expires. If a trader not wishing to take physical delivery is holding a 21-day contract and the 21-day notice period expires, the trader is referred to as being 5 o'clocked or simply clocked, as the notice period expires at 5pm GMT 21 days before 3 day loading window. The trader then has a wet barrel Dated Brent contract.

The 21-day BFOE market used to be known as the 15-day Brent market. However, in 2002 as Brent blend production was declining to under 500,000 barrels per day, Platts, which is one of the primary oil price trade journals, decided, in consultation with industry, to include Forties and Oseberg North Sea crudes in assessing the daily settlement price of Brent crude. The addition of Forties and Oseberg made the Brent market less prone to squeezes as a market player would have to buy a huge amount of not only Brent, but also now Forties and Oseberg crudes in order to corner or squeeze the market. The notification period for Forties and Oseberg lifting windows is typically a few days longer than that for Brent blend and therefore the 15-day notice period was widened to 21 days notice.

Brent blend is deliverable at the Sullom Voe terminal on the Shetland Islands in Scotland. Shell manages the Brent crude loading schedule at Sullom Voe which is where the Brent and Ninian pipeline systems bring ashore a blend of the crudes from the approximately 20 North Sea oil fields which make up Brent blend crude.

Forties crude lifting schedule is managed by BP and cargoes are loaded FOB at Hound Point and Kinneil terminals in Scotland. Oseberg crude lifting schedule is managed by StatoilHydro and cargoes are loaded FOB at Sture terminal in Norway. Ekofisk crude is delivered to Teeside in England which is managed by ConocoPhilips.

The four crude oils are deliverable on a FOB (Free On Board) basis, and therefore BFOE prices do not normally fluctuate due to freight prices.

Dated Brent

Once a 21-day BFOE has been assigned dates for its 3-day loading window, the crude becomes what is referred to as dated Brent. Dated Brent can be bought or sold, but is no longer a paper derivative contract, it is instead a physical wet barrel contract. In other words, the buyer has legal ownership of physical oil; he simply has to wait a few days for delivery. There are four grades of crude which are deliverable against a 21-day BFOE obligation: Brent blend, Forties, Oseberg, and Ekofisk. The buyer is notified which grade will be delivered when he is given 21-days notice of the 3-day laytime.

Platts, the oil trade journal, assesses prices at the end of each trading day based on trade of FOB cargoes of dated Brent blend, dated Forties, dated Oseberg, and dated Ekofisk, loading within a 10-21 day timeframe (10-23 day time frame on Fridays.) Cargoes within 10 days of loading usually have found their end user and are no longer of value in determining market prices.

Although dated Brent is considered the spot, or cash, market for Brent crude, it is in reality a short-term (up to 21-days) forward market until loading actually occurs. This differs from other oil spot markets, where physical oil can be deliverable the day after the spot trade has been completed providing there is transportation and storage available.

Dated Brent - Contract for Differences (CFDs)

Dated Brent CFDs are an OTC market which allows traders to hedge or speculate on the price of dated Brent over short periods such as an individual week, ten day period (called a decade in the oil market), half month, or month. A CFD is the difference between the average dated Brent price assessment for a specific week or half-month, and a 21-day BFOE contract for a month in advance (remember 21-day BFOE trades for up to six or so months in advance.) As an example of a CFD, the week of July 10-14 dated Brent may be traded as a spread to the month of August 21-day BFOE. Usually up to eight individual weeks ahead of CFDs are traded in the market.

The Brent CFD market developed in order to allow traders who were given specific loading windows for lifting, i.e. dated Brent, a mechanism of hedging the price movement during that window.

Brent - Dated to Front Line (DFL)
Just as the Brent CFD contract is the difference between the price of dated Brent and a forward month of 21-day BFOE, the dated to front line contract is an OTC swap for the difference between Platts dated Brent assessments over an entire month and the ICE Futures prompt Brent futures contract.

Why are CFDs and DFLs traded? Both the CFD and DFL contracts allow a trader to get exposure to, or hedge, small dated Brent lifting windows. For example if a trader is lifting Brent crude during a certain week or month, a CFD or DFL can be used to hedge flat price movements by linking the price of dated Brent during that time period to more liquid either ICE Futures Brent futures contracts or 21-day BFOE contracts.

EFS and EFP
The ICE futures exchange has two additional features: Exchange Futures for Swaps (EFS) and Exchange Futures for Physical (EFP). EFS and EFP are all about changing risk from flat price risk to less volatile spread risk.

EFS (Exchange Futures for Swaps):
An EFS transaction allows one to trade the spread between an OTC swap, such as a swap on Dubai or Tapis crudes, and an exchange traded contract, such as ICE Brent futures. EFS trades are done in two stages: first with an OTC trade and then calling a futures broker.

For example, suppose you buy, through the OTC market, a Brent/Dubai EFS for August paying $8.50/bbl and say you agree to use August Brent of $150 (you have to agree with your OTC counterparty the Brent side price reference). In this example, you would be buying an August Brent Futures contract paying $150 and selling an August Dubai OTC Swap at $142.50. The trade is usually done with an OTC broker but can also be transacted directly with a counterparty. After dealing through the OTC broker each trader then enters the futures side of the trade electronically, or calls a futures broker on the futures exchange and tells them to book the futures leg of the trade. The months of an EFS trade don't necessarily have to coincide, for example one can trade a September ICE Brent futures contract with an August OTC Dubai Swap.

EFS allows traders to get exposure, or hedge, relatively illiquid OTC swaps, such as Dubai or Tapis crudes, while simultaneously hedging the flat price risk with a more liquid futures contract, such as OTC Brent futures. Basically, the liquidity of Brent futures is used to enhance the liquidity of OTC swaps such as Dubai or Tapis.

EFP (Exchange Futures for Physical):
An EFP transaction allows traders of physical crude cargoes to trade the spread between physical crude and an exchange traded futures contract. For example, a seller of a crude oil cargo becomes the buyer of futures and the buyer of a crude oil cargo becomes a seller of futures. The EFP trade is similar to an EFS trade in that the physical side is done OTC and both the buyer and seller need to enter the futures electronically, or call their futures broker to book the futures on an exchange. EFP could be used, for example, if I am a refinery trader, I want to buy crude for physical delivery and I already have a long (I have bought) futures position. The refinery trader could simply exchange its long futures position for physical crude with an OTC counterparty having the offsetting physical position.

APPENDIX TWO
CONVERSION FACTORS

The two most common methods of comparing various fuels are by density and energy content (Table A2-1).

Table A2-1

Approximate density and energy values

	Density					Energy Content		
	mass/volume	volume /mass	Relative to Water			Gravimetric (energy/mass)	Volumetric (energy/volume)	
	Kg/M³ liquid	lb/US Gal liquid	bbl/MT liquid	Specific Gravity	API°	MJ/Kg	Btu/lb	Btu/cu ft STP gas
Methane	424	3.54	14.83	0.424	202.2	49.7	21,433	910
Natural gas (Avg. stats for US delivered)	465	3.88	13.52	0.465	172.8	49.6	21,375	1,030
Ethane	356	2.97	17.67	0.356	266.0	47.1	20,295	1,630
Propane	508	4.24	12.38	0.508	147.0	45.9	19,770	2,365
Isobutane	562	4.69	11.18	0.562	120.3	46.3	19,975	2,975
Normal butane	584	4.87	10.78	0.584	110.8	47.0	20,257	3,250
							Btu/gal liquid	MJ/L liquid
Ethanol (E100)	794	6.63	7.92	0.794	46.7	29.7	12,800	76,100	21.21
E85 (15% by vol. CG and 85% by vol. ethanol)	780	6.51	8.06	0.780	49.9	29.2	12,590	82,000	22.85
E10 (90% by vol. CG and 10% by vol. ethanol)	720	6.01	8.73	0.720	65.0	42.5	18,297	110,000	30.66
Aviation Gasoline	715	5.97	8.79	0.715	66.4	43.7	18,844	112,500	31.35
Motor Gasoline (US Conventional Winter)	715	5.97	8.79	0.715	66.4	43.7	18,843	112,500	31.35
Motor Gasoline (US Conventional Summer)	735	6.14	8.55	0.735	61.0	43.6	18,787	115,300	32.13
Jet Fuel - Naphtha/Wide cut Type (B)	762	6.36	8.25	0.762	54.2	43.5	18,711	119,000	33.17
Diesel (US No. 1-D diesel fuel)	796	6.65	7.89	0.796	46.2	43.4	18,650	124,000	34.56
Jet Fuel - Kerosene/Narrow cut Type (A/A-1)	810	6.76	7.76	0.810	43.2	43.3	18,609	125,800	35.06
Diesel (US No. 2-D diesel fuel)	850	7.10	7.40	0.850	35.0	42.6	18,316	130,000	36.23
Distillate Fuel Oil (Heating Oil / No. 2 Fuel Oil)	870	7.26	7.23	0.870	31.1	42.1	18,102	131,500	36.65
Biodiesel (B100)	886	7.40	7.10	0.886	28.2	40.0	17,250	133,000	37.07
Residual Fuel Oil (US No. 6 Fuel Oil)	990	8.27	6.35	0.990	11.4	39.4	16,936	140,000	39.02

Standardized conversion factors used when trading oil (shown in Table 17-1 on page 336) are not the same as the more exact figures shown above.

Energy content, also known as the heat of combustion, is the energy released as heat when a fuel undergoes a complete reaction with oxygen. Energy content is measured using a calorimeter. Calorimeters measure gross (high) and net (low) heat values. The distinction between high and low heat values occur as water vapor is produced during combustion. The high heating value includes all energy released; including the energy contained in the vaporized water until it is condenses. Low heating value excludes energy contained in vaporized water. Almost all oil burning engines exhaust water vapor produced during combustion and so the low heating value is used to measure energy. The table above just shows approximate low heat values for energy content.

High heat values are especially important for natural gas power generation facilities. Older natural gas electrical power generation plants would release water vapor into the atmosphere. Modern efficient natural gas power generation facilities use the water vapor produced during combustion to generate further power.

References

ASTM International (ASTM): www.astm.org

Association for the Study of Peak Oil and Gas (ASPO): www.peakoil.net

Baltic Exchange: www.balticexchange.com

BP: www.bp.com

California Air Resources Board (CARB): www.arb.ca.gov

IntercontinentalExchange (ICE): www.theice.com

New York Mercantile Exchange (NYMEX): www.nymex.com

Organization of the Petroleum Exporting Countries (OPEC): www.opec.org

Organization for Economic Cooperation and Development (OECD), International Energy Agency (IEA): www.iea.org

Platts: www.platts.com

SAE International (SAE): www.sae.org

Society of Petroleum Engineers (SPE): www.spe.org

United Nations (UN): www.un.org

United Nations (UN), International Maritime Organization (IMO): www.imo.org

U.S. Central Intelligence Agency (CIA): www.cia.gov

U.S. Department of Agriculture (USDA): www.usda.gov

U.S. Department of Commerce (DOC), National Oceanic and Atmospheric Administration (NOAA), National Weather Service (NWS): www.nws.noaa.gov and National Hurricane Center (NHC): www.nhc.noaa.gov

U.S. Department of Energy (DOE), Energy Information Administration (EIA): www.eia.doe.gov

U.S. Department of the Interior (DOI), Minerals Management Service (MMS): www.mms.gov

U.S. Department of Transportation (DOT), Federal Highway Administration (FHWA): www.fhwa.dot.gov

U.S. Environmental Protection Agency (EPA): www.epa.gov

Bibliography

Oil histories

Tarbell, I.M. 1904. *The History of the Standard Oil Company.* McClure, Phillips and Co.

Longhurst, H. 1959. *Adventures in Oil – The story of British Petroleum.* Sidgewick and Jackson.

Ferrier, R.W. 1982. *The History of the British Petroleum Company, Vol. 1: The Developing Years, 1901-1932.* Cambridge University Press.

Yergin, D. 1991. *The Prize: The Epic Quest for Oil, Money and Power.* Simon & Schuster.

Bamberg, J.H. 1994. *The History of the British Petroleum Company, Vol. 2: The Anglo-Iranian Years, 1928-1954.* Cambridge University Press.

Dolin, E.J. 2007. *Leviathan: The History of Whaling in America.* W. W. Norton.

Technical oil books focusing on specific areas of the industry

Levorsen, A.I. 1967. *Geology of Petroleum, 2nd Edition.* W H Freeman & Co.

North, F.K. 1985. *Petroleum Geology.* Springer.

Tiratsoo, E.N. 1986. *Oilfields of the World, 3rd Edition.* Gulf Publishing Company.

Gold, T. 1992. *The Deep Hot Biosphere.* Proceedings of the National Academy of Sciences of the United States of America.

Magoon, L.B. and Dow, W.G. 1994. *The Petroleum System: From Source to Trap.* American Association of Petroleum Geologists.

Hunt, J. 1995. *Petroleum Geochemistry and Geology, 2nd Edition.* W. H. Freeman.

Barber, R., et al. 1996. *Motor Gasolines Technical Review.* Chevron Corporation.

Tusiani, M.D. 1996. *The Petroleum Shipping Industry: Operations and Practices.* Pennwell.

Myers, P.E. 1997. *Above Ground Storage Tanks.* McGraw-Hill Professional.

Selley, R.C. 1997. *Elements of Petroleum Geology, 2nd Edition.* Pennwell.

Conaway, C.F. 1999. *The Petroleum Industry: A Nontechnical Guide.* Pennwell.

Devereux, S. 1999. *Drilling Technology in Nontechnical Language.* Pennwell.

Leffler, W.L. 2000. *Petroleum Refining in Nontechnical Language, 3rd Edition.* Pennwell.

Burdick, D.L. and Leffler, W.L. 2001. *Petrochemicals in Nontechnical Language, 3rd Edition.* Pennwell.

Hyne, H.J. 2001. *Nontechnical Guide to Petroleum Geology, Exploration, Drilling, and Production, 2nd Edition.* Pennwell.

Leffler, W.L., Pattarozzi, R.A., and Sterling, G. 2003. *Deepwater Petroleum Exploration & Production: A Nontechnical Guide.* Pennwell.

Parkash, S. 2003 *Refining Processes Handbook.* Gulf Professional Publishing.

Dake, L.P. 2004. *The Practice of Reservoir Engineering, Revised Edition.* Elsevier Science.

Gluyas, J. and Swarbrick, R. 2003. *Petroleum Geoscience.* Wiley-Blackwell.

Raymond, M.S. and Leffler, W.L. 2005. *Oil & Gas Production in Nontechnical Language.* Pennwell.

Hemighaus, G., et al. 2006. *Aviation Fuels Technical Review.* Chevron Corporation.

Ahmed, T. 2006. *Reservoir Engineering Handbook, 3rd Edition.* Gulf Professional Publishing.

Leffler, W.L., and Miesner, T.O. 2006. *Oil & Gas Pipelines in Nontechnical Language.* Pennwell.

Bacha, J. et al. 2007. *Diesel Fuels Technical Review.* Chevron Corporation.

Books describing the global peak and decline in oil production

Deffeyes, K.S. 2001. *Hubbert's Peak: The Impending World Oil Shortage.* Princeton University Press.

Heinberg, R. 2003. *The Party's Over: Oil, War and the Fate of Industrial Societies.* New Society Publishers.

Heinberg, R. 2004. *PowerDown: Options And Actions For A Post-Carbon World.* New Society Publishers.

Campbell, C. J. 2004. *The Coming Oil Crisis*. Multi-Science Publishing.

Kunstler, J.H. 2005. *The Long Emergency: Surviving the Converging Catastrophes of the Twenty-First Century*. Atlantic Monthly Press

Simmons, M.R. 2005. *Twilight in the Desert: The Coming Saudi Oil Shock and the World Economy*. John Wiley & Sons.

Deffeyes, K.S. 2005. *Beyond Oil: The View from Hubbert's Peak*. Hill and Wang.

Roberts, P. 2005. *The End of Oil: On the Edge of a Perilous New World*. Mariner Books.

Leeb, S. and Strathy, G. 2006. *The Coming Economic Collapse: How You Can Thrive When Oil Costs $200 a Barrel*. Business Plus.

Tertzakian, P. 2006. *A Thousand Barrels a Second: The Coming Oil Break Point and the Challenges Facing an Energy Dependent World*. McGraw-Hill.

McBay, A. 2006. *Peak Oil Survival: Preparation for Life After Gridcrash*. Lyons Press.

Books on reducing oil consumption and alternatives.

Goodstein, D. 2004. *Out of Gas: The End of the Age of Oil*. W. W. Norton & Company.

Sandalow, D. 2007. *Freedom From Oil: How the Next President Can End the United States' Oil Addiction*. McGraw-Hill.

Oil novels

Sinclair, U. 1927. *Oil!* Robert Bentley, Inc.

Margonelli, L. 2007. *Oil on the Brain: Adventures from the Pump to the Pipeline*. Nan A. Talese.

Other books

Commodity Research Bureau, 2008. *The CRB Commodity Yearbook: 2008*. John Wiley & Sons.

Index